Essential Oils - Oils of Nature

Edited by Hany A. El-Shemy

Published in London, United Kingdom

IntechOpen

Supporting open minds since 2005

Essential Oils - Oils of Nature
http://dx.doi.org/10.5772/intechopen.77673
Edited by Hany A. El-Shemy

Contributors
Zakaria Hazzoumi, Youssef Moustakime, Khalid Amrani Joutei, Jason Jerry Atoche Medrano, Jean Baptiste Hzounda Fokou, Pierre Michel Jazet, Fekam Boyom Fabrice, Razzagh Mahmoudi, Marzieh Moosavi-Nasab, Armin Mirzapour-Kouhdasht, Najme Oliyaei, Swee Hua Erin Lim, Kok-Song Lai, Mariam Nasser Aljafari, Maryam Sultan Alhosani, Aisha Abushelaibi, Muhammad Irshad, Muhammad Ali Subhani, Amjad Hussain, Saqib Ali, Elena Stashenko, Jairo René Martínez, Abdelwahed Fidah, Mohamed Rahouti, Bouselham Kabouchi, Abderrahim Famiri, Fernando Almeida De Souza, Amanda Mara Teles, Adenilde Nascimento Mouchrek, Gustavo Oliveira Everton, Ana Lucia Abreu-Silva, Kátia Da Silva Calabrese, Juan Francisco Pérez-Sabino, Antonio Jorge Ribeiro Da Silva, Max Samuel Mérida-Reyes, Manuel Alejandro Muñoz-Wug, Isabel Cristina Gaitán-Fernández, Daniel Luiz Reis Simas, Bessie Evelyn Oliva Hernández, José Vicente Martínez-Arévalo, Aml ElAwady, Filomena Nazzaro, Florinda Fratianni, Antonio D'Acierno, Raffaele Coppola, J. Fernando Ayala-Zavala, Adriano Gomes Da Cruz, Vincenzo De Feo

First published in London, United Kingdom, 2020 by IntechOpen
IntechOpen is the global imprint of INTECHOPEN LIMITED, registered in England and Wales, registration number: 11086078, 7th floor, 10 Lower Thames Street, London, EC3R 6AF, United Kingdom
Printed in Croatia

British Library Cataloguing-in-Publication Data
A catalogue record for this book is available from the British Library

Additional hard and PDF copies can be obtained from orders@intechopen.com

Essential Oils - Oils of Nature
Edited by Hany A. El-Shemy
p. cm.
Print ISBN 978-1-78984-640-9
Online ISBN 978-1-78984-641-6
eBook (PDF) ISBN 978-1-83880-851-8

For EU product safety concerns:
IN TECH d.o.o., Prolaz Marije Krucifikse Kozulić 3, 51000 Rijeka, Croatia, info@intechopen.com or visit our website at intechopen.com.

Meet the editor

Prof. Hany A. El-Shemy received his two PhD degrees in Biochemistry and Genetic Engineering from the University of Cairo, Egypt, and the University of Hiroshima, Japan. He has filed two patents, written 13 international books, published more than 90 SCI journal papers and 50 conference presentations, and served as a technical committee member as well as chair in many international conferences; he is also the editor of *PLoS ONE*, *BMC Genomics*, and *Current Issues in Molecular Biology*. He has received several awards, including state prizes awarded from the Academy of Science, Egypt (2004, 2012, and 2018), Young Arab Researcher Prize awarded from the Shuman Foundation, Jordan (2005), State Excellence Prize from the Academy of Science, Egypt (2011), State Recognition Prize, Egypt (2017), and Cairo University Prizes (2007, 2010, and 2014). He served as an expert for the African Regional Center of Technology, Dakar, Senegal, and as a visiting professor at the Pan African University, African Union. He was appointed Vice President of the Academy of Science and Technology from 2013 to 2014, Egypt. From 2014 to December 2017 he worked as the Dean of the Faculty of Agriculture, Cairo University.

Contents

Preface

This book provides an overview as well as in-depth knowledge on essential oils. It consists of 13 chapters.

Many of the chapters attempt to review the recent research on essential oils, such as resources, chemical contents, etc. Other chapters aim to present a number of focal issues related to applications of essential oils in medical fields as natural treatments and traditional medicines.

Historically, essential oils have been used by consumers for several reasons, such as their aromas, medical applications, etc. Many international pharmaceutical companies produce essential oil derivatives for medical applications.

Therefore, this book contains information on the benefits and properties of essential oils as well as major compounds with their chemical structures.

It seems that essential oils will continue to contribute positively to pharmaceutical markets globally because of their multifunctional effects.

This book will be useful for researchers and other academics who wish to enhance their scientific backgrounds and abilities for teaching.

Hany A. El-Shemy
Professor,
Cairo University,
Giza, Egypt

Section 1

Medical Applications of Essential Oils

Chapter 1

Applications of Medicinal Herbs and Essential Oils in Food Safety

Razzagh Mahmoudi, Ata Kaboudari and Babak Pakbin

Abstract

In the last few years, more and more studies on the biological properties of essential oils (EOs) especially antimicrobial and antioxidant properties in vitro and food model have been published in all parts of the world. But so far no comprehensive reports of these studies have been reported in food model from Iran. The focus of this overview lies in the using of EOs from some indigenous medicinal plants of Iran (including *Mentha longifolia*, *Cuminum cyminum*, *Teucrium polium*, *Pimpinella anisum* and *Allium ascalonicum*) in probiotic dairy products (especially cheese, yoghurt and Aryan) in recent years. Recently, consumers have developed an ever-increasing interest in natural products as alternatives for artificial additives or pharmacologically relevant agents. Among them, EOs have gained great popularity in the food, cosmetic as well as pharmaceutical industries. Despite the reportedly strong antimicrobial activity of EOs against food-borne pathogens and spoilage microorganisms, their practical application as preservatives is currently limited owing to the undesirable flavour changes they cause in food products. Nonetheless, more studies are necessary to the applicability of various EOs on other food models in Iran and other countries.

Keywords: EOs, functional dairy foods, natural preservative, sensory quality

1. Introduction

Today, due to the adverse effects of chemical preservatives, regarding the carcinogenic potential and toxicity to humans, as well as the high levels of antimicrobial agents present in plants, there is a growing interest in the use of natural preservatives derived from natural sources [1]. Food storage methods which maintain the quality and extend the shelf life of food because of improved production, supply and trading are important. Humans are familiar in keeping foods by different methods such as the use of heating, cooling, salting, etc. a long time ago, but for inhibition of pathogen growth and also prevention of food spoilage, a new method is more in need; therefore one of these methods is the use of essential oils and natural materials as antimicrobial additives in food [2].

Essential oils are composed of lipophilic and highly volatile secondary plant metabolites [3]. As defined by the International Organisation for Standardisation (ISO), the term "EOs" is reserved for a product obtained from vegetable raw material, either by distillation with water or steam, or from the epicarp of citrus fruits by a mechanical process, or by dry distillation, that is, by physical means only. EOs have been proposed as natural preservatives and are used as alternatives for the control of pathogenic microorganisms. Herbal EOs are aromatic oil liquids, extracted from various parts of plants, and are used as flavouring agents in foods; thus the importance of the use of medicinal plants in food products can be multiple times [4].

2. How plants work

Preservatives are used to limit the growth and microbial activity in pharmaceutical products, food and cosmetics, and by interfering with cell membranes, enzymes and genetic structure of microorganisms have a preventive effect. To apply the essential oils as chemical preservatives in food, investigating their antibacterial activities alone and in combination with other factors affecting the growth of microorganisms in food and nutrition is essential in laboratory models [2]. The use of natural antimicrobial compounds such as essential oils, herbal extracts and spices for the protection of food against microbial spoilage has led to the identification of some of their unique features such as taste effects and antioxidant activity [5]. EOs have been used in human health as functional food, food additives, medicine, nutritional supplements and cosmetic manufacturing [4, 6, 7].

Essential oils and plant extracts with various biological compounds have very high potential for using as new drug combinations, healthcare and human and animal diseases as well due to the presence of anti-microbial compounds especially against Gram-Positive and Gram-Negative pathogens, Anticancer, Antioxidant and Free Radical Removal Factors as one of the most important natural sources for the using of them in medicines and foods [1]. Essential oils and extracts from medicinal herbs with antimicrobial, anticancer and antioxidant compounds (due to the presence of free radicals eliminating agents) have importance as new and natural drug combinations, both in the field of health and disease management and in the protection of raw and processed foods [8].

3. Importance of food-borne pathogens

Food-borne diseases recognised as one of the major public health problems worldwide, especially in developing countries, and, on the other hand, increasing incidence of food-borne disease along with its social and economic consequences have led to conducting extensive research in order to produce safer food and develop new antimicrobial agents; among them, the extensive use of probiotics and bacteriocins as biological additives is of considerable importance. With increase of urban population, tourism, immigration, a variety of food with different components, improve technology in the food industry, changes in food consumption culture and approach to food consumption, food preparation, and finally international trade in food, overburdened the more food illness in the present age, so that about 30 per cent of people in developed countries at least once a year to develop food-borne diseases [4]. Despite the reportedly strong antimicrobial activity of EOs against food-borne pathogens and spoilage microorganisms, their practical application as preservatives is currently limited owing to the undesirable flavour changes they cause in food products [6, 9–11].

4. Some of the most important plants that can be used in foods

4.1 *Mentha longifolia* L.

Mentha longifolia L. from *Lamiaceae* essentially grows in wet river banks of temperate areas of Central and South Europe, Australia, South-West Asia and Iran.

The EOs of this plant varying in quantity according to variety and characteristic of the growing site are composed of cationic compounds especially pulegone, 15–40% total alcohols, 7–12% limonene and Dilantin. This plant bears medicinal characteristics and has proven to be of benefit for digestive system disorders, vomiting and loss of appetite, ulcerative colitis and liver malfunctions. Other reported inhibitory effects have been reported towards microorganisms causing food-borne diseases, for example, *S. aureus*, *E. coli*, *Bacillus subsp*., *Salmonella subsp*. and *Aspergillus subsp*. [8, 12].

4.2 *Cuminum cyminum* L.

C. cyminum with the vernacular name of "Zireh e Sabs" (in Iran) is a plant belonging to the *Apiaceae* family applied in Iranian folk medicine since more than 200 years ago. Major constituents in *C. cyminum* essential oil (EO) are cumin aldehyde, cuminic alcohol, gamma-terpinene and ß-pinene [11, 13]. This plant has inhibitory effects on *E.coli*, *L. monocytopenia* and *S. aureus* [14].

4.3 *Teucrium polium* L.

This plant is belonging to the mint family, plateau, a height of 10 to 30 cm, with a white cottony appearance, usually in poor areas (nutrients and organic matter), rocky areas and sand dunes in Europe, the Mediterranean region, north of Africa and south west of Asia, including Iran, especially Khorasan province [2]. The studies have shown that this herb has antioxidant effects and antipyretic, antimicrobial and antispasmodic effects [15]. It has been reported that the ethanol extract of this herb has also an antibacterial activity against Gram-positive and Gram-negative microorganisms of the show itself [16]. *Bacillus cereus* in the food samples is one of the ingredients that is inhibited by the essential oil of this plant [2].

4.4 *Pimpinella anisum*

Pimpinella anisum L. is a plant with white leaves and small green yellowish seeds and is from the *Umbelliferae* family. This plant grows in countries such as Iraq, Turkey, Iran, India, Egypt and many tropical areas of the world [14, 17]. EOs of some species of this plant are used in treating diseases such as epilepsy [18].

4.5 *Echinophora orientalis*

Echinophora is a plant of the family *Apiaceae* that includes 10 species that have been distributed from the Mediterranean area to Iran. *E. orientalis* is a common species in Iran [19]. Two species of 10, including *E. sibthorpiana* and *E. orientalis*, are also growing in Anatolia, Turkmenistan, Armenia, Russia, Syria, the Balkans, Cyprus and Afghanistan [20, 21]. *Echinophora* EO contains alkaloid compounds and flavonoids [19]. γ-Decalactone, β-cis-ocimene and linalool L are the most important compounds in the EO of this plant [19]. This plant and its oil have antiseptic, antibacterial, antioxidant and antifungal effects and can inhibit human platelet aggregation and are also used in folk medicine to heal wounds and have carminative and digestive properties [22–25]. In the result of a study, different concentrations of *E. orientalis* EO significantly affected the growing of *S. aureus* bacterial in food model [19]. In another study, *E. tenuifolia* EO showed strong antimicrobial activity against *B. cereus* and *Staphylococcus spp* [26].

4.6 *Aloe vera* gel (*Aloe barbadensis Miller*)

Aloe vera (*Aloe barbadensis Miller*) is a plant, which belongs to the family of *Liliaceae*. *Aloe vera* grows in arid climates and is widely distributed in India, Africa and other arid areas [27]. The 0.7% of gel of leaves is made up of solids mainly carbohydrates [28]. Activity against a variety of infectious agents has been attributed to *Aloe vera* such as antiviral, antibacterial and antifungal effects [29–31]. Some specific plant's compounds have been proposed to have direct antimicrobial activity, for example, anthraquinones, dihydroxyanthraquinones and saponins [32–35]. The antibactericidal activity of *Aloe vera* gel may be attributed to active compounds such as alkaloids, tannins, flavonoids as well as saponins which have a direct antimicrobial activity [33, 36]. In the results of a study, the antimicrobial potency of *Aloe vera* gel aqueous extract against *E. coli* has been shown in yoghurt [37]. In the study of Agarry et al. [38], they reported that leaf extracts had antibacterial activity against bacterial species such as *S. aureus*, *Klebsiella pneumoniae* and *E. coli*. In another study, the ethanol extract of *Aloe vera* gel inhibited the growth of *E. coli* and *S. aureus* [39].

4.7 *Ferula sharifi*

The genus *Ferula* belongs to the family *Apiaceae* that comprises about 170 species in the world. These genera are produced from central Asia to northern Africa [40]. These plants are well documented as a good source of biologically active compounds such as sesquiterpenoids and sulphur-containing compounds [41]. Species of this genus have been used in traditional medicine for the treatment of various organ disorders, for example, *F. assa-foetida* used as anticonvulsant, carminative, antispasmodic, diuretic, aphrodisiac, antihelmintic, tonic, laxative and alterative or *F. persica* used as laxative, carminative, antihysteric and for treatment of lumbago, diabetes and rheumatism [22, 40, 42–45].

5. The sensitivity of some important food-borne pathogens to plant extracts and EOs

5.1 *Listeria monocytogenes*

Food-borne diseases are one of the major public health problems worldwide, and recent reports indicate that *Listeria monocytogenes* is a major concern. *Listeria monocytogenes* can cause food intoxication, meningitis and encephalitis [6]. Control of these bacteria and its diseases are very important. So, various studies have been carried out on the effects of different essential oils and extracts on growth and control of these bacteria.

In the study of Ehsani et al. [46], the results have shown that treatments of 0.1% *Allium ascalonicum* and *Pimpinella anisum* essential oils at the end of cheese ripening period showed the highest decrease in the mean bacterial colony counts. The results of Mahmoudi et al. study have shown that *Cuminum cyminum L.* essential oil on *Listeria monocytogenes* has effects at different concentrations [8]. These results, as well as the results from other studies, have shown that essential oils and plant extracts could help the control of the bacteria in the food industry.

5.2 *Salmonella typhimurium*

Salmonella typhimurium is a pathogenic food-borne bacterium. *Salmonella* is widespread worldwide and found sporadically in water, soil, animal food, meat, faeces and vegetables and can infect many mammals and birds [47]. Considering that this bacterium can cause disease through food, including dairy products in humans, its control through essential oils and plant extracts is very much considered. *Salmonella* infection may occur in one of three clinical forms of self-sustaining gastroenteritis, then septicaemia with local lesions or an enteric fever or typhoid fever [48].

For example, *Teucrium polium* EO has the best *Salmonella* growth inhibition at 60 ppm and 80 ppm concentrations. In this research, no *Salmonella* was isolated during the 28 days of preservation of probiotic yoghurt [49].

5.3 *Staphylococcus aureus*

Staphylococcus aureus is one of the most important pathogenic food-borne bacteria. These bacteria can cause diarrhoea and vomiting intoxication [5]. Due to the importance of these bacteria and its toxicity, as well as due to the health hazards of chemical preservatives, researchers have used various essential oils and herbs to control these bacteria in their experiments [6].

In the result of one study, essential oil of *Mentha longifolia L.* with a concentration of 150 ppm has an inhibitory effect against *Staphylococcus aureus* [6]. In a study, 21 essential oils were used for antibacterial effects. As a result, the essential oils of *Corydothymus capitatus*, *Cinnamomum cassia*, *Origanum heracleoticum*, *Satureja montana* and *Cinnamomum verum* were effective against *Staphylococcus aureus* [50, 51].

Today, the importance of biofilm formation in the food industry is also high, so studies have also been carried out and are expanding. In these studies, vegetable oils are used to prevent and eliminate biofilms. For example, the compounds in essential oils such as SAB, C3 and C4 are highly effective against biofilms created by *Staphylococcus aureus* [52].

6. Herbal medicines appear relatively safe

Traditional medicine has brought the foundation of health care around the world from the earliest days of human beings. Medicinal plants have been known for many years as a rich source of well-known therapeutic agents for the treatment and prevention of various diseases, the most important of which is the social, cultural, spiritual and medicinal fields. Over the past centuries, severe changes in human lifestyle and dietary habits have led to the emergence of various chronic pathologies. Recently, "herbal renaissance" is a visible phenomenon worldwide, and two-thirds of the plant species in the world may have medicinal value. The World Health Organisation believes that 80% of the population in Africa and Asia uses traditional medicine as the first source for their health-care needs. Also, in the United States, more than 40% of the population has recently been identified with complementary and alternative supplements, including herbal supplements [53].

7. Conclusion

Herbal drugs appear to be relatively safe, but human research or prospective data on adverse effects and plant and drug interactions are limited. They generally have fewer drugs than their pure relatives because they contain a mixture of chemicals that are in low amounts. According to studies, the importance of edible oral pathogens is not covered for everyone. According to studies, the importance of food-borne pathogen bacteria is not covered for everyone. On the other hand, due to the harmful effects of chemical preservatives and also the increase of drug resistance, the use of plants and their essential oils is very important. Essences and their effective compounds can be used to prevent poisoning and disease and to prevent the transmission of bacteria from food and food industry like dairy.

Author details

Razzagh Mahmoudi[1]*, Ata Kaboudari[2] and Babak Pakbin[3]

1 Medical Microbiology Research Center, Qazvin University of Medical Sciences, Qazvin, Iran

2 Faculty of Veterinary Medicine, Urmia University, Urmia, Iran

3 Faculty of Veterinary Medicine, University of Tehran, Tehran, Iran

*Address all correspondence to: r.mahmodi@yahoo.com

References

[1] Mahmoudi R, Kazeminia M, Kaboudari A. Review on composition and antimicrobial effects of Teucrium (*Teucrium polium* L.) grown in Iran and comparison with the around the world. Journal of Babol University of Medical Sciences. 2017;**19**(2):54-64

[2] Keykavousi M, Ghiassi Tarzi B, Mahmoudi R, bakhoda H, Kaboudari A, Pir Mahalleh SFR. Study of antibacterial of effects of *Teucrium polium* oil on *Bacillus cereus* in cultural laboratory and commercial soup. Carpathian Journal of Food Science and Technology. 2016;**8**(2):176-183

[3] Sell C. Chemistry of essential oils. In: Baser KH, Buchbauer G, editors. Handbook of Essential Oils. Science, Technology, and Applications. Boca Raton, FL: CRC Press; 2010. pp. 121-150

[4] Burt S. Essential oils: Their antibacterial properties and potential applications in foods—A review. International Journal of Food Microbiology. 2004;**94**(3):223-253

[5] Salehi P, Sonboli A, Eftekhar F, Ebrahimi S, Yousefzadi M. Effect of essential oils from certain *Ziziphora* species on swimming performance in mice. Phytotherapy Research. 2005;**9**:225-227

[6] Ehsani A, Mahmoudi R. Effects of *Mentha longifolia* L. essential oil and *Lactobacillus casei* on the organoleptic properties, and on the growth of *Staphylococcus aureus* and *Listeria monocytogenes* during manufacturing, ripening and storage of Iranian white brined cheese. The International Journal of Dairy Technology. 2012;**66**:77-82

[7] Thabet HM, Nogaim QA, Qasha AS, Abdoalaziz O, Alnsheme N. Evaluation of the effects of some plant derived essential oils on shelf life extension of Labneh. Merit Research Journal of Food Science and Technology. 2014;**2**(1):8-14

[8] Mahmoudi R, Ehsani A, Zare P. Phytochemical, antibacterial and antioxidant properties of *Cuminum Cyminum* L. essential oil. Journal of Food Industry Research. 2012;**22**(3):311-321

[9] Basti AA, Misaghi A, Khaschabi D. Growth response and model- ling of the effects of *Zataria multiflora* Boiss. essential oil, pH and temperature on *Salmonella Typhimurium* and *Staphylococcus aureus*. Food Science & Technology. 2007;**40**:973-981

[10] Yamazaki K, Yamamoto T, Kawai Y, Inoue N. Enhancement of antilisterial activity of essential oil constituents by nisin and diglycerol fatty acid ester. Food Microbiology. 2004;**21**:283-289

[11] Mahmoudi R. Improvement the hygienic quality and organoleptic properties of bioyoghurt using *Cuminum cyminum* L. essential oil. Journal of Agroalimentary Processes and Technologies. 2013;**19**(4):405-412

[12] Gulluce M, Sahin F, Sokmen M, Ozer H, Daferera D, Sokmen A. Antimicrobial and antioxidant properties of the essential oils and methanol extract from *Mentha longifolia* L. ssp. longifolia. Food Chemistry. 2007;**103**:1449-1456

[13] Charlier C, retenet M, Even S. Interactions between *Staphylococcus aureus* and lactic acid bacteria: An old story with new perspectives. International Journal of Food Microbiology. 2009;**131**:30-39

[14] Singh G, Kapoor IP, Pandey SK. Studies on essential oils: Part 10; antibacterial activity of volatile oils of some spices. Phytotherapy Research. 2002;**16**:680-682

[15] Zare P, Mahmoudi R, Ehsani A. Biochemical and antibacterial properties of essential oil from *Teucrium polium* using resazurin as the indicator of bacterial cell growth. Pharmaceutical Sciences. 2011;**17**(3):183-188

[16] Darabpour E, Motamedi H, Seyyed Nejad SM. Antimicrobial properties of *Teucrium polium* some clinical pathogens. Asian Pacific Journal of Tropical Medicine. 2010:124-127

[17] Pourgholami MH, Majzoob S, Javadi M, Kamalinejad M, Fanaee GHR, Sayyah M. The seeds essential oil of *Pimpinella anisum* exerts anticonvulsant effects in mice. Journal of Ethnopharmacology. 1999;**66**:211-215

[18] Al-Bayati FA. Synergistic antibacterial activity between *Thymus vulgaris* and *Pimpinella anisum* essential oils and methanol extracts. Journal of Ethnopharmacology. 2008;**116**:403-406

[19] Farzanehnia E, Ghajarbeygi P, Mahmoudi R, Mardani K. Phytochemical and antibacterial properties of *Echinophora orientalis* essential oil against *Staphylococcus aureus* in soup. Journal of Biology and Today's World. 2016;**5**(8):150-156

[20] Georgiou C, Koutsaviti A, Bazos I, Tzakou O. Chemical composition of *Echinophora tenuifolia* subsp. *sibthorpiana* essential oil from Greece. Records of Natural Products. 2010;**4**:167-170

[21] Mileski K, Dzamic A, Ciric A, Grujic S, Ristic M, Matevski V, et al. Radical scavenging and antimicrobial activity of essential oil and extracts of *Echinophora sibthorpiana* Guss. From Macedonia. Archives of Biological Sciences. 2014;**66**(1):401-413

[22] Hashemi M, Ehsani A, Jazani NH, Aliakbarlu J, Mahmoudi R. Chemical composition and in vitro antibacterial activity of essential oil and methanol extract of *Echinophora platyloba* DC against some of food-borne pathogenic bacteria. Veterinary Research Forum. 2013;**4**(2):123-127

[23] Hadjmohammadi M, Karimiyan H, Sharifi V. Hollow fibre-based liquid phase microextraction combined with high-performance liquid chromatography for the analysis of flavonoids in *Echinophora platyloba* DC. And *Mentha piperita*. Food Chemistry. 2013;**141**(2):731-735

[24] Lv J, Huang H, Yu L, Whent M, Niu Y, Shi H, et al. Phenolic composition and nutraceutical properties of organic and conventional cinnamon and peppermint. Food Chemistry. 2012;**132**(3):1442-1450

[25] Genç İ, Ecevit-Genç G. The synopsis of the genus *Echinophora* L. (*Apiaceae*) in Turkey. Journal of Faculty Pharmacy of Istanbul University. 2014;**44**(2):233-240

[26] Gokbulut I, Bilenler T, Karabulut I. Determination of chemical composition, total phenolic, antimicrobial, and antioxidant activities of *Echinophora tenuifolia* essential oil. International Journal of Food Properties. 2013;**16**(7):1442-1451

[27] Nemati Niko Z, Ghajarbeygi P, Mahmoudi R, Mousavi S, Mardani K. Inhibitory effects of *Aloe vera* gel aqueous extract and *L. casei* against *E.coli* in yoghurt. Journal of Biology and Today's World. 2015;5(9):157-162

[28] Foster S. *Aloe vera*: The Succulent with Skin Soothing Cell Protecting Properties. Herbs for Health Magazine Health World [Online]. 1999. Available from: http://www.healthy.net/library/articles/hfh/aloe.htm

[29] Ferro VA, Bradbury F, Cameron P, Shakir E, Rahman SR, Stimson WH. In vitro susceptibilities of *Shigella flexneri* and *Streptococcus pyogenes* to inner gel of

Aloe barbadensis Miller. Antimicrobial Agents and Chemotherapy. 2003;**47**(3):1137-1139

[30] Kawai K, Beppu H, Shimpo K, Chihara T, Yamamoto N, Nagatsu T, et al. In vivo effects of *Aloe arborescens* Miller var. natalensis Berger (*Kidachi aloe*) on experimental tinea pedis in Guinea-pig feet. Phytotherapy Research. 1998;**12**(3):178-182

[31] Antonisamy JMA, Beaulah N, Laju R, Anupriya G. Anti-bacterial and antifungal activity of *Aloe vera* gel extract. International Journal of Biomedical and Advance Research. 2012;**3**(3):184-187

[32] García-Sosa K, Villarreal-Alvarez N, Lübben P, Chrysophanol P-RM. An antimicrobial anthraquinone from the root extract of *Colubrina greggii*. Journal of the Mexican Chemical Society. 2006;**50**(2):76-78

[33] Dabai Y, Muhammad S, Aliyu B. Antibacterial activity of anthraquinone fraction of *Vitex doniana*. Pakistan Journal of Biological Sciences. 2007:1-3

[34] Wu YW, Ouyang J, Xiao XH, Gao WY, Liu Y. Antimicrobial properties and toxicity of anthraquinones by microcalorimetric bioassay. Chinese Journal of Chemistry. 2006;**24**(1):45-50

[35] Reynolds T, Dweck A. *Aloe vera* leaf gel: A review update. Journal of Ethnopharmacology. 1999;**68**(1):3-37

[36] Nusrat SI, Ljber K, Gul A. Commercial extraction of gel from *Aloe vera* (L) leaves. Journal of The Chemical Society of Pakistan. 2000;**22**(1):47

[37] Hasani P, Yasa N, Vosough-Ghanbari S, Mohammadirad A, Dehghan G, et al. In vivo antioxidant potential of *Teucrium polium*, as compared to α-tocopherol. Acta Pharmaceutica. 2007;**57**:123-129

[38] Agarry O, Olaleye M. Comparative antimicrobial activities of *Aloe vera* gel and leaf. African Journal of Biotechnology. 2005;**4**(12):1413

[39] Stanley M, Ifeanyi O, Eziokwu O. Antimicrobial effects of *Aloe vera* on some human pathogens. International Journal of Current Microbiology and Applied Sciences. 2014;**3**(3):1022-1028

[40] Sahebkar A. Biological activities of essential oils from the genus Ferula (Apiaceae). Asian Biomedicine. 2011;**4**(6):835-847

[41] Iranshahi M, Hassanzadeh-Khayyat M, Sahebkar A, Famili A. Chemical composition of the fruit oil of *Ferula flabelliloba*. Journal of Essential Oil Bearing Plants. 2008;**11**:143-147

[42] Zargari A. Medicinal Plants. Vol. 2. Tehran: Tehran University Press; 1992

[43] Aboabrahim Z. Zakhirah Kharazmshahi. Vol. 2. Teheran: Nacional Works; 1970. p. 141

[44] Abdul-Ghani A-S, El-Lati S, Sacaan A, Suleiman MS, Amin RM. Anticonvulsant effects of some Arab medicinal plants. International Journal of Crude Drug Research. 1987;**25**:39-43

[45] Shahraki MR, Arab MR, Mirimokaddam E, Palan MJ. The effect of *Teucrium polium* (Calpoureh) on liver function, serum lipids and glucose in diabetic male rats. Iranian Biomedical Journal. 2007;**11**:65-68

[46] Ehsani A, Mahmoudi R, Zare P, Hasany A. Biochemical properties and antimicrobial effects of *Allium ascalonicum* and *Pimpinella anisum* essential oils against *Listeria monocytogenes* in white brined

cheese. Journal of Food Research. 2011;**21**(3):317-328

[47] Mahmoudi R, Amini K, Kaboudari A. Pir Pir Mahalleh SFR, Babak Pakbin Detection of ESBL genes in *Salmonlla enteritudis* isolated from clinical samples. International Journal of Food Nutrition and Safety. 2016;7(1):10-25

[48] Razavi-Rohani SM, Griffits MW. The effect of mono and poly glycerollaurate on spoilage and pathogenic bacteria associated with foods. Journal of Food Safety. 1994;**14**:131-151

[49] Mahmoudi R, Zare P, Nosratpour S, Mardani K, Safari A. Hygienic effects of *Teucrium polium* essential oil against *Salmonella typhimorium* LT2 in probiotic yoghurt. Urmia Medical Journal. 2014;**25**(8):769-777

[50] Oulahal N, Brice W, Mrtial A, Degraeve P. Quantitative analysis of *Staphylococcus aureus* or *Listeria monocytogenes* on two types of surfaces: Propylene and stainless steel in contact with three different dairy products. Food Control. 2008;**19**:178-185

[51] Oussalah M, Caillet S, Saucierc L, Lacroix M. Inhibitory effects of selected plant essential oils on the growth of four pathogenic bacteria: *E. coli* O157:H7, *Salmonella Typhimurium*, *Staphylococcus aureus* and *Listeria monocytogenes*. Food Control. 2007;**18**(5):414-420

[52] Borges A, Lopez-Romero JC, Oliveira D, Giaouris E, Simões M. Prevention, removal and inactivation of *Escherichia coli* and *Staphylococcus aureus* biofilms using selected monoterpenes of essential oils. Journal of Applied Microbiology. 2018;**123**(1):104-115

[53] Sanghi DK, Tiwle R. Herbal drugs an emerging tool for novel drug delivery systems. Research Journal of Pharmacy and Technology. 2013;**6**:962-966

Chapter 2

Essential Oil's Chemical Composition and Pharmacological Properties

Jean Baptiste Hzounda Fokou, Pierre Michel Jazet Dongmo and Fabrice Fekam Boyom

Abstract

Essential oil, sent by nature, is a complex mixture of volatile secondary metabolites. Its composition varies along with many parameters that can lead to misunderstanding of its wonderful pharmacological property. In fact, from post-harvest treatment to the compound's identification through extraction approaches, the original chemotype of essential oils can be misdescribed. The pharmacological potentials of these oils are well known in the traditional system since time immemorial. Nowadays, some chemotypes of these oils had shown the effect against WHO's top 10 killer diseases. But the misuses of these essential oils are in part due to the lack of robust and easy analysis strategy that can allow the quality of the essential oils.

Keywords: essential oils, chemotype, Kovats indexes, FTIR, 10 leading death diseases

1. Introduction

Essential oils are a complex mixture of plant volatile compounds. Those compounds are essentially composed of terpenoids and phenolic compounds. The biosynthesis of these flavoring volatile compounds is done in dedicated cell types present in almost all parts of the plant, from the leaves or flower to the roots depending on the plant's genus. These cells are glandular trichome, adduct cavities and osmophores were the compounds are biosynthesized and accumulated [1]. According to the scent of these compounds, the plant that can produce those compounds that when extracted are called essential oils are then called aromatic plants. Aromatic plants are not specific to a given taxonomic group, but they are present widely across the plant kingdom. It must be mentioned that the composition of the essential oils is different from plant taxonomic group to another [2–5]. It should be noted as well that, in the plant, compounds that will later form the essential oils are considered as secondary metabolites with a volatile characteristic. The variation in chemical composition of the essential oil can change from plant to plant even in the same species. These changes in the chemical profile are associated with many factors such as abiotic and biotic factors, postharvest treatment, extraction methods, and conservation conditions.

Abiotic factors group all the nonliving factors that affect the plant's secondary metabolite production. This includes the soil hydrology, pH and salinity, and the climate in general but more interestingly the microclimate in which plant is growing [6–10].

Biotic factors group the living organisms that can affect the plant's metabolite productions. These factors include, in one part, soil organism and microorganism. In fact, secondary metabolites present in essential oils are produced to help the plant to fight against plant invaders, interact with the symbiotic organism, and attract insects for pollination, among others. In another part, biotic factors are inherent in the plant itself.

Postharvest treatment include all the procedure that occurs between the plant collection and essential oil extraction. There are numerous reports that highlight the fact that drying plant material before extraction increases drastically the yield of extraction [5, 11, 12]. The counterpart of this method is the fact that after plant collection, there is a biochemical reaction that occurs between secondary metabolites. The output of this biochemical interactions is the changes in the chemistry of the obtained essential oil in comparison with that originally present in plant. At that point, the balance is between biological activity and the yield [5, 11, 12].

The extraction method can also lead to significant modification of the chemical profile of the essential oil. There are a huge number of approaches that can help to obtain an extract from the aromatic plant, including distillation methods, expression, solvent extraction, enfleurage, and microwave-assisted extraction. Apart from hydrodistillation, microwave assisted, and expression (specifically the citrus pericarp expression), all the other methods lead to a product that is not recognized as essential oil sensu stricto. In fact, with all the other methods, the extract obtained usually contains nonvolatile compounds that are extractable by the process and must be further removed. Also, the hydrodistillation method can lead to the chemical transformation of the compounds such as ester and other compounds that are highly hydrolyzable due to the long stay of the essential oil in water in the extraction system [13–17].

Regarding the conservation method that group all the process between the extraction and chemical analysis, the nature of the oil can lead to some modifications. In fact, most of the compounds that are present in essential oils are unsaturated (contain double bound) and most of them are sensitive to light (photosensitive) and to oxygen (oxidable).

Essential oils are used for their wonderful biological properties. The biological properties include the effect on humans, animals, plants, insects, and microorganisms. In humans, every single part of the human life can be affected by essential oils. In fact, essential oil is employed in nutrition as a food preservative or flavoring, in cosmetics as an odorant part, and in pharmacology as an active ingredient. The essential oil's pharmacological properties comprise their effect on transmissible and non-transmissible diseases.

This chapter aims at going through the mechanism of essential secondary metabolite biosynthesis, essential oil extraction, essential oil chemical profile, and its pharmacological potential against the top list of human killer diseases as presented by WHO.

2. Chemical composition

2.1 Biosynthesis

Terpenoid and phenylpropanoid derivatives are the main components found in essential oils. In most plants, their essential oils contain terpenoids at around 80%. But the presence of phenylpropanoid derivatives affords the essential oils significant flavor, odor, and piquant. These two groups of compounds are derived from two different pathways from different primary metabolites [18].

2.1.1 Biosynthesis of terpenoids

The name terpene was first attributed to the compounds with molecular formula C10H16 by Kekule, a German chemist, due to its abundance in turpentine oil. The derivatives C10H16O and C10H18O were named generically camphor and related to terpenes. Wallach, Kekule's assistant, characterized pinenes, limonene, dipentene, phellandrene, fenchone, terpinolene, and sylvestrene, which at that period were recognized as artifacts for turpentine oil [19]. But nowadays, they are considered as compounds of some essential oils.

Terpenoids are a heterogeneous group of terpenes (compounds with double bonds) and their oxygenated derivative. Sensu stricto, an essential oil terpene, is a group of compounds derived from isoprene. Isoprene in its part is an organic scaffold with 5 carbon units with 1 double bond. Terpenoids are also called isoprenoids.

To start at the beginning, there are two pathways that lead to the synthesis of isoprene. These two pathways occur in the different parts of the upstate specialized cell. In the cytosol, the so-called mevalonic acid pathway is used, while in the plastid, the Rohmer or 1-deoxy-D-xylulose-5-phosphate is used. As this part is the most important part to understand the differences between essential oils' chemistry and thereby their pharmacological properties, let us walk you through some chemistry (**Figure 1**).

According to plant species and foremost to the plant enzymatic ability, there is interconversion between certain compounds. The most observed case is that of conversion of thymol into carvacrol or vice versa. This depends on what plants need and environmental conditions [22, 23]. This leads in some case to a huge difference between chemotype of the same plants collected in different areas or different period of time during the same day or during different maturity stages. This has been reported for *Allium sativum* [24–26] and for *Lavandula angustifolia* [14, 27–30] and *Melissa officinalis* [4, 12, 13, 16, 31–36].

2.1.2 Biosynthesis of phenylpropanoids

The synthesis of this class of compounds in aromatic plant leads to a wide variety of compounds, but in this chapter, we will focus on the pathway that leads to volatile compounds.

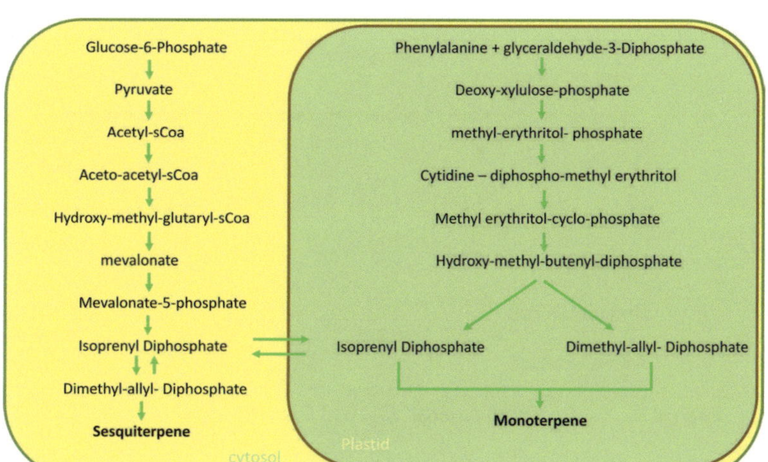

Figure 1.
Biosynthesis of terpenoids summarized from [20, 21].

In comparison to isoprenoids, volatile phenylpropanoid compound synthesis occurs less often. But this is not the case for almost all plants; in fact for clove oil, eugenol is the major compounds [37].

The enzymatic arsenal involved in the synthesis of volatile compounds in the plant is not well known. Therefore, there are many approaches to explain the synthesis of those groups of compounds. But from a metabolite point of view, the starting point of these volatile phenols is the phenyl-alanine that is transformed into cinnamic acid by the phenylalanine amino lyase. The cinnamic acid in turn is transformed into para-COUMARIC acid. This latter compound depending on the enzymatic ability of the plants can undergo two different ways of synthesis (**Figure 2**).

2.2 Essential oil extraction

Essential oil extraction is one of the critical points that can affect the chemical profile of the essential oil. Sensu stricto, essential oils are a volatile odorant complex mixture obtained by distillation. Many techniques have been developed to obtain essential oil such as microwave-assisted extraction, expression, enfleurage, and solvent extraction. Most of these extraction methods lead most of the time to artefactual products as well as transformed products. To better understand, the next paragraphs will present the most used methods and their principal limit in the way of modification of the original chemical profile of the essential oil.

2.2.1 Distillation methods

Distillation methods are a group of methods using steam as compound vector or transporter. In fact, in distillation method, the plant material may be immersed or not in water, and after heating to water boiling point, the impression created in the reactor by steam as well as the high temperature will create the vaporization of these volatile compounds from their stockade cell to the environment of the reactor. The gas is pouched throughout a cooler. The condensation of the water and volatilized compounds from their vapor to water phase form a mixture that can be separated according to their density. There are two varieties of distillation methods: the hydrodistillation and the so-called steam distillation methods [38–40].

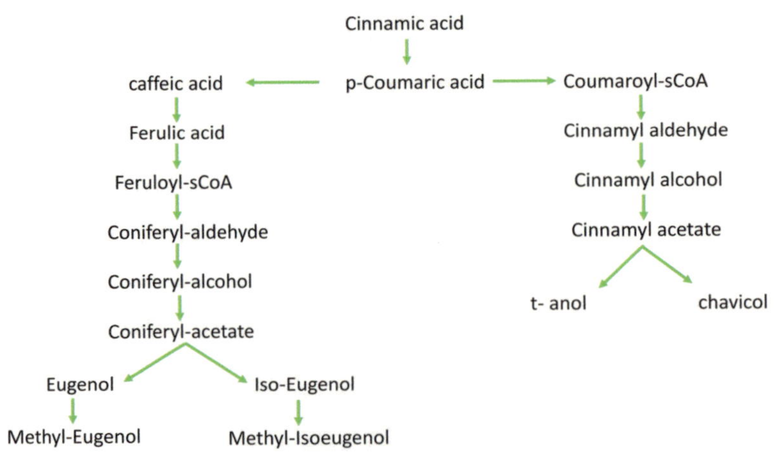

Figure 2.
Biosynthesis of phenylpropanoids summarized from [20].

Hydrodistillation is an essential oil extraction method in which the material is immersed in water, but in the steam distillation, the plant material is not in direct contact with water but will go through, for example, steam flow before entering the cooler.

The principal limits of this method regarding the impact on the chemical profile of the essential oil are as follow:

- The thermosensitive compounds will undergo a transformation or simply degradation. In fact, the high temperature can catalyze some chemical reactions that the normal cell will not, and this will lead to the chemical entities that were not present in the plants.

- Lasting contact with water increases hydrolysis. Esterified compounds are highly sensitive to water as they will be broken down into compounds that do not exist in the essential oil storage cells.

- When the glass Clevenger is used, the essential oil is exposed to light and can, therefore, undergo photo-oxidation and the chemical profile will reveal a high number of oxygenated derivatives that do not really exist in the plant.

Many reports have highlighted these observations and one of them is that of Kamii et al. [14]. In fact when analyzing the chemical profile of *Lavandula angustifolia* extract using two different technics including hydrodistillation, they obtained 25.3% of linalyl acetate, 16.4% of terpene-4-on, 13% of linalool, and 13.6% of ocimene for hydrodistillation and 30.6% of linalyl acetate, 14.1% of terpene-4-on, 8.4% of lavandulyl acetate, 7.3% of β-caryophyllene, and 7.2% of β-farnesene for supercritical fluid extraction as well as lavandulol and phytol that were solely present in that later method. Many other reports are available in the literature for many other plants [1, 15–17].

2.2.2 Microwave-assisted essential oil extraction

Microwave-assisted essential oil extraction is a variant of the distillation method where the heating source has been changed from the normal electric heating cap by the microwave. The plus here is the hypothetic increasing in extraction yield: hypothetic because the increase in yield is not as spectacular as tough [41–43]. It is true that it is better to crush the plant material, but in comparison to the classic distillation method, the essential oil yield is systematically the same. The principle of this method is based on the change of the polarity of water by the waves and of course the heating that will play the same role as in classic distillation method. This method has in addition the limit of the normal distillation method, the fact that the microwave can lead to chemical stereo switching from one isomer to another.

2.2.3 Expression

This method is specific for citrus pericarp. In fact, to avoid the thermal destruction, cold pressing of the citrus fruit rinds as the essential oil is store at that part of the plant has led to a good quality oil. The limit of this technique is the fact that it is not applicable to other parts of the plant [38–40].

2.2.4 Enfleurage

Enfleurage is an old technique of essential aroma extraction. It is based on the solubilization of the essential oil's component on a greasy wax. When using cold extraction, it can take more than a month, but this can be reduced to a few hours by heating. The aromatized wax is then called concrete. This later undergoes solubilization in a polar solvent and then partitioned with absolute ethanol. The product obtained after this process is at most a part of the essential oil as present in plant, and it contains many other terpenes that can be solubilized in fats used; that is, while at the end, the product is not called essential oil but absolute.

2.2.5 Solvent extraction

This is the oldest method for obtaining crude extract from plants. The principle is based on the solubilization of the compounds in the cells by the solvent. This method also has two variants as it can be run in room temperature or in high temperature. But no matter whether the extraction was performed in room temperature or not, the solvent will be separated by Rotavapor regarding the volatility of the solvent. This method does not really lead to obtaining the essential oil as all the nonvolatile compounds are also extracted by this approach. Therefore, the chemical profile will not be that of the volatile fraction of the plants but for the compound that is soluble in the solvent used in the process [13–17].

2.2.6 Supercritical fluid extraction

This method is the most modern and sophisticated. It uses gases at their supercritical stage. The gas at supercritical stage is liquid due to the high pressure applied to it. Many solvents can be used as the method brings the solvent at its temperature and pressure above its thermodynamic critical point, but the most used is CO_2 for the reason that it needs less pressure to be liquified, it is less reactive than other, it is noninflammable, it is nontoxic and available at low cost with high purity, and most importantly it can be removed from the plant material using just the press release. This method is based on the fact that gas at the supercritical state can enter throughout the plant material like a gas and dissolve component like a liquid. After the extraction procedure, the essential oils compounds are mixed with the supercritical fluid (in liquid form). The separation is performed by reducing temperature and increasing the pressure up to room conditions [13–17]. The principal limit of this method is the complexity of the system.

2.3 Method for chemical analysis

There is a myriad of technics and methods for essential oil chemical profiling. All the methods used in organic chemistry can be used here. Due to their volatility nature, the compounds that constitute essential oil are preferably analyzed by gas chromatography (GC). Gas chromatography alone does not provide enough data for good chemical proofing. Therefore, many other analytical tools have been used such as mass spectrometry (MS), infrared spectroscopy (IR), and nuclear magnetic resonance (NMR). As well, many technics have been used to make the GC a better tool for chemical profiling, and these include chiral selective GC and multidimensional GC. Advances in liquid chromatography have highlighted the usefulness of high-pressure liquid chromatography (HPLC) as a tool for essential oil analysis. Many variants are available to date as multidimensional HPLC, HPLC-MS, and HPLC-GC. The following section will walk you through gas chromatography and gas chromatography coupled with mass spectrometry for their popularity and the Fourier transform infrared spectroscopy for its simplicity and for being environment-friendly and long-term cost-effectiveness.

2.3.1 Gas chromatography and variant in essential oil analysis

Gas chromatography is an old essential oil chemical profiling method. As all the chromatographic methods, it is based on the single compound separation from a complex matrix between two phases regarding their affinity between the phases, their shape, and mass. The mobile phase is the one which moves by capillarity or pressure transporting compounds faster or slower toward the fixed phase. In GC, the mobile phase is a gas and the fixed phase is a solid at room temperature, but at a high temperature, this phase will be slightly melted but will remain fixed on the column.

For analysis, the essential oil is mixed with a solvent (mostly hexane or pentane) and introduced into the injector room set at 200°C with the help of a syringe. The mixture is carried by the mobile phase into the column. Here the temperature drops down to about 50°C. All the compounds settle down at different distance in the column. This column itself is placed in an oven set with gradient increasing temperature (1–5°C/min). Each compound has it vaporization point and therefore when the oven temperature will reach this point, it will be volatilized thereby take off from the stationary (fixed) phase and carry and spit off out of the column.

Usually, the column is connected to a detector, which in the case of essential oil analysis is the flame ionization detector (FID). In this case, when the compound comes out of the column, it is burned and the increasing temperature in the flame, proportional to the compound scaffold, is detected and transformed into a peak. The compound identification is done by using the retention time and/or the Kovats index. The following formula as presented in Adams is used to calculate the Kovats index.

$$KI_{(x)} = 100\,P_z + 100\left(\frac{logRT_x - logRTP_z}{logRTP_z - logRTP_{z+1}}\right) \quad (1)$$

$KI_{(x)}$ is the Kovats indices of compounds x, $P_{(z)}$ is the paraffin with z carbon atoms, $P_{(z+1)}$ is the paraffin with $z + 1$ carbon atom, and RT is the retention time.

The linearization of this formula presented in Adams is as follows:

$$AI_{(x)} = 100\,P_z + 100\left(\frac{RT_x - RTP_z}{RTP_z - RTP_{z+1}}\right) \quad (2)$$

$AI_{(x)}$ is the arithmetic index of compounds x, $P_{(z)}$ is the paraffin with z carbon atom, $P_{(z+1)}$ is the paraffin with the $z + 1$ carbon atom, and RT is the retention time.

After calculation of index, the value is compared with that of the online libraries or the Adams library [45].

When the GC is connected to MS, the compound is accelerated and ionized into the different compartment of its scaffold and detected as bands.

Table 1 summarizes few retention or Kovats indexes reported in the literature for randomly selected plant and randomly selected papers. The diversity of chemotype obtained per essential oil is interesting. But the question about the accuracy of this profiling method has come up. Moreover, the Internet-based data bank proposes a huge range of retention indexes for the same compounds. For illustration mater, let us take the eucalyptol also called cineol-1,8. In the NIST database in the slightly polar column, here dimethyl-silicone with 5% phenyl group (DIMS 5P), this compound have a retention indexes ranging from 1021 to 1044 [44]. This range represents 23 units in the retention indexes. Taking into consideration the Kovats or Van Den Dool and Kartz formula, this range (8 min 57 s to 9 min 44 s) represents a gap of 47 s. For the same column, in terms of polarity, in Adams Database, during this period, 24 single compounds can be identified [45]. More complexly, there are overlappings in

Retention indexes								
Allium sativum								
Compounds	AP	SAPC	PC	[26]	[25]	[24]	[85]	[86]
Diallyl sulfide	835–872	831–872	1118–1177	855	784	—	846	854
Allyl methyl disulfide	891–915	887–928	1241–1322	916	918	—	908	915
Diallyl disulfide	1054–1078	1048–1095	1436–1526	1080	1111	1085	1079	1084
Allyl methyl trisulfide	1100–1132	1123–1165	1587–1605	1138	1116	1145	1126	1131
Diallyl trisulfide	1266–1292	1277–1320	1775–1822	1301	1207	1311	1301	1305
Lavandula angustifolia								
Compounds	AP	SAPC	PC	[5]	[28]	[29]	[30]	[14]
Ocimene	1027–1050	1028–1047	1211–1251	1030	—	1043	—	1043
Linalool	1074–1098	1088–1109	1507–1564	1092	—	1107	1100	1101
Terpineol	1148–1180	1178–1203	1655–1687	1167	1179	—	—	1181
Linalyl acetate	1234–1254	1238–1268	1532–1570	1246	—	1256	—	1258
Borneol	1134–1172	1152–1177	1653–1717	1155	1167	1170	1144	1169
Melissa officinalis								
Compounds	AP	SAPC	PC	[4]	[13]	[16]	[12]	[33]
Neral	1211–1240	1231–1269	1641–1706	—	—	1246	1245	1240
Nerol	1206–1239	1216–1250	1752–1832	—	—	1228	—	—
Geranial	1236–1260	1252–1291	1680–1750	—	—	1278	1271	1270
Geraniol	1231–1256	1238–1269	1795–1865	1185	1252	1258	—	—
Caryophyllene	1400–1442	1405–1440	1569–1632	—	1417	1405	1421	1418
Germacrene D	1458–1491	1446–1493	1676–1726	—	1484	1480	1489	1480
Cadinene	1506–1542	1503–1541	1734–1803	—	1523	1538	1514	—
Ocimum basilicum								
Compounds	AP	SAPC	PC	[87]	[88]	[89]	[90]	[17]
Eucalyptol	1013–1039	1021–1044	1186–1231	1031	1031	1027	1029	1026
Ocimene	1027–1050	1028–1047	1211–1251	1038	1045	—	1048	1034
Linalool	1074–1098	1088–1109	1507–1564	1111	1120	1085	1116	1095
Eugenol	1323–1372	1345–1375	2100–2198	—	—	1377	1376	1330
Humulene	1439–1459	1436–1456	1637–1689	1447	1455	1430	1459	1445
Cadinene	1506–1542	1503–1541	1734–1803	1519	—	1497	—	1505
Peper nigrum								
Compounds	AP	SAPC	PC	[3]	[89]	[91]	[92]	[15]
a-Pinene	924–951	921–944	1008–1039	938	930	931	931	939
b-Phellandrene	1005–1036	995–1013	1148–1186	1040	978	1024	1021	1003
g-Terpinene	1035–1062	1049–1069	1222–1266	—	1049	1051	1050	1060
a-Cubebene	1345–1359	1438–1480	1438–1480	1357	—	1377	1352	1351
Farnesene	1484–1509	1488–1493	1627–1668	1455	1518	1453	1445	1443
b-Caryophyllene	14,000–1442	1405–1440	1569–1632	1437	1395	1419	1419	1419

Retention indexes								
Rosmarinus officinalis								
Compounds	AP	SAPC	PC	[6]	[7]	[8]	[10]	[93]
a-Pinene	924–951	921–944	1008–1039	931	936	1075	961	939
Myrcene	975–991	980–995	1140–1175	980	991	1174	991	991
Camphene	936–965	936–959	1043–1086	944	952	1102	943	954
Eucalyptol	1013–1039	1021–1044	1186–1231	1021	1031	1221	1032	1031
Camphor	1106–1153	1127–1155	1481–1537	1122	1148	1547	1088	1144
Verbenone	1167–1198	1190–1224	1696–1735	1183	1209	—	1119	1207
Bornyl acetate	1259–1209	1264–1297	1549–1597	1272	1292	1612	1277	—
Saliva officinale								
Compounds	AP	SAPC	PC	[89]	[36]	[9]	[94]	[95]
Myrcene	975–991	980–995	1140–1175	981	988	—	980	992
a-Thujone	1076–1104	1099–1117	1385–1441	1104	1101	1105	—	1102
Camphor	1106–1153	1127–1155	1481–1537	1122	1141	1143	1108	1142
Eucalyptol	1013–1039	1021–1044	1186–1231	1027	1026	1034	1191	1032
b-Caryophyllene	1400–1442	1405–1440	1569–1632	1395	1417	1418	—	1418
Humulene	1439–1459	1436–1456	1637–1689	1430	1452	—	1430	1454
Viridiflorol	1561–1598	1569–1604	2041–2110	—	—	—	1587	—
Syzygium aromaticum								
Compounds	AP	SAPC	PC	[96]	[97]	[98]	[99]	[89]
Eugenol	1323–1372	1345–1375	2100–2198	1392	1353	1370	1354	2098
b-Caryophyllene	14,000–1442	1405–1440	1569–1632	1458	1428	1426	1421	—
Eugenyl acetate	1472–1493	1514–1531	2252–2277	1552	1538	—	1522	2107
Humulene	1439–1459	1436–1456	1637–1689	1579	—	1460	1455	—
Thymus vulgaris								
Compounds	AP	SAPC	PC	[22]	[100]	[23]	[98]	[101]
a-Terpinene	1001–1024	1007–1026	1154–1195	—	1010	1016	1019	1019
p-Cymene	1004–1029	1011–1033	1246–1291	1029	1014	1024	1026	1033
Limonene	1012–1038	1019–1039	1178–1219	—	1022	1028	1031	—
Terpineol	1148–1180	1178–1230	1655–1687	—	1162	1194	1190	1177
g-Terpinene	1035–1062	1049–1069	1222–1266	1064	1049	1057	1062	1060
Linalool	1074–1098	1088–1109	1507–1564	—	1084	1099	1100	1107
Thymol	1260–1289	1272–1304	2100–2205	—	—	1288	1296	1315
Carvacrol	1272–1300	1291–1344	2140–2246	1308	1286	1296	1305	—
b-Caryophyllene	14,000–1442	1405–1440	1569–1632	—	—	1419	1419	1423
Zingiber officinale								
Compounds	APC	SPC	PC	[102]	[103]	[104]	[105]	[106]
a-Pinene	1001–1024	1007–1026	1154–1195	939	—	926	943	935
Camphene	936–965	936–959	1043–1086	953	—	944	954	950
Linalool	1074–1098	1088–1109	1507–1564	—	1112	—	—	1103

Retention indexes								
Eucalyptol	1323–1372	1345–1375	2100–2198	1033	1060	1015	1027	1032
Neral	1211–1240	1231–1269	1641–1706	—	1265	1249	1227	1247
Geranial	1236–1260	1252–1291	1680–1750	—	1292	—	1252	1379
Germacrene D	1458–1491	1446–1493	1676–1726	1481	1532	—	1469	—
b-Farnesene	14,000–1442	1405–1440	1569–1632	—	1518	—	—	1458
Zingiberene	1463–1494	1485–1509	1696–1743	1495	1521	1492	1487	1508

APC, apolar column; SPC, slightly polar column; PC, polar column.

Table 1.
Retention indexes of important compounds from 10 randomly selected essential oils.

the range proposed by the NIST as reported by Babushok [44] as it can be observed in **Table 1**; for the same DIMS5P column, α pinene (921–944) is overlapping with camphene (936–959) and eucalyptol (1021–1044) is overlapping with ocimene (1028–1047) and camphor (1127–1155). These observations are the real pitfalls of these identification approaches and can lead to misuses of the valuable data from those data banks. In fact, these overlappings lead to compound identification oriented to the chemotype available in the literature rather than the real chemotype that is analyzed. And the previous paragraph had highlighted that the chemical profile thereby the pharmacological profile can change because of a myriad of factors. All these lead to questioning the robustness of this actual gold essential oil chemical profiling strategy not in the view of the provided database or analytic tools but in the view of data interpretation by the users.

2.3.2 Infrared spectroscopy and chemometric analysis

The idea besides the spectroscopy analysis is the fact that, in the electromagnetic radiation, there is a range of a tiny part of visible light that exists as a wave. This light moves in a straight line if the part of this light is not reflected, refracted, or absorbed by the matter. In biology and chemistry, this technique is used to produce an infrared spectrum in the case of infrared spectroscopy by passing an infrared radiation through a given sample knowing that the atom constituting the sample will lead to absorption of a part of the light energy. When a part of the molecule present in the sample absorbs this energy, it will become unstable and will react by twisting, stretching, bending, rocking, wagging, or scissoring depending on the bond linking these atoms with the rest of the molecule. The energy absorbed, the different reactions of the atom lead to the appearance of a specific peak on the spectrum [46–49].

The spectrum can be at least qualitatively interpreted without prior or additional chemometric algorithm as it provides a high level of specific information on the molecular aspect of the essential oil [50]. In fact, the spectrum regions are known and correspond to established characteristic group absorbances [47] and therefore the assignment can lead to specific chemical identification (**Figure 3**). But this is when the sample is a single compound. When the mixture is analyzed, the peak cannot be simply attributed to the compounds as it is most of the time the result of the overlapping small vibration of different closely related groups from a different molecule in the mixture.

For the quality purpose, the fingerprint region is most important. In fact, the fingerprint region is a unique vibrational signature that a given essential oil can have. It is as specific as the fingerprint is specific to a human. But to be more

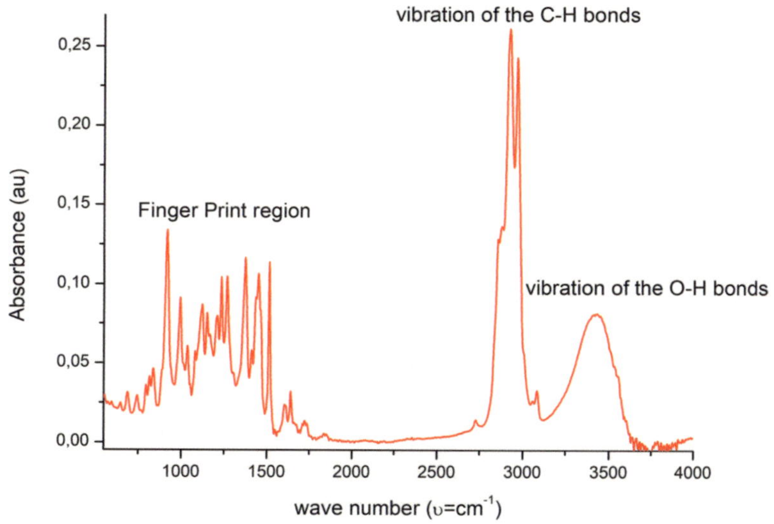

Figure 3.
Infrared spectrum of Ocimum basilicum essential oil obtained from Fourier transform infrared spectroscopy.

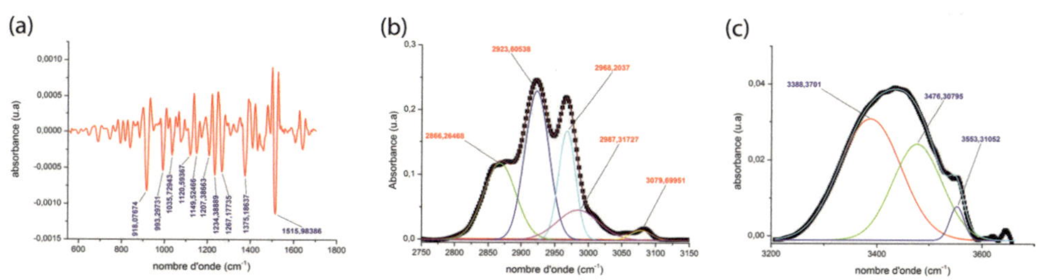

Figure 4.
Spectral processing of fingerprint (a), C-H (b) and O-H (c) regions of Ocimum basilicum essential oil's FTIR spectrum [51].

efficient in the analysis of the area, chemometric tool should be used such as derivation. This approach is a simple application of the derivate formula to the equation of the fingerprint region or a given region. This technique provides with key benefits. In fact, it enhances the resolution for the first derivative and the second derivative gives negative peaks of each band in the examined region [47] an example from **Figure 3** is given in **Figure 4A**. It is also possible to apply other data processing such as baseline correction, smoothing, or curve fitting [47].

Using this approach, an increasing number of authors have succeeded in discriminating the essential oil of various plants based on their chemotype [46, 51–56]. Those authors have suggested that for a known essential oil, the fingerprint of the oil should be used as a tool for rapid essential oil quality control.

3. Pharmacological property of essential oils

This part of the chapter is huge and covers various aspects of human and animal care. It will be easier to find a disease that no essential oil as a whole or its component can exert any effect. Essential oils are used to manage even diseases that are not scientifically measurable these days. These include protecting against misfortune and bad spirit and protecting the soul after death. This aspect of the therapeutic use of the essential oil will not be described here. The following paragraphs will be devoted to the review of essential oil use against top 10 human killers. Regarding the classification of 10 global causes of death around the world as published by WHO [57], ischemic heart disease, stroke, chronic obstructive pulmonary disease, lung cancer, diabetes, lower respiratory infection, diarrheal disease, tuberculosis, HIV/AIDS, and road injuries were respectively the most important human killers in 2016. Unfortunately, this chapter could not review the potential of essential oils against road accident.

3.1 Essential oil against ischemic heart disease and stroke

During the last 18 years, ischemic heart disease had won the gold medal in killing a human. In the second position with silver medal comes the stroke. In 2016, these affections were incriminated in 15.2 million human death [57]. This is clear, known, and localizable but a silent terrorist against which all efforts should be focused on. These two affections share the same origin but different target. In fact, they are both cardiovascular diseases. Cardiovascular disease is known as a disease that affects the heart and blood vessels of the circulatory system [58]. In this case, ischemic heart disease and stroke are specific to blood vessels. Ischemic heart disease is a class of cardiovascular disease that appears on the vessels that supply the heart cells with blood in terms of nourishing rather than pumping for other organs [58]. Stroke on its part can be defined as neurological shortfall caused by crucial injury of the central nervous system. The reason behind is the cerebral infarction or intracerebral hemorrhage. Ischemic stroke accounts for 80% [59] of all types of stroke this chapter will focus on these later on.

Death happens when cells of these two vital organs are not supplied with nutrients and oxygen and are not free from their metabolic wastes. The main cause is the obstruction of the vessels by a clot (thrombus) from venal or upper artery atherosclerosis in coronary (in the heart) or arterioles (in the brain) and/or atherosclerosis on these two vessels [58, 59]. It is therefore clear that targeting specific clot dissolution or arteriosclerosis can help to manage ischemic heart and stroke disease.

In that regards, a massive number of essential oils have been studied for their effect on cardiovascular diseases. These reports were previously reviewed by Grenier et al. [60] for the vasodilatation effect. The major compounds from these essential oils were eucalyptol (*Croton nepetaefolius*); piperitenone oxide (*Mentha* × *Villosa*); terpineol (*Alpinia zerumbet*); eugenol and methyl eugenol (*Ocimum gratissimum* and *Aniba canelilla*); anethole and estragole (*Croton zehntneri*); α pinene, caryophyllene, and eucalyptol (*Hyptis fruticose*); geraniol, citronellol, and citronellal (*Cymbopogon winterianus*); and many other essential oils containing linalool, bisabolol, and β pinene [60]. For the anti-arteriosclerotic effect, these same authors reviewed 17 essential oils that have been reported for their anti-arteriosclerotic effect both *in vitro* and *in vivo*. These plants include *Allium sativum*, *Ocimum sanctum*, *Melissa officinalis*, and *Lavandula angustifolia*.

3.2 Chronic obstructive pulmonary disease and lower respiratory infection other than tuberculosis

Chronic obstructive pulmonary disease and lower respiratory infection other than tuberculosis are respectively the third and the fourth leading causes of death worldwide. These diseases caused cumulatively 6 million deaths in 2016. The lower respiratory infection disease owns the gold medal in communicable disease. It is important to notice that, in the developing world, the lower respiratory infectious disease is the first leading cause of death [57].

The chronic obstructive pulmonary disease (COPD) can be defined as a progressive respiratory condition clinically characterized by dyspnea, cough, and sputum production. This is the manifestation of alveolar enlargement and destruction as well as inflammation of lung parenchyma airways [61]. *In fine*, this disease can be categorized as the inflammation of the lung. Essential oils or their components have been reported to have a positive effect on lung inflammation. In fact, eucalyptol, a main component of eucalyptus essential oil, has shown evidence on the resolution of pulmonary inflammation in humans [62, 63] and in rats [64]; myrtle, eucalyptus, and citrus essential oils have significantly reduced the inflammatory response *ex vivo* [65].

Lower respiratory infection disease is mainly caused by microorganisms including bacteria (other than Mycobacterium), fungi and viruses. A myriad of reports has proved the potential of essential oils as a possible source or as a drug against pathogens causing respiratory infection. Essential oils from *Pimpinella anisum*, *Foeniculum vulgare*, *Eucalyptus globulus*, *Mentha piperita*, *Melaleuca alternifolia*, and *Thymus vulgaris* were reviewed by Horvath and Acs from European Pharmacopeia [66] and had an effect *in vitro*, *in vivo,* and on a human. *Syzygium aromaticum*, *Cinnamomum zeylanicum* Nees., Batch, *Eucalyptus globulus*, *Thymus vulgaris* L., *Pinus sylvestris*, and *Mentha piperita* have been reported for their inhibitory effect on bacteria that cause lower and upper respiratory infections [67].

3.3 Lung cancer

Lung cancer is the fifth leading cause of death worldwide, and in 2016, WHO estimated that 1.7 million persons were dying from lung cancer. It is established that there is an urgent need for anticancer drugs due to the high selectivity that anticancer drug should possess. There are an important number of publications in the literature on the anticancer effect of the essential oils. More specifically, a certain number of essential oils have been claimed to possess anticancer effect against lung cell line; all these experiments are as per now *in vitro* or in an animal model [68–70].

3.4 Diabetes

Diabetes is the sixth leading cause of death worldwide. In 2016, this affection caused 1.6 million deaths with an increase of 0.6 million compared to 2000. As the number of affected persons is increasing, this means that the management method is not that efficient, and alternatives are needed. Essential oils have been analyzed for their antidiabetic effect. *Lavandula angustifolia*, *Melissa officinalis*, *Ocimum basilicum*, *Peper nigrum*, *Rosmarinus officinalis*, *Saliva officinale*, and *Thymus vulgaris* were analyzed of their stimulation of glucose consumption. And *Melissa officinalis* essential oils present better effect compared to insulin [71]; moreover, *in vivo* studies have also been reported [72–76].

3.5 Diarrheal disease

Diarrhea is a disease caused by infection of the gastrointestinal tract by bacteria, fungi, virus, and ameba. The manifestation includes the urination of more than 3 liquid stools per day. The epidemiology of diarrheal disease has decreased from 200 with 2.4 million death cases to 1.4 million cases in 2016 [57]. This reduction is mostly made by hygiene management. Essential oils have been analyzed for their beneficial effect in the therapeutic management of diarrheal infection [77–80]. Some components as linalool, eucalyptol, terpineol, geraniol, thymol, carvacrol, spathulenol, caryophyllene, elemene, viridiflorol, bisabolol, eugenol, t-anetol, cinnamaldehyde, and allyl isothiocyanate were used to optimize the effect of the available drugs against some of the pathogens with successful results [78].

3.6 Tuberculosis

Tuberculosis is the infection of the lung by *Mycobacterium tuberculosis*, and it caused 1.3 million deaths in 2016. Many drugs are available, and control by quarantining the affected person had caused a reduction in this affection during the last 19 years [57]. Essential oils can also play an important role in the management of this affection. In fact, *Salvia aratocensis*, *Lippia Americana*, *Turnera diffusa* [81], *Cuminum cyminum*, *Eugenia caryophyllata*, *Cinnamomum verum*, *Laurus nobilis*, *Pimpinella anisum* [82], and *Hyptis suaveolens* [83], among others have been analyses successfully for their effect against *Mycobacterium tuberculosis*.

3.7 HIV/AIDS

HIV/AIDS is the acquired immunodeficiency syndrome, caused by human immunodeficiency virus. The disease is the ninth more important terrorist worldwide, killing in 2016 1 million persons [57]. The mortality due to this virus has decreased along the years due to the generations of antiretrovirals as well as the protective measures adopted by the populations. There are only a few publications available on the effect of the essential oils on this virus. The reason can be the fact that this virus is very dangerous, and the testing equipment must be sophisticated to allow this test. However, *Ridolfia segetum* and *Oenanthe crocata* have been analyzed successfully for their effect on RNA-dependent DNA polymerase and ribonuclease H [84].

4. Conclusion

Essential oils are the oil of nature, sent by nature, the spirit of nature, and the soul of the forest. Their scent depends on their composition that in turn depends on the enzymatic package of the plant species. This enzymatic package can be influenced by external factors that can be biotic, depending on a living organism, or abiotic depending on molecules or climate and geography. The pharmacological potential of this spirit of the forest is a fact. But the scientific proof of that fact is subject to misunderstanding or misinterpretation. That may be why it is difficult to have essential oil recommended as an official drug for a disease. In fact, there are many variations that could affect the chemical composition not in the plant but during essential oil processes from plant collection time to the analytic method and the data interpretation. As well, Aromatic plants containing these scent of the nature are mostly present in tropical part of the word, and most of those tropical countries are developing countries and as such they do not have robust technology equipment to test and prove the whole efficiency of that essential oils, spirit of nature. To solve these small but crucial issues, scientists should come along with a robust analytical tool (FTIR, for example) and standard postharvest and extraction protocols for standardization of all essential oils not only in few local pharmacopeias but on a world level.

Conflict of interest

The authors declare no conflict of interest.

Author details

Jean Baptiste Hzounda Fokou[1*], Pierre Michel Jazet Dongmo[2]
and Fabrice Fekam Boyom[3]

1 Antimicrobial and Biocontrol Agent Unit, Department of Biochemistry, University of Yaoundé I, Yaounde, Cameroon

2 Department of Biochemistry University Douala, Douala, Cameroon

3 Anti-Microbial and Biocontrol Agent Unit, University of Yaounde I, Yaounde, Cameroon

*Address all correspondence to: hzoundafokou@gmail.com

References

[1] Rehman R, Asif Hanif M. Biosynthetic Factories of Essential Oils: The Aromatic Plants. Natural Products Chemistry & Research. 2016;**04**(04):227

[2] Tchoumbougnang F et al. Activité larvicide sur Anopheles gambiae Giles et composition chimique des huiles essentielles extraites de quatre plantes cultivées au Cameroun. Biotechnologie, Agronomie, Société et Environnement. 2009;**13**(1):77-84

[3] Sahari IS, Assim ZBIN, Ahmad FB, Jusoh IBIN. Essential oils from fresh fruits, fruit stalks and stem barks of four Piper nigrum varieties from Sarawak. Borneo Journal of Resource Science and Technology. 2013;**3**(1):43-51

[4] Mahmoudi R, Amini K, Hosseinirad H, Valizadeh S, Kabudari A, Aali E. Phytochemistry and insecticidal effect of different parts of *Melissa officinalis* on *Tetranychus urticae*. Research Journal of Pharmacognosy. 2017;**4**(4):49-56

[5] Smigielski K, Prusinowska R, Stobiecka A. Biological properties and chemical composition of essential oils from flowers and aerial parts of lavender (*Lavandula angustifolia*). Journal of Essential Oil-Bearing Plants. 2018;**21**(5):1303-1314

[6] Pintore G et al. Chemical composition and antimicrobial activity of *Rosmarinus officinalis* L . oils from Sardinia and Corsica. Flavour and Fragrance Journal. 2002;**17**:15-19

[7] Ngioni ALA, Arra ANB, Ereti ELC, Arile DAB, Sanita D, Porcell V. Chemical composition, plant genetic differences, antimicrobial and antifungal activity investigation of the essential oil of *Rosmarinus officinalis* L. Journal of Agricultural and Food Chemistry. 2004;**52**:8-10

[8] Hcini K, Sotomayor JA, Jordan MJ, Bouzid S. Chemical composition of the essential oil of rosemary (*Rosmarinus officinalis* L.) of Tunisian origin. Asian Journal of Chemistry. 2013;**25**(5):2601-2603

[9] Russo A et al. Chemical composition and anticancer activity of essential oils of Mediterranean sage (*Salvia officinalis* L.) grown in different environmental conditions. Food and Chemical Toxicology. 2013;**55**:42-47

[10] Asressu KH, Tesema TK. Chemical and antimicrobial investigations on essential oil of *Rosmarinus officinalis* leaves grown in Ethiopia and comparison with other countries. Journal of Applied Pharmacy. 2014;**6**(2):132-142

[11] SádeCká J. Influence of two sterilisation ways, gamma-irradiation and heat treatment, on the volatiles of black pepper. Czech Journal of Food Sciences. 2010;**28**(1):44-52

[12] Mirahmadi S, Norouzi R, Nohooji MG. The influence of drying treatments on the essential oil content and composition of *Melissa officinalis* L. compared with the fresh. Journal of Medicinal Plants. 2017;**16**:61

[13] Rehman S, Latief R, Bhat KA, Khuroo MA, Shawl AS, Chandra S. Comparative analysis of the aroma chemicals of *Melissa officinalis* using hydrodistillation and HS-SPME techniques. Arabian Journal of Chemistry. 2013;**10**(2):S2485-S2490

[14] Kamiie Y, Sagisaka M, Nagaki M. Essential oil composition of *Lavandula angustifolia* 'Hidcote': Comparison of hydrodistillation and supercritical fluid extraction methods. Transactions of the Materials Research Society of Japan. 2014;**39**(4):485-489

[15] Song G, Hu Y. GC–MS analysis of the essential oils of *Piper nigrum* L. and

Piper longum L. Chromataographia. 2014;**66**:785-790

[16] Abdellatif F, Hassani A. Chemical composition of the essential oils from leaves of *Melissa officinalis* extracted by hydrodistillation, steam distillation, organic solvent and microwave hydrodistillation. Journal of Materials and Environmental Science. 2015;**6**(1):207-213

[17] Chenni M, El Abed D, Rakotomanomana N, Fernandez X, Chemat F. Comparative study of essential oils extracted from microwave extraction. Molecules. 2016;**21**:113-129

[18] Sangwan NS, Farooqi AHA, Shabih F, Sangwan RS. Regulation of essential oil production in plants. Plant Growth Regulation. 2001;**34**(1):3-21

[19] Husnu can Baser K, Buchbauer G. Handbook of Essential Oils: Sciences Technologies and Applications. New york: Taylor & Francis Web; 2010

[20] Lange BM, Rujan T, Martin W, Croteau R. Isoprenoid biosynthesis: The evolution of two ancient and distinct pathways across genomes. Proceedings of the National Academy of Sciences. 2000;**97**(24):13172-13177

[21] Akhila A. Metabolic engineering of biosynthetic pathways leading to isoprenoids: Mono- and sesquiterpenes in plastids and cytosol. Journal of Plant Interactions. 2007;**2**(4):195-204

[22] Zantar S et al. Effect of harvest time on yield, chemical composition, antimicrobial and antioxidant activities of *Thymus vulgaris* and *Mentha pulegium* essential oils. European Journal of Medicinal Plants. 2015;**8**(2):69-77

[23] Satyal P, Murray BL, Mcfeeters RL, Setzer WN. Essential oil characterization of *Thymus vulgaris* from various geographical locations. Food. 2016;**5**(70):1-12

[24] Mallet ACT et al. Chemical characterization of the *Allium sativum* and *Origanum vulgare* essential oils and their inhibition effect on the growth of some food pathogens. Revista Brasileira de Plantas Medicinais. 2014;**16**(4):804-811

[25] Boubechiche Z, Chihib N, Jama C, Hellal A. Comparison of volatile compounds profile and antioxydant activity of *Allium sativum* essential oils extracted using hydrodistillation, ultrasound-assisted and sono-hydrodistillation processes. Indian Journal of Pharmaceutical Education and Research. 2017;**51**(3):281-285

[26] Satyal P, Craft JD, Dosoky NS, Setzer WN. The chemical compositions of the volatile oils of garlic (*Allium sativum*) and wild garlic (*Allium vineale*). Food. 2017;**6**(63):1-10

[27] Verma RAMS, Rahman LU, Chanotiya CS. Essential oil composition of *Lavandula angustifolia* Mill. Available online at www.shd.org.rs/JSCS/. Journal of the Serbian Chemical Society. 2010;**75**(3):343-348

[28] Mantovani ALL et al. Chemical composition , antischistosomal and cytotoxic effects of the essential oil of *Lavandula angustifolia* grown in Southeastern Brazil. Revista Brasileira de Farmácia. 2013;**23**:877-884

[29] Hamad KJ, Al-Shaheen SJA, Kaskoos RA, Ahamad J, Jameel M, Mir SR. Essential oil composition and antioxidant activity of lavandula angustifolia from Iraq. International Journal of Pharmaceutics. 2013;**4**(4):117-120

[30] Adaszyńska-skwirzyńska M, Swarcewicz M, Dobrowolska A. The potential of use lavender from vegetable waste as effective antibacterial and sedative agents. Medicinal Chemistry. 2014;**4**(11):734-737

[31] De Sousa AC, Alviano DS, Blank AF. *Melissa officinalis* L. essential oil: Antitumoral and antioxidant activities. The Journal of Pharmacy and Pharmacology. 2004;**5**(2002):677-681

[32] Jalal Z, El Atki Y, Lyoussi B, Abdellaoui A. Phytochemistry of the essential oil of *Melissa officinalis* L. growing wild in Morocco: Preventive approach against nosocomial infections. Asian Pacific Journal of Tropical Biomedicine. 2015;**5**(6):458-461

[33] De Menezes CP, Queiroga F, Guerra S, Pinheiro LS, Trajano VN, Pereira FDO. Investigation of *Melissa officinalis* L. essential oil for antifungal activity against *Cladosporium carrionii*. International Journal of Tropical Disease & Health. 2015;**8**(2):49-56

[34] Popova A, Mihaylova D, Hristova I, Alexieva I. *Melissa officinalis* L.—GC profile and antioxidant activity. International Journal of Pharmacognosy and Phytochemical Research. 2016;**8**(4): 634-638

[35] Efremov AA, Zykova ID, Gorbachev AE. Composition of the essential oil from the lemon balm growing in the neighborhood of krasnoyarsk as indicated by gas chromatography – mass spectrometry data. Russian Journal of Bioorganic Chemistry. 2016;**42**(7):726-729

[36] Couladis M, Koutsaviti A. Chemical composition of the essential of saliva officinalsm saliva fructicosam Melissa officinalis and their infusions. Ratarstvo i Povrtarstvo. 2017;**54**(1):36-41

[37] Bakkali F, Averbeck S, Averbeck D, Idaomar M. Biological effects of essential oils nigelle.pdf. Foud and Chemical Toxicology. 2008;**46**:446-475

[38] Wasserscheid AJ, Wasserscheid P. A brief review on essential oil extraction and equipment. Chemical Technology. 2010;**5**(1):19-24

[39] Rassem HHA, Nour AH, Yunus RM. Techniques for extraction of essential oils from plants: A review. Australian Journal of Basic and Applied Sciences. 2016;**10**(1016):117-127

[40] Khan MF, Dwivedi AK. A review on techniques available for the extraction of essential oils from various plants. International Research Journal of Engineering and Technology. 2018;**5**(5):5-8

[41] Fadel O, Ghazi Z, Mouni L, Benchat N, Ramdani M, Amhamdi H. Comparison of microwave-assisted hydrodistillation and traditional hydrodistillation methods for the *Rosmarinus eriocalyx* essential oils from Eastern Morocco. Journal of Materials and Environmental Science. 2011;**2**(2):112-117

[42] Norfatirah M, Tajuddin S, Chemat F, Rajan J, Yusoff M. Comparison of microwave-assisted extraction and hydrodistilation method in the extraction of essential oils from *Aquilaria malaccensis* (Agarwood) oil. In: Proceedings of the ICNP. Vol. 4. 2013. p. 227. DOI: 10.2174/2210289201304010227

[43] Jeyaratnam N, Nour AH, Akindoyo JO. Comparative study between hydrodistillation and microwave-assisted hydrodistillation for extraction of cinnamomum cassia oil. Journal of Engineering and Applied Science. 2016;**11**(4):2647-2652

[44] Babushok VI, Linstrom PJ, Zenkevich IG. Retention indices for frequently reported compounds of plant essential oils. Journal of Physical and Chemical Reference Data. 2011;**40**(4):043101-1 043101-30. DOI:10.1063/1.3653552

[45] Adams RP. Identification of Essential Oil Components by Gas Chromatography/ Mass Spectroscopy, 4. Carol Stream: Allured Publishing; 2017

[46] Schulz H, Schrader B, Quilitzsch R, Pfeffer S, Kruger H. Rapid classification of basil chemotypes by various vibrational spectroscopy methods. Journal of Agricultural and Food Chemistry. 2003;**51**(9):2475-2481

[47] Stuart B. Infrared spectroscopy: Fundamentals and applications. West Sussex: John Wiley & Sons; 2004. p. 221

[48] Schulz H, Baranska M. Identification and quantification of valuable plant substances by IR and Raman spectroscopy. Vibrational Spectroscopy. 2007;**43**:13-25

[49] Dufour É. Principles of infrared spectroscopy. In: SUN D, editor. Infrared Spectroscopy for Food Quality Analysis and Control. 1st ed. Oxford, UK: Elsevier; 2009. pp. 3-27

[50] Schulz H. Rapid analysis of medicinal and aromatic plants by non-destructive vibrational spectroscopy methods. Acta Horticulturae. 2005;**679**(5):181-187

[51] Hzounda Fokou JB et al. Spectral and chemometric analyses reveal antioxidant properties of essential oils from four Cameroonian Ocimum. Industrial Crops and Products. 2016;**80**:101-108

[52] Schulz H, Baranska M, Belz H-H, Rösch P, Strehle MA, Popp J. Chemotaxonomic characterisation of essential oil plants by vibrational spectroscopy measurements. Vibrational Spectroscopy. 2004;**35** (1-2):81-86

[53] Schulz H, Özkan G, Baranska M, Krüger H, Özcan M. Characterisation of essential oil plants from Turkey by IR and Raman spectroscopy. Vibrational Spectroscopy. 2005;**39**(2):249-256

[54] Baranska M et al. Investigation of eucalyptus essential oil by using vibrational spectroscopy methods. Vibrational Spectroscopy. 2006. Available from: http://linkinghub.elsevier.com/retrieve/pii/S0924203106001809 [Accessed: 06 February 2014]

[55] Argyropoulou C, Daferera D, Tarantilis PA, Fasseas C, Polissiou M. Chemical composition of the essential oil from leaves of *Lippia citriodora* H.B.K. (Verbenaceae) at two developmental stages. Biochemical Systematics and Ecology. 2007. Available from: http://www.sciencedirect.com/science/article/pii/S0305197807001512 [Accessed: 11 February 2014]

[56] Almeida MR, Fidelis CHV, Barata LES, Poppi RJ. Classification of Amazonian rosewood essential oil by Raman spectroscopy and PLS-DA with reliability estimation. Talanta. 2013;**117**:305-311

[57] WHO. Deaths by Cause, Age, Sex, by Country and by Region, 2000-2016. Geneva: World Health Organization; 2018. p. 1

[58] Gaze DC. Introduction to ischemic heart disease. Intech Open. 2016;**54**(June):713-727

[59] Minnerup J, Schmidt A, Albert-Weissenberger C, Kleinschnitz C. Stroke: Pathophysiology and Therapy. California: Morgan & Claypool Life Sciences; 2013:1500-1516

[60] Greiner R et al. The effects of food essential oils on cardiovascular diseases: A review. Critical Reviews in Food Science and Nutrition. 2017;**58**(10):1688-1705

[61] Sharafkhaneh A, Hanania NA, Kim V. Pathogenesis of emphysema: From the bench to the bedside. Proceedings of the American Thoracic Society. 2008;**5**(4):475-477

[62] Juergens UR, Dethlefsen U, Steinkamp G, GIllissen A, Repges R, Vetter H. Bronchial asthma: A double-blind placebo-controlled trial. Respiratory Medicine. 2003;**97**:250-256

[63] Fischer J, Dethlefsen U. Efficacy of cineole in patients suffering from acute bronchitis: A placebo-controlled double-blind trial. Cough. 2013;**9**(1):1-5

[64] Chunzhen Z et al. Protective effect of eucalyptus oil on pulmonary destruction and inflammation in chronic obstructive pulmonary disease (COPD) in rats. Journal of Medicinal Plant Research. 2017;**11**(6):129-136

[65] Rantzsch U, Vacca G, Gillissen A. Anti-inflammatory effects of myrtol standardized and other essential oils on alveolar macrophages from patients with chronic obstructive pulmonary disease. European Journal of Medical Research. 2009;**14**:205-209

[66] Horváth G, Ács K. Essential oils in the treatment of respiratory tract diseases highlighting their role in bacterial infections and their anti-inflammatory action: A review. Flavour and Fragrance Journal. 2015;**30**(5):331-341

[67] Ács K, Balázs VL, Kocsis B, Bencsik T, Böszörményi A, Horváth G. Antibacterial activity evaluation of selected essential oils in liquid and vapor phase on respiratory tract pathogens. BMC Complementary and Alternative Medicine. 2018;**18**(1):1-9

[68] Khan I, Bahuguna A, Kumar P, Bajpai VK, Kang SC. In vitro and in vivo antitumor potential of carvacrol nanoemulsion against human lung adenocarcinoma A549 cells via mitochondrial mediated apoptosis. Scientific Reports. 2018;**8**(1):712-714

[69] Sehgal K, Singh M. Essentials to kill the cancer. Cancer Therapy & Oncology International Journal. 2017;**4**(5):4-7

[70] Lesgards JF, Baldovini N, Vidal N, Pietri S. Anticancer activities of essential oils constituents and synergy with conventional therapies: A review. Phytotherapy Research. 2014;**28**(10):1423-1446

[71] Yen HF, Hsieh CT, Hsieh TJ, Chang FR, Wang CK. In vitro anti-diabetic effect and chemical component analysis of 29 essential oils products. Journal of Food and Drug Analysis. 2015;**23**(1):124-129

[72] Boukhris M, Bouaziz M, Feki I, Jemai H, El Feki A, Sayadi S. Hypoglycemic and antioxidant effects of leaf essential oil of *Pelargonium graveolens* LHér. in alloxan induced diabetic rats. Lipids in Health and Diseases. 2012;**11**(1):1

[73] Bharti SK, Kumar A, Prakash O, Sharma NK, Krishnan S, Gupta AK. Vivo Experiments and Computational Studies. Open Access Scientific Reports. 2013;**2**(3):1-9

[74] Al-hajj NQM, Sharif HR, Aboshora W, Wang H. In vitro and in vivo evaluation of antidiabetic activity of leaf essential oil of *Pulicaria inuloides*—Asteraceae. 2016;**4**(7):461-470

[75] Kaur H, Richa R. Antidiabetic activity of essential oil of hedychium spicatum. International Journal of Pharmacognosy and Phytochemical Research. 2017;**9**(6):853-857

[76] Aa Y, Oa E, Sa H, Za E, Hm I. Antidiabetic effects of essential oils of some selected medicinal lamiaceae plants from Yemen against a-glucosidase enzyme phytochemistry & biochemistry. Journal of Phytochemistry & Biochemistry. 2018;**2**(1):1-5

[77] Aljarallah KM. Conventional and alternative treatment approaches for *Clostridium difficile* infection. International Journal of Health Sciences. 2017;**11**(1):50-59

[78] Miron A, Aelenei P, Aprotosoaie A, Bujor A, Gille E, Trifan A. Essential oils

and their components as modulators of antibiotic activity against gram-negative bacteria. Medicines. 2016;**3**(3):19

[79] Jalilzadeh-Amin G, Maham M. Antidiarrheal activity and acute oral toxicity of *Mentha longifolia* L. essential oil. Avicenna Journal of Phytomedicine. 2015;**5**(2):128-137

[80] Hawrelak JA, Cattley T, Myers SR. Essential oils in the treatment of intestinal dysbiosis: A preliminary in vitro study. Alternative Medicine Review. 2009;**14**(4):380-384

[81] Bueno J, Escobar P, Martínez JR, Leal SM, Stashenko EE. Composition of three essential oils, and their mammalian cell toxicity and antimycobacterial activity against drug resistant-tuberculosis and nontuberculous mycobacteria strains. Natural Product Communications. 2011;**6**(11):1743-1748

[82] Andrade-Ochoa S, Chacón-Vargas F, Nevarez-Moorillon G, Rivera-Chavira B, Hernández-Ochoa L. Evaluation of of antimycobacterium activity of the essential oils of cumin (*Cuminum cyminum*), clove (*Eugenia caryophyllata*), cinnamon (*Cinnamomum verum*), laurel (*Laurus nobilis*) and anis (*Pimpinella anisum*) against *Mycobacterium tuberculosis*. Advances in Biological Chemistry. 2013;**3**:480-484

[83] Runde M, Kubmarawa D. Compositional analysis and antimycobacterium tuberculosis activity of essential oil of hyptis suaveolens lamiceae obtained from North-East. Research Journal of Chemical Sciences. 2015;**4**(9):45-49

[84] Bicchi C et al. HIV-1-inhibiting activity of the essential oil of ridolfia segetum and oenanthe crocata. Planta Medica. 2009;**75**(12):1331-1335

[85] Dziri S, Casabianca H, Hanchi B, Hosni K. Composition of garlic essential oil (*Allium sativum* L .) as influenced by drying method. Journal of Essential Oil Research. 2014;**26**(3):91-96

[86] Mnayer D et al. Chemical composition, antibacterial and antioxidant activities of six essentials oils from the Alliaceae family. Molecules. 2014;**19**:20034-20053

[87] Hzounda Fokou JB et al. Optimized combinaition of Ocimum essential oils Inhibit growth of four *Candida albicans*. International Journal of Drug Discovery. 2014;**6**(1):198-206

[88] Özcan M, Chaltat J-C. Essential oil composition of *Ocimum basilicum* L . and *Ocimum minimum* L. in Turkey. Czech Journal of Food Sciences. 1995;**20**(6):223-228

[89] Politeo O, Juki M, Milo M. Chemical composition and antioxidant activity of essential oils of twelve spice plants. Croatica Chemica Acta. 2006;**79**(4):11-13

[90] Kumar A, Shukla R, Singh P, Prakash B, Dubey NK. Original article chemical composition of *Ocimum basilicum* L . essential oil and its efficacy as a preservative against fungal and aflatoxin contamination of dry fruits. International Journal of Food Science and Technology. 2011;**46**:1840-1846

[91] Orav A, Stulova I, Kailas T, Muurisepp M. Effect of storage on the essential oil composition of *Piper nigrum* L. fruits of different ripening states. Journal of Agricultural and Food Chemistry. 2009;**52**:2582-2586

[92] François T et al. Comparative essential oils composition and insecticidal effect of different tissues of *Piper capense* L., *Piper guineense* Schum. et Thonn., *Piper nigrum* L. and *Piper umbellatum* L. grown in Cameroon. African Journal of Biotechnology. 2009;**8**(3):424-431

[93] Hussain AI, Anwar F, Chatha SAS, Jabbar A, Mahboob S, Nigam PS. *Rosmarinus officinalis* essential oil: Antiproliferative, antioxidant and antibacterial activities. Brazilian Journal of Microbiology. 2010;**41**:1070-1078

[94] Mehdizadeh T, Hashemzadeh MS, Nazarizadeh A, Tat M. Chemical composition and antibacterial properties of *Ocimum basilicum*, *Salvia officinalis* and *Trachyspermum ammi* essential oils alone and in combination with nisin. Research Journal of Pharmacognosy. 2016;**3**(4):51-58

[95] Porte A, Godoy RLO, Maia-Porte LH. Chemical composition of sage (*Salvia officinalis* L.) essential oil from the Rio de Janeiro State (Brazil). Revista Brasileira de Plantas Medicinais. 2013;**15**(3):438-441

[96] Safrudin I, Maimulyanti A, Prihadi AR. Effect of crushing of clove bud (*Syzygium aromaticum*) and distillation rate on main constituents of the essential oil. American Journal of Essential Oils and Natural Products. 2015;**2**(3):12-15

[97] Taroq A et al. Research article phytochemical screening of the essential oil of *Syzygium aromaticum* and antibacterial activity against nosocomial infections in neonatal intensive care 1. International Journal of Pharmaceutical Sciences Review and Research. 2018;**48**(14):58-61

[98] Viuda-martos M, Ruíz-navajas Y, Fernández-lópez J, Pérez-álvarez JA. Chemical composition of the essential oils obtained from some spices widely used in mediterranean region. Acta Chimica Slovenica. 2007;**54**:921-926

[99] Faraco A, Araujo DO, Ribeiro-paes JT, De Deus JT. Larvicidal activity of *Syzygium aromaticum* (L.) Merr and *Citrus sinensis* (L.) Osbeck essential oils and their antagonistic effects with temephos in resistant populations of *Aedes aegypti*. Memórias do Instituto Oswaldo Cruz. 2016;**111**(7):443-449

[100] El Hattabi L, Talbaoui A, Amzazi S, Bakri Y, Harhar H. Chemical composition and antibacterial activity of three essential oils from south of Morocco. (*Thymus satureoides*, *Thymus vulgaris* and *Chamaelum nobilis*). Journal of Materials and Environmental Science. 2016;**7**(9):3110-3117

[101] Hay Y-O, Abril-Sierra MA, Sequeda-Castañeda LG, Bonnafous C, Raynaud C. Evaluation of combinations of essential oils and essential oils with hydrosols on antimicrobial and antioxidant activities. Journal of Pharmacy & Pharmacognosy Research. 2018;**6**(3):216-230

[102] Pino JA, Marbot R, Rosado A, Batista A. Chemical composition of the essential oil of zingiber chemical composition of the essential oil of *Zingiber officinale* Roscoe L. from Cuba. Journal of Essential Oil Research. 2011;**16**(3):186-188

[103] Ohlmuth HANSW, Mith MIKEKS, Rooks LYOB, Yers STPM, Each DANL. Essential oil composition of diploid and tetraploid clones of ginger (*Zingiber officinale* Roscoe) grown in Australia. Journal of Agricultural and Food Chemistry. 2006;**54**:1414-1419

[104] Sukari MA et al. Chemical constituents variations of essential oils from rhizomes of four zingiberaceae species. Malaysian Journal of Analytical Sciences. 2008;**12**(3):638-644

[105] Sasidharan I, Menon AN. Comparative chemical composition and antimicrobial activity fresh & dry ginger oils (*Zingiber officinale* roscoe). International Journal of Current Pharmaceutical Research. 2010;**2**(4):40-47

[106] Kizhakkayil J, Sasikumar B. Characterization of ginger (*Zingiber officinale* Rosc). Germplasm based on volatile and non-volatile components. African Journal of Biotechnology. 2012;**11**(4):777-786

Chapter 3

Biological Importance of Essential Oils

Muhammad Irshad, Muhammad Ali Subhani, Saqib Ali and Amjad Hussain

Abstract

Essential oils are the volatile compounds having the oily fragrance. Essential oils are obtained from the different plant parts, and they are extracted from the different techniques and the most preferable method of extraction is the hydrodistillation which is cheap and easy to use. Plant parts including the flowers, leaves, stem, bark and roots are used for the isolation of essential oils. Essential oils are used in almost every field of life and because of these characteristics, the market of essential oils is growing rapidly. Essential oils are used in the aromatherapy and act as antioxidant, antimicrobial, antifungal, pain relievers, anxiety, depression. In the field of cosmetics and industries, the essential oils are used rapidly and mostly used in the perfume industries which are growing increasingly. Essential oils are used in the food preservations and many food items. Essential oils are used as the folk herbal medicines and their fragrance is used for the improvement of the mood and as the depression release.

Keywords: acts as the antioxidant, antimicrobial, antifungal, pain reliever, anxiety and depression

1. Introduction

Essential oils have been used in the folk medicines throughout the history. Essential oils are called the ethereal or volatile oils, which are fragrant oily liquid that are extracted from the various parts of the plants and mostly used as the food flavors. An essential oil is "essential" in sense that it contains the essence of the different fragrance, and the properties of the plants from which they are derived. These volatile oils showed the different kinds of biological activities including the antibacterial, antioxidant, antiviral, insecticidal, etc. [1]. These oils are also used for cancer treatment, while some other has been used for the food preservations, aromatherapy, and in the perfumery industries [2]. The antimicrobial and antioxidant screening of essential oil acts as the root of numerous applications including the processed and fresh food preservations, natural therapies, pharmaceuticals, and alternative medicines [3]. Essential oils are used in aromatherapy as an alternative source of wound healing because of the aromatic compounds that are present in the essential oils. It is also used as a relaxation process, but this evidence is not under consideration [4].

Numerous efforts are made to explore the essential oils usage as the treatment of various infectious diseases that supernumerary to the pharmaceutical's remedies. Medicinal and aromatic plants are extensively used as natural organic compounds and as medicines [5]. Previously, essential oils have been used for the treatment of various sorts of infectious diseases in the whole world. Now, in this era, the importance of essential oils is increasing day by day, because they are mostly used in the beverage and food industries, cosmetics and fragrance industries for making valuable perfumes, and with lot of biological activities [6].

Various essential oils have been used for the insecticidal activities against the different pests, but in detail, studies showed that they do not show the repellence, avicidal, phytochemistry, antifungal, and oviposition. The essential oils do not show the abovementioned characteristics, but there is still urgent need to work on this side of research and study the in vivo and in vitro studies to control the pests, and most of the oils have shown good antioxidant activities [7]. Essential oils that showed good antioxidant activates and acts as the defensive role for the unsaturation of lipids in the tissue of the animal and they also act as hepatoprotective negotiators in mammals. The antioxidant substances are most important for human being because of the oxygen which is a toxic element and has the ability to change the metabolic activities into the most reactive form of oxygen just like the super oxide, hydrogen peroxide, hydroxyl free radicals, and the singlet oxygen which are collectively called as active oxygen [8]. Essential oils are best known for their action as the antispasmodic, antiviral negotiators, antimicrobial, and carminative, and the essential oil composition is variable; they also show different sorts of activities and mostly depend upon the chemo types [9].

2. Sources and isolation of essential oils

Essential oils were extracted from different aromatic plants. These plants are distributed in the tropical countries and Mediterranean. These plants got importance because local people use them for the treatment of diseases. The essential oil is produced in every part of the plant including the leaves, seeds, buds, stem, flowers, leaves etc as shown in **Figure 1**. Essential oil is accumulated from the epidermic cell, cavities, secretary cells, and channels [10]. The odor that is produced in plants is because of essential oils. The essential oils were extracted from the dried, fresh, or partially dehydrated materials of plant. The extraction rate depends upon the diffusion via plant tissues that directly involve the surface from which the essential oil was removed by different processes. The extraction of essential oil depends upon the stability of the essential oil. The two most important method that are used for the extraction of essential oil was used are steam distillation method and the hydro distillation process as shown in **Figure 2**. These are the most suitable and effective techniques for the extraction processes [11]. Some other methods were also used for extraction but they are not too much suitable for this process these are the microwave or liquid carbon dioxide, high- or low-pressure distillation with the help hot water or steam water (**Figures 1** and **2**) [12].

The essential oil extracted from the steam distillation method is mostly used in pharmacological activities and food items, while the essential oil that are used in the fragrance industry or perfume industry are extracted from the lipophilic solvents and sometime with the supercritical carbon dioxide going more attractive [13]. The quality of the essential oil depended on the basis of the age of plants, parts that are used for extraction, vegetative cycle stage, effect of climate, etc. The chromatographic and the spectroscopic techniques fully changed the chemical analysis of the essential oils. The chemical composition of the essential oils was studied with the help of IR-spectroscopy, UV-Vis spectroscopy, gas chromatography, NMR

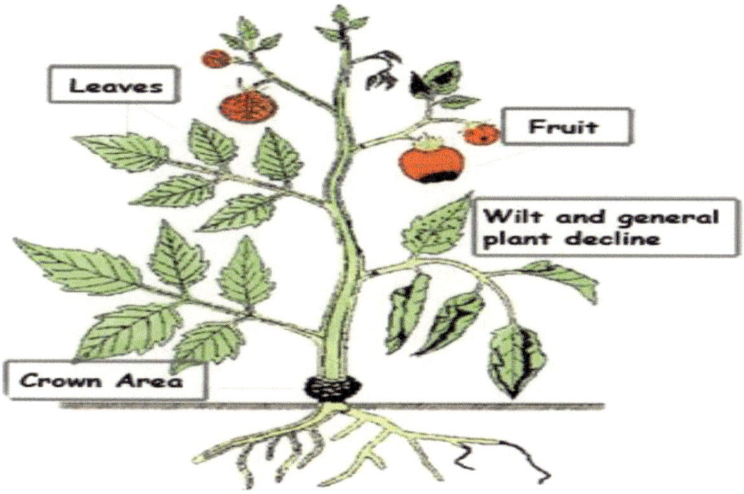

Figure 1.
Plants and their parts used for the isolation of essential oils.

Figure 2.
Hydrodistillation apparatus used for the extraction of essential oils.

spectroscopy [14]. The enhanced demand for the essential oil in various fields of life provoked us to access the reliable methods for the essential oil analysis, and the techniques used are the GC-MS and GC analyses [15]. The characterization of the essential oil was carried out by using the gas chromatography. The compounds that are present in the essential oil was confirmed by using the GC and GC-MS analysis [16]. The storage and handling of the essential oil also affect its yield and quality, ad essential oil was deposited in the oil glands that are present in the organization of the plant material [16].

3. Essential oil industry

The worldwide essential oil market demand was 226.8 kilotons in year 2018. It is expected to expand at a CAGR of 8.6% from the 2019 to 2025. Usage of essential oils in industries are increases day by including the beverage, food, personal care, aromatherapy, and cosmetics. Various sorts of the health-related benefits are offered by essential oils and they are reported as the anticipated fuel and their demand is increasing in the medical and pharmaceuticals applications. Most of the conventional drugs have no side effects. The growing inclination of the consumers toward the organic and natural products is leading to increase the use of essential oils in the beverage, food, and cosmetics industries. Worldwide essential oil market will cross USD 13 billion in the year of 2024 the latest report of the Global Market Insights, Inc. The increase in the World population are suffering from the different kinds of health-related issues and essential oils are used in aromatherapy products and due to this reason, the Worldwide market of essential oils are increasing day by day [17].

The period when essential oils were utilized first on a commercial scale is hard to recognize. The nineteenth century is for the most part viewed as the beginning of the cutting-edge period of commercial utilization of essential oils. Notwithstanding, the extensive scale use of essential oils goes back to antiquated Egypt. In 1480 BC, Queen Hatshepsut of Egypt sent a campaign to the nation of Punt (presently Somalia) to gather fragrant plants, tars, and oils, as elements for medicaments, scents, and flavors and for the preservation of bodies. Valuable scents have been found in numerous Egyptian archeological unearthing, as an image of riches and social position. The huge global exchange of fundamental oil-based items is the standard for modern use; "Ruler of Hungary Water" was the primary alcoholic scent ever. This aroma, in view of rosemary basic oil distillate, was made in the mid-fourteenth century for the Polish-conceived Queen Elisabeth of Hungary. Following an uncommon introduction to King Charles V, The Wise of France in 1350, it ended up prevalent in all medieval European courts. The start of the eighteenth century saw the presentation of "Eau de Cologne," in light of bergamot and different citrus oils, which remains broadly used right up 'til the present time. This crisp citrus aroma was the making of Jean Maria Farina, a relative of Italian perfumers who came to France with Catherine de Medici and settled in Grasse in the sixteenth century. As indicated by the city of Cologne files, Jean Maria Farina and Karl Hieronymus Farina, in 1749, built up a processing plant (Fabriek) of this water, which sounds exceptionally "mechanical." The "Kolnisch Wasser" turned into the main unisex aroma as opposed to one basically for men, known and utilized all over Europe, and it has been rehashed in this manner in incalculable countertypes as a scent for men. The essential oil market was extended day by day because of increase in demand for the essential oil products including the soap, cosmetics, and food industries. The international companies are the major contributors of the development of the essential oil industries in the mid-nineteenth century [18].

Changing the standards of the living led to the occurrence of different sorts of mental issues including the depression, anxiety, insomnia, and stress that led to grow the market of essential oils because they are used for the treatment of such kinds of diseases. There are more than 300 industries in the Pakistan which industrialized various human resources. These industries used unprocessed material especially essential oils that are imported from the western countries. Pakistan imports more than Rs. 1526.8 million to buy essential oils and perfumes and isolates [19]. Pakistan is an agricultural country that is rich in aromatic sorts of plants, which are used as natural medicines and are used in local areas to cure common diseases. The environment of Pakistan is much more suitable for the growth of essential oils crops. And from these plants the essential oils obtained, and they are used in essential oil industries, but this industry is not much more attractive in Pakistan.

4. Modern trends of essential oils

The essential oil has been large number of usages in worldwide products including the ice creams perfumes, backed food stuff, beverage, and cosmetics as shown in **Figure 3**. Newly, at least 300 kinds of essential oils out of 3000 are commercially important in various kinds of industries including the perfume and sanitary industries, cosmetics, food, beverage, agronomics, and pharmaceuticals [20]. Some of the bioactive components that are present in essential oils are the limonene, geranyl acetate, carvone, etc., and these are the important components of the hygienic products and tooth pastes. Essential oils are used for the preservation of the food additives; for the treatment of common diseases and folk medicines; and used by aromatherapist. Essential oils are used as the natural antioxidant. The usage of natural antioxidant is prominent in the defensive medicines and food items, and because of this reason, essential oils are getting popular day by day. Recently, the growth explores the applications of the volatile essential oils for remedial usage and in the treatment of some infectious diseases [21].

Essential oils are widely used in perfumes, personal hygiene products, and in aromatherapy including the inhalation, massage, masking agent to avoid the unpleasant odor in the textile industries, paint and plastic industries, and pharmaceuticals formulations. Essential oils are also used as the natural antifungal and antibacterial agents in the food safety items; essential oil also used in the various kinds of cereals, antimicrobial packing of the food items, edible thin film, nanoemulsion, preservation of the fruits and vegetables, soft drinks, as the flavoring agents in the carbonated drinks, as the major ingredients in soda/citrus concentrates, seafood preservations, fish, etc. (**Figure 3**) [22].

Figure 3.
Modern trends of essential oils.

5. Growing trends of essential oils adaptation

The essential oils are the products that are obtained from the plant extracts and have been used for large-scale industrial and homemade products. The major usages of essential oils are pest control products, cleaning actions, and counter medications among the other products and personal care products. Essential oils have various advantages in wound healing, rejuvenation, and relaxation. Alongside their applications in the betterment of the health issues, the most common health issues such as migraines and nausea are cured from the essential oils. It is also used in the food industries because of their preservative potential in contrast to the foodborne pathogens, antibacterial, antimicrobial, and antifungal characteristics. The use of aromatherapy as the harmonizing care is speedup due to their unique characteristics which include the coping with some of side effects of cancer and to promote the wound healing [23].

6. Uses of essential oils in perfumery

The essential oils that are used in the perfume industries are classified according to their diffusion rate in air and volatility:

Base note: these are the least volatile essential oils and last for a longtime period. These remain for longtime duration including several hours. Some essential oils that are used as the base notes are the Myrrh, vanilla, sandalwood, and frankincense.

Middle note: these sorts of essential oils are tending to be spicy or floral and give body to blends; their time duration is less and remain up-to 1 hour. These include Ylang-ylang, jasmine, geranium, clove, and lavender.

Top note: these are the most volatile and the first perceptible odors from the perfume. Their time duration is too much less and remains maximum for 30 minutes. These include berry, bergamot, cinnamon, juniper, and gardenia.

Perfumes are formulated mostly using alcohol, though these may contain the cloudy solutions. Eau de types of perfumes are mostly formulated using the essential oils generally amber color because of their natural oils color but normally they are clear.

6.1 Percentage of essential oils in different perfumes products

- Eau de perfume usually contains 8–15% amount of essential oils or sometimes their fragrance, and 80–90% alcohol.

- Splash cologens usually contain 1–3% fragrance or essential oil, and 80% alcohol.

- Eau de cologne usually contains 3–5% of fragrance or essential oil, and up to 70% alcohol.

- Eau de toilette usually contains the essential oil between 4 and 8% or its fragrance, and 80–90% alcohol [24].

6.2 Increasing the sales of essential oils to the home appliances

All over the world, people are shifting toward the herbal products for the treatment of skin diseases compared to medicines and synthetic drugs. The essential oil is pure and does not have any side effects. The demand for essential oil is increasing because of their usage in daily life and it is mostly used for the relaxation purpose and people prefer it because of its no side effects. Aura Cacia that is manufacturer of Iowa-based care products said that the essential oil sale was increased 90% between the 2009 and 2012, and the sale of household items that contains the essential oil was increased from 6 to 12%.

Essential oils play a key role in treating the dermatological issues including the rashes, acne, hives, eczema, and psoriasis which made the essential oil suitable for the skin treatment care products that enhance the growth of skin industry. The market of essential oils is growing because it has no side effects, and other synthetic chemicals have side effects, so they are less preferred. Essential oils market of home care products and cleaning products will be increased to 550 million USD by 2024. The growth in essential oil market along with the companies that are introducing the products with supplementary benefits such as better cleaning, easy fragrance, and germ fighting.

The essential oils market of France will be increased up to 8.5% by 2024. Major cosmetics industries used essential oils in cosmetics and imported these oil products worldwide. Companies used the marketing strategies to spread the awareness to the people regarding the usage and benefits of essential oils, and the aromatherapy markets gets more enhanced customers to buy these products. The essential oil market of India will be exceeding up to 790 million USD by 2024. Since being a large country, India used the large-scale agricultural techniques to grow crops of essential oil plants including lemon, mint, and spices, and its aromatherapy market are growing day by day.

Lavender oil market will be reached up to 20 kilotons by 2024. It is used in fighting the serious health conditions, including the chronic anxiety, relieves pain, cancer, stress with reverse sign of the ageing, headache, cosmetics applications, pharmaceuticals applications, aromatherapy etc. as shown in (**Figure 4**). The major companies that share large market size of essential oil-based products are Firmenich, Frutarom, Flaex, Rock Mountain Moksha Lifestyle, and Florihana Falcon Young living (**Figure 4**) [25].

6.3 Some major essential oils and their applications

6.3.1 Bergamot

The essential oil of bergamot obtained from the peel of the fruits of the *Citrus bergamia* is known as the bitter orange tree. The extract of the bergamot is used in flavoring

Figure 4.
Applications of essential oils in daily life.

in Earl Grey tea and essential oil of this is used also for the same purpose. It is applicable for the treatment of skin diseases, and it improves the mood, relieves pain, reduces fever, treats the digestive system problems, and breaks up chest congestion [26].

6.3.2 Clove

It is extracted from the aromatic flower buds of *Syzygium aromaticum* tree that is native to Maluku, Indonesia. The essential oil of Clove provides the strong fragrance, used in cooking spice foods; medically, it is used as pain relief, used for the treatment of dental disorders, for nausea treatment, to reduce inflammation, for the treatment of the digestive ailment, and to clear up acne [26].

6.3.3 Eucalyptus

It is extracted from the different species of genus *Eucalyptus*. Every type of species contains different and unique usage in every field. The most familiar essential oil obtained from the *Eucalyptus globulus* has a mint-like fragrance. It is used for decongestant chest rub, as pain relievers, as an antimicrobial agent, immunostimulant, and for the treatment of the flu and cold cough. It is used in aromatherapy and it provides mental clarity; it also boosts up energy and used as a natural insect repellent [26].

6.3.4 Frankincense

The earliest known and the most useful essential oil is Frankincense and it is obtained from the resin of the four species of the generous Bowwellia and the most known from this genus is the Bowsellia carterii hard tree which grow in the arid land of Arabian Peninsula and north eastern. The old African people used the essential oil of Frankincense in the religious and spiritual ceremonies. The Frankincense essential oil is unique from all other obtained essential oils because of the perfect combination of wood, balsam, earth, and citrus. It is used as the mood enhancer, antimicrobial, stress reducer, for faster wound healing, aid in digestion, anti-inflammatory, fades scars, reduces swelling of insect bites, for the treatment of skin diseases, and eases itching [26].

6.3.5 Lavender

The most effective essential oil obtained from the *Lavandula angustifolia* is the most popular garden herb English lavender. Its odor is same as the flowers from which they are obtained having the sweet smell, floral, and green, and the health benefits are greater as compared to their fragrance. The best purpose of essential oil of lavender is their sleep-inducing properties and calmness. It showed good antioxidant, anti-inflammatory, antibacterial, and antifungal properties, and it is also used for the treatment of various sorts of skin diseases including eczema or ringworm and acne. Lavender essential oil is used to enhance the digestive system, to reduce the swelling of sore muscles, and to relieve pain. Due to its attractive smell, it attracts butterflies, bees, and some pollinators, and it also acts as a natural repellent for many flying six-legged pests [26].

6.3.6 Lemon

Essential oils obtained from the lemon are mostly used. The essential oil obtained from the *Citrus limon* is used worldwide. The essential oil of lemon is used as antimicrobial agents, in household items including soaps, polishes, furniture, fresheners, and in most of the cleaning products. Some other uses of these essential oils are that they are the pain relievers, show antifungal activity, help for the loss of weight, and alleviate the severe nausea; the essential oil of lemon is used in aromatherapy to reduce the anxiety and stress and simultaneously enhances the concentration and mood. It is also used for cleaning the hair and enhancing the natural growth of hairs [26].

6.3.7 Oregano

The essential oil of Oregano was obtained from the kitchen spice *Origanum vulgare*. It is the perfect combination of the earth, spice, and warmth. When the bottle of essential oil is opened, their fragrance has an effect on the senses. The usage of the essential oil of Oregano is increasing day by day and it is mostly used for the skin care treatment like eczema, rosacea, and psoriasis. It is used to alleviate the menstrual problems or painful menstrual cramp, used to cure stomach problems, and helps to control the flu and cold infections [26].

6.3.8 Peppermint

The essential oil of peppermint is used worldwide and it is obtained from the *Mentha piperita*. This mint hybrid is the most favorite between the essential oil and gardeners. It is the most famous type of essential oil because of its unique applications, and it is mostly used in preventing flu and cold, alleviating headache, relieving pain in muscles and joints, clearing the skin conditions, relieving nausea, and improving the digestive system processes [26].

6.3.9 Rosemary

The essential oil of rosemary is obtained from the evergreen shrub of *Rosmarinus officinalis* and is famous albeit common kitchen herb has the extraordinary healing potential in its natural oil. Just like the rosemary, the essential oil of this herb has the crisp woody, herbal, and somewhat balsamic odor just like the camphor. Due to its unique fragrance of rosemary oil, it is used for cleaning the inside and outside of the body. It is further used for the treatment of various diseases, especially skin care, dandruff treatment, to improve the scalp health, to boost up the immune system, flu infections, and ward off cold. Although this oil is used to alleviate the pain, swelling in joints and muscles, for treatment of the digestive tissues, soothe tension headaches, to promote the mental clarity, to enhance the memory, and improve mood, it is also the best natural insecticide and the bug repellent [26].

6.3.10 Tea tree

The essential oil of the tea tree is obtained from the leaves and stem of *Melaleuca alternifolia* and shrub of *Camellia sinensis.* The oil is toxic if ingested directly and it is used mostly for the external purposes and has the herbal, fresh, and slightly camphorates aroma. Melaleuca claims that essential oil of tea tree act as an antimicrobial agent, treating antifungal infections, and cleansing wounds. It is used in cosmetics products including the shampoo to clear some scalp conditions and dandruff and used for the treatment of insect bite to reduce itching and inflammation [26].

6.3.11 Some plant species essential oils and their usage

Some plant species essential oils and their usage are shown in **Table 1**.

S. no.	Plant species	Essential oil	Uses
1	*Pimenta racemosa*	Bay oil	Aches, muscle circulation, improve dandruff [27]
2	*Callitris intratropica*	Blue cypress oil	Asthma [27]
3	*Daucus carota*	Carrot seed oil	Detoxification, eczema [27]
4	*Apium graveolens*	Celery seed oil	Treat of gout, antifungal diuretic, blood pressure, antiseptic reduces sedative, urinary antirheumatic [27]
5	*Stellaria species*	Chickweed infusion	Wound healing, antirheumatic, astringent [27]
6	*Cinnamon species*	Cinnamon oil	Antifungal, uterine stimulant, antibacterial [27]
7	*Artemisia pallens*	Davana oil	Coughs, including menstruation, anxiety, healing of wounds, antiseptic [27]
8	*Canarium luzonicum*	Elemi oil	Coughs, healing wounds, stress [27]
9	*Eucalyptus citriodora*	*Eucalyptus citriodora* oil	Fever and flu, to improve blood circulation and sinusitis, arthritis, bronchitis, catarrh, cold stores, colds and coughs [27]
10	*Eucalyptus globulus*	*Eucalyptus* oil	Antiseptic, antispasmodic, treatment of scarlet fever, influenza, measles and typhoid, infusion reduces blood sugar levels [27]
11	*Alpinia galanga*	Galanga oil	Aphrodisiac, easing heart pain and angina, dizziness and fatigue. Stomach, spleen, relief of pain, treatment of flu and colds, travel sickness [27]
12	*Pelargonium graveolens*	Geranium oil	Acne, cellulites, lice treatment, menopause [27]
13	*Zingiber officinalis*	Ginger oil	Promotes sweating, expectorant, prevents vomiting, antiseptic, anti-spasmodic, carminative, antibacterial, circulatory stimulant, nausea, relaxes peripheral blood vessels, promotes sweating [27]
14	*Hyssopus officinalis*	Hyssop oil	Nervous exhaustion, anxiety, used topically as an anti-inflammatory, bruises and anti-viral, increases alertness, uplifting and relaxing nerves [27]
15	*Ammi visnaga*	Khella oil	Antiasthmatic, diuretic, antispasmodic, relaxant [27]
16	*Citrus Limon*	Lemon oil	Blemishes, varicose veins, warts, chilibains, colds, corns, flu, skin, athletes foot [27]
17	*Backhousia citriodora*	Lemon myrtle oil	Insect repellent, stress, athletes foot, colds, flu, skin blemishes [27]
18	*Citrus reticulata*	Mandarin oil	Blemishes, stress and wrinkles, acne, insomnia, scars, skin [27]
19	*Mentha species*	Mint oil	Analgesic, calming, cooling for migraines, anti-bacterial, clear nasal congestion, prevents vomiting, relaxes peripheral blood vessels, promotes bile flow [27]
20	*Myrtus communis*	Myrtle oil	Sore throat, asthma, coughs [27]
21	*Piper nigrum and other species*	Pepper oil	Aches and pains, coughs, chills, cramps, digestion, antiseptic, anti-bacterial, topical use increases blood flow around area [27]
22	*Zingiber cassumunar*	Plai oil	Uterine relaxant, inflammatory [27]

S. no.	Plant species	Essential oil	Uses
23	*Rosa species*	Rose oil	Astringent, sedative, digestive stimulant, increase bile production, expectorant, anti-bacterial, antiseptic, kidney tonic, blood tonic, anti-depressant, anti-spasmodic, aphrodisiac [27]
24	*Mentha spicata*	Spearmint oil	Flu and fever, nausea, vertigo, asthma, exhaustion [27]
25	*Citrus sinensis*	Sweet orange oil	Constipation, cough relief, flu, gum treatment, calms nerves, digestive stimulant, aids energy
26	*Tagetes glandulifera and patula*	Tagetes oil	Warts and corns [27]
27	*Vetiveria zizanoides*	Vetivert oil	Insomnia, muscle aches, sores and stress, acne, cuts, anti-depressant, exhaustion [27]
28	*Viola odorata*	Violet leaf absolute	Poor blood circulation, sore throat, bronchitis, head ache, insomnia, rheumatism [27]
29	*Melaleuca alternifolia*	Tea tree oil	Fungal, antiseptic, anti-viral, candida, cold sores, corns, cuts, flu, anti-bacterial [27]
30	*Cananga odorata*	Ylang oil	Hypertension, palpitations, stress, anxiety, anti-depression, frigidity, hypertension [27]
31	*Zanthoxylum alatum*	Leaf, stem, root oil	Antioxidant, antifungal, antibacterial [28]
32	*Skimmea laureola*	Leaf oil	Antioxidant, antifungal, antibacterial and perfumery [29]
33	*Angelica glauca*	Root oil	Antioxidant, antifungal, antibacterial, and phytotoxicity [30]
34	*Thymus serpyllum*	Whole plant oil	Antioxidant, antifungal, antibacterial, and phytotoxicity [31]
35	*Plectranthus rugosus*	Leaf oil	Antifungal, antioxidant, antibacterial [32]

Table 1.
Some plant species essential oils and their uses.

7. Conclusion

Essential oils are the natural volatile compounds having loveable odor. The essential oils are isolated mostly from the hydrodistillation method which is more suitable for this process and easy to carry. Whole parts of the plants are used for the extraction of plants. Steam distillation method is expensive than the hydrodistillation, so it is less preferred. Essential oils have good medicinal applications and used in the treatment of different diseases including the infectious diseases, depression, anxiety, act as the antifungal, antimicrobial, anticancer, and wound healing; they are also used in cosmetics and perfume industries. In the field of heath, essential oils are used more frequently and are mostly applied to the external body parts to relieve the pain. In the field of fragrance, essential oils are used in the perfume industry and due to attractive odor, the essential oils are used mostly in this industry. It is used worldwide and due to their better usage, the world essential oil market is growing rapidly and getting more importance day by day.

Acknowledgements

The author wishes to thank University of Kotli for providing the facilities to write this chapter.

Conflict of interest

Author has no conflict of interest.

Author details

Muhammad Irshad[1*], Muhammad Ali Subhani[1], Saqib Ali[1] and Amjad Hussain[2]

1 Department of Chemistry, University of Kotli Azad Jammu and Kashmir, Pakistan

2 Department of Zoology, University of Kotli Azad Jammu and Kashmir, Pakistan

*Address all correspondence to: chemist_q2005@yahoo.com

References

[1] Abu-Shanab B, Adwan GM, Abu-Safiya D, Jarrar N, Adwan K. Antibacterial activities of some plant extracts utilized in popular medicine in Palestine. Turkish Journal of Biology. 2005;**28**(2-4):99-102

[2] Kelen M, Tepe B. Chemical composition, antioxidant and antimicrobial properties of the essential oils of three salvia species from Turkish flora. Bioresource Technology. 2008;**99**(10):4096-4104

[3] Celiktas OY, Kocabas EH, Bedir E, Sukan FV, Ozek T, Baser KH. Antimicrobial activities of methanol extracts and essential oils of Rosmarinus officinalis, depending on location and seasonal variations. Food Chemistry. 2007;**100**(2):553-559

[4] Lee MS, Choi J, Posadzki P, Ernst E. Aromatherapy for health care: An overview of systematic reviews. Maturitas. 2012;**71**(3):257-260

[5] Tepe B, Daferera D, Tepe AS, Polissiou M, Sokmen A. Antioxidant activity of the essential oil and various extracts of Nepeta flavida hub-Mor. from Turkey. Food Chemistry. 2007;**103**(4):1358-1364

[6] Rios JL, Recio MC. Medicinal plants and antimicrobial activity. Journal of Ethnopharmacology. 2005;**100**(1-2):80-84

[7] Dorman HJ, Deans SG. Antimicrobial agents from plants: Antibacterial activity of plant volatile oils. Journal of Applied Microbiology. 2000;**88**(2):308-316

[8] Pérez Gutiérrez RO, Hernández Luna H, Hernández Garrido S. Antioxidant activity of Tagetes erecta essential oil. Journal of the Chilean Chemical Society. 2006;**51**(2):883-886

[9] Mimica-Dukić N, Božin B, Soković M, Mihajlović B, Matavulj M. Antimicrobial and antioxidant activities of three Mentha species essential oils. Planta Medica. 2003;**69**(05):413-419

[10] Gilani AH, Khan AU, Jabeen Q, Subhan F, Ghafar R. Antispasmodic and blood pressure lowering effects of Valeriana wallichii are mediated through K+ channel activation. Journal of Ethnopharmacology. 2005;**100**(3):347-352

[11] Kulisic T, Radonic A, Katalinic V, Milos M. Use of different methods for testing antioxidative activity of oregano essential oil. Food Chemistry. 2004;**85**(4):633-640

[12] Bousbia N, Vian MA, Ferhat MA, Petitcolas E, Meklati BY, Chemat F. Comparison of two isolation methods for essential oil from rosemary leaves: Hydrodistillation and microwave hydrodiffusion and gravity. Food Chemistry. 2009;**114**(1):355-362

[13] Donelian A, Carlson LH, Lopes TJ, Machado RA. Comparison of extraction of patchouli (*Pogostemon cablin*) essential oil with supercritical CO_2 and by steam distillation. The Journal of Supercritical Fluids. 2009;**48**(1):15-20

[14] Hussain AI, Anwar F, Sherazi ST, Przybylski R. Chemical composition, antioxidant and antimicrobial activities of basil (*Ocimum basilicum*) essential oils depends on seasonal variations. Food Chemistry. 2008;**108**(3):986-995

[15] Daferera DJ, Ziogas BN, Polissiou MG. The effectiveness of plant essential oils on the growth of *Botrytis cinerea*, *Fusarium* sp. and *Clavibacter michiganensis* subsp. michiganensis. Crop Protection. 2003;**22**(1):39-44

[16] Van Vuuren SF, Viljoen AM, Őzek T, Demirci B, Başer KH. Seasonal and geographical variation of Heteropyxis

natalensis essential oil and the effect thereof on the antimicrobial activity. South African Journal of Botany. 2007;**73**(3):441 8

[17] Grand Review Research Inc. US. Available from: https://www.grandviewresearch.com/

[18] Market Research Reports, Consulting: Global Market Insights Inc. [Internet]. Available from: https://www.gminsights.com

[19] Pakistan Statistical Year Book. Government of Pakistan, Statistical Division, Federal Bureau of Statistical; 2008

[20] Burt S. Essential oils: Their antibacterial properties and potential applications in foods—A review. International Journal of Food Microbiology. 2004;**94**(3):223-253

[21] Hajhashemi V, Ghannadi A, Sharif B. Anti-inflammatory and analgesic properties of the leaf extracts and essential oil of *Lavandula angustifolia* mill. Journal of Ethnopharmacology. 2003;**89**(1):67-71

[22] Mahato N, Sharma K, Koteswararao R, Sinha M, Baral E, Cho MH. Citrus essential oils: Extraction, authentication and application in food preservation. Critical Reviews in Food Science and Nutrition. 2019;**59**(4):611-625

[23] Aromatherapy Market Trend 2019: Increasing Sales of Essential Oils for Home Usage. 2019. Available from: https://reportsuptodate.us/1390 [Accessed: April 20, 2019]

[24] Use Essential Oils for Perfume to Improve your Emotional and Physical Well Being [Internet]. 2015. Available from: www.experience-essential-oils.com/essential-oils-for-perfume.html

[25] Aromatherapy Market Trend 2019: Increasing Sales of Essential Oils for Home Usage. 2019. Available from: https://reportsuptodate.us/1390/aromatherapy-market-trend-2019-increasing-sales-of-essential-oils-for-home-usage/

[26] Taylor J. The 10 Most Popular Essential Oils and 174 Magical Ways to Use Them. 2016. Available from: https://www.naturallivingideas.com/author/jan/ [Accessed: April 12, 2016]

[27] Murry H. Essential Oils: Art, Agriculture, Science, Industry and Entrepreneurship (A Focus on The Asia-Pacific Region). Nova; 2009. pp. 626-633

[28] Irshad M, Aziz S, Ahmed MN, Asghar G, Akram M, Shahid M. Comparisons of chemical and biological studies of essential oils of stem, leaves and seeds of Zanthoxylum alatum Roxb growing wild in the state of Azad Jammu and Kashmir, Pakistan. Records of Natural Products. 2018;**12**(6):638

[29] Irshad M, Aziz S, Shahid M, Ahmed MN, Minhas FA, Sherazi T. Antioxidant and antimicrobial activities of essential oil of Skimmea laureola growing wild in the state of Jammu and Kashmir. Journal of Medicinal Plant Research. 2012;**6**(9):1680-1684

[30] Irshad M, Shahid M, Aziz S, Ghous T. Antioxidant, antimicrobial and phytotoxic activities of essential oil of Angelica glauca. Asian Journal of Chemistry. 2011;**23**(5):1947

[31] Aziz S, Habib-ur-Rehman, Irshad M, Asghar SF, Hussain H, Ahmed I. Phytotoxic and antifungal activities of essential oils of *Thymus serpyllum* grown in the state of Jammu and Kashmir. Journal of Essential Oil-Bearing Plants. 2010;**13**(2):224-229

[32] Irshad M, Aziz S, Habib-ur-Rehman, Hussain H. GC-MS analysis and antifungal activity of essential oils of *Angelica glauca, Plectranthus rugosus*, and *Valeriana wallichii*. Journal of Essential Oil-Bearing Plants. 2012;**15**(1):15-21

Chapter 4

Essential Oils: Partnering with Antibiotics

Mariam Aljaafari, Maryam Sultan Alhosani,
Aisha Abushelaibi, Kok-Song Lai and Swee-Hua Erin Lim

Abstract

Essential oils (EO) are volatile, non-lipid-based oils produced as a plant defense mechanism. Studies from our group have validated the potential usefulness of EOs to synergistically and additively work with antibiotics. In this book chapter, we aim to outline some background on the EOs and their uses and applications, to discuss the different mechanisms of action in partnering with antibiotics, and, finally, to explore their potential use against multidrug-resistant bacteria. Applications of EO in therapy will enable the revival of previously sidelined antibiotics and enhance the development of new drug regimens to better mitigate what may be the biggest health challenge by year 2050.

Keywords: lavender oil, cinnamon bark oil, peppermint oil, multidrug-resistant bacteria, synergistic interaction, antimicrobial

1. Introduction

Essential oil (EO) is a concentrated mixture of organic compounds. EOs are produced by plants as a form of defense in addition to being an attractant to insects for dispersion of pollens and seeds [1, 2]. These oils are formed by the glandular trichomes and specialized secretory structure like secretory hairs, ducts, cavities, and glands; they then diffuse to the surface organs of plant such as leaves and flowers [3, 4]. The process of EOs formation involves three pathways which are the methyl-D-erythritol-4-phosphate (MEP), mevalonate, and malonic acid pathways [5]. The MEP and mevalonate pathways contribute in the biosynthesis of isoprenoids, whereas the malonic acid pathway will form the phenolic compounds [6, 7].

EOs have been used for many years for different purposes, such as to preserve raw and processed food because it can inhibit the growth of microorganisms like bacteria, viruses, and fungi [1, 8, 9]. Besides food, EO was also utilized in the area of perfumery for many years especially for ancient civilizations of India, Greece, Egypt, and Rome [10, 11].

In addition, EOs also serve as an alternative medicine that is important for local populations to treat severe burns to accelerate healing [11] and also for diseases such as leishmaniasis, schistosomiasis, and malaria [12, 13]. To date, approximately 10% of all EOs have been analyzed and commercially used as an insect repellent, attributed by its low toxicity to mammalian cells and the environment [10, 14]. However, certain EOs may cause toxicity or allergies which results in health and safety problems. Hence, national and international organizations have set standards to control the use of EOs [15].

Part	Name of essential oil	References
Flowers	Lavender, jasmine	[18]
Leaves	Mint, lemongrass	[19, 20]
Wood	Sandal, cedarwood	[21, 22]
Roots	Sassafras, valerian	[23, 24]
Seeds	Fennel, nutmeg	[25, 26]
Rhizomes	Ginger, orris	[27, 28]
Fruits	Orange, juniper	[18, 29]

Table 1.
EOs extracted from plant parts.

EOs can be found in various plants species, in particular those that belong to the Coniferae, Myrtaceae, Rutaceae, Labiatae, Umbelliferae, Alliaceae, and Zingiberaceae families [16, 17]. EOs are derived from different plant parts, such as flowers, leaves, wood, roots, seeds, rhizomes, and fruits [18]. See **Table 1** for examples of EOs found in each of the plant parts.

2. Classification of essential oils

In general, EOs can be classified based on their chemical composition, aroma created by the oil, evaporation speed, taxonomy or the families they belong to, their therapeutic uses, consistency, their origin, and the alphabetical order [16, 30]. Classification based on consistency, for example, can be divided into essences, balms, and resins [16, 31]. See **Table 2** for definition and examples of each.

Furthermore, there are three classifications of EOs based on their origin which are natural, artificial, and synthetic [16]. The natural EOs are taken from the plant without physical or chemical modifications, while the artificial oils are obtained by

Based on consistency	Definition	Examples
Essences	Volatile liquid at room temperature [16]	Lavender, jasmine, geranium, rose [32, 33]
Balsams	Thick very volatile natural extract from tree or bush [16]	Copaiba balsam, Peruvian balsam, Canada balsam, Tolu balsam, Cabreuva balsam, Bangui balsam [16, 34]
Resins	Solid or semisolid products that comprise of derivates and abietic acid [16]	Patchouli, sandalwood, frankincense [33]

Table 2.
Classification of EOs based on consistencies.

Types of essential oil	Disadvantages	Advantages
Natural	Expensive, need a lot of natural sources to create, can cause burns if not diluted [15, 35, 36]	Great smell, helpful for physical and mental health [36]
Synthetic	No therapeutic properties, damaging the skin and respiratory system [36]	Cheap, commonly used as fragrance and taste enhancers, long lasting [15, 36]

Table 3.
Classification of EOs based on their origin.

enriching the essence with extra components (can be one or more). The synthetic EOs, however, are obtained by combining many chemical substances together [16]. See **Table 3** for comparison between natural and synthetic EOs.

3. Essential oil extraction

Five thousand years ago, the ancient civilizations have already incorporated the use of machines for EO extraction [11]. However, there has been an expansion of the different extraction methods today. One of the important methods is the hydro-distillation which is divided into water distillation, water-steam distillation, and steam distillation [37, 38]. Hydro-distillation method involves hydro-diffusion, hydrolysis, and decomposition by heat [18]. In addition, steam distillation is performed by using the Clevenger system to extract oil from both fresh and dried plants, and it takes about 3 h [1, 11]. Another method is the expression method which utilizes the machines to compress the EO out of the plant [9, 11]. Additionally, solvent extraction and ultrasonic extraction methods are also routinely used [17].

Throughout the distillation process, water is separated by gravity, and at the end it leaves the volatile liquid behind; this liquid is the EO [16, 39]. EOs that are extracted by the use of chemical solvents cannot be called true EOs according to the National Cancer Institute, because they can cause changes in the clarity, scent, and fragrance of the oil [40]. The four criteria that affect the amount of essential oils produced are (1) time of distillation, (2) temperature, (3) pressure, and (4) plant quality.

3.1 Hydro-distillation

Hydro-distillation is the most commonly used method of extraction of EOs in which the plant is boiled in water [41, 42]. This method takes 1 h of distillation for fresh samples and 1 h and 15 min for dried samples. In the hydro-distillation method, a round-bottomed flask is used to place the plant material in with distilled water; if the plant material is dry, 1000 ml of distilled water should be used for 75 g of plant material, and if it is fresh material, 400 ml of distilled water should be used with 200 g of plant material; if the sample of plant is smaller, however, they can adjust the amount of water using this ratio: 13.3 ml of distilled water for each gram of dry plant. For water distillation, the modified Clevenger trap should be used to extract EO, and at the end the volume of the oil should be determined, and the EO should be analyzed immediately [43–45]. An advanced distillation method which is the microwave-assisted hydro-distillation can be used to shorten extraction time [46, 47].

3.2 Steam distillation

Steam distillation is the traditional method of extraction of EOs from plants [37]. The fundamental principle of steam distillation is that the mixture is allowed to be distilled at a temperature that is lower than the boiling point of the component; EO substances have a high boiling point that can reach 200°C; however, these substances will be volatile when steam or boiling water is present which is in 100°C; then the hot gas mixture will be condensed to form oil if it passes through a cooling system [48]. In steam distillation, the steam is first passed into a flask that contains the plant material; after that the condensate at the bottom of the flask should be collected which will be the water and oil; then the extract is condensated three times with ethyl ether to ensure that the essential oil is fully extracted; then the moisture should be removed by adding sodium sulfate to the ethyl ether, followed by rotary evaporation to remove ethyl ether; and finally the volume of the EO is determined [43]. The advantages of this method of extraction are that it is rapid and can be controlled by the operator and it gives an acceptable quality than EOs extracted with other methods [48].

3.3 Solvent extraction

Solvent extraction method or liquid-liquid method is done by separating compounds based on their part solubility [49]. The basic principle of the solvent extraction method is that between two immiscible solvents, the solute distributes itself in a fixed ratio, whereby one is usually water and the other is an organic solvent [50]. In this method, the plant material will be grinded in a mortar that contains anhydrous hexane Na_2SO_4, followed by four rounds of extraction with hexane to obtain the yellow extract, then this is followed by adding a sufficient amount of Norite A charcoal for all extracts to remove the yellow color after low-speed centrifugation, and eventually the solution will be concentrated under air stream at room temperature [37, 43, 49]. A newer method of solvent extraction, called the microwave-assisted simultaneous distillation-solvent extraction (MW-SDE), is faster and simpler and uses fewer solvents to determine volatile compounds than conventional methods [51].

4. Composition of essential oils

4.1 Physical properties of EOs

EOs are volatile and become liquid at room temperature; they might be colorless or slightly yellow in color when extracted. Moreover EOs are lower in density than water, except for sassafras and clove essences [16]. EOs can be either liposoluble or soluble in alcohol and organic solvents, but they are only slightly soluble in water [4, 16, 32].

4.2. Chemical properties of EOs

Plants metabolites are divided into primary and secondary metabolites. The primary metabolites include proteins, DNA, and compounds that are important for cellular function. Secondary metabolites are produced by plants as a response of stress to deter herbivores or animals that would feed on them [52, 53]. Of the secondary metabolites, plant terpenes are the most numerous and diverse natural products of plant secondary metabolites which can be found in EOs [53]. They are found in monoterpene and diterpene oils and may be aliphatic, cyclic, or aromatic depending on the functional group [16]. According to the functional group, they can be alcohols, esters, ethers, hydrocarbon, and aldehydes [16].

The composition varies due to the place of origin, harvesting moment, extraction method, planting time, mineral fertilization, and climate [5, 16, 54]. For example, in warm places there will be more EOs than the cold or hot areas [16]. The concentration of EOs is extremely high due to the extraction methods used [23]. The simplest unit of EOs is the isoprene units that are composed of five carbons which can be assembled to form terpenes [16, 52]. EOs are composed of hydrocarbon molecules. Terpenes, for example, are hydrocarbon molecules that comprise of 10, 15, 20, and 30 carbon atoms and are made out of five-carbon isoprene units [55, 56].

EOs' main components are divided into terpenoid and non-terpenoid groups present in different concentrations [4]. The non-terpenoid group contains short-chain aliphatic, aromatic, nitrogenated, and sulfated substances [16, 57]. The terpenoid group contains a different composition of hydrocarbon terpenes, terpenoids, and sesquiterpenes which is responsible for the special aroma [5, 58]. In general, the non-terpenoid group is less important than the terpenoid in terms of applications [53].

5. Use of essential oils against multidrug-resistant bacteria

Antibiotics are effective drugs that play an important role in treating infections and decreasing morbidity and mortality rates [59, 60]. In general, antibiotics kill multidrug-resistant (MDR) bacteria through various mechanisms. Examples include the β-lactam antibiotics that inhibit the bacterial cell wall synthesis, fluoroquinolones that inhibit DNA synthesis, tetracycline which is an inhibitor of protein synthesis, sulfonamides as a metabolic pathway or folic acid synthesis inhibitor, and polymyxin B which interferes with cell membrane integrity [60–63]. Antibiotic resistance develops naturally but is accelerated when the antibiotics are misused in both human and animals; the bacteria will evolve and develop resistance toward antibiotics, preventing the antibiotic from killing the bacteria [59, 64]. The bacteria subsequently become resistant by many mechanisms depending on the selective pressure incurred by the antibiotic used; for example, if the penicillin is used, the bacteria will become resistant to it by producing enzymes that will act against the antibiotic which is in this situation penicillinase enzyme [39]. For instance, a study conducted in 173 hospitals in Europe showed that high antibiotic consumption hospitals have a higher number of methicillin-resistant *Staphylococcus aureus* (MRSA) [65].

Antibiotic resistance in microorganisms is increasing at a worrisome rate [66]. Hence, over the years, researchers are exploring possible alternative sources that will be helpful to mitigate MDR bacteria. Of all the potential sources, EO was identified as one of the good alternative sources, because of their effectiveness in folk medicine [67]. Bacteria can be divided into two main types: the gram-positive and the gram-negative. The gram-positive have a thicker peptidoglycan layer than the gram-negative bacteria [68]. Besides that, the gram-negative bacteria also have an outer membrane that is absent in the gram-positive bacteria (**Figure 1**).

Generally, the gram-positive bacteria are less resistant to EOs than gram-negative bacteria [69, 70]. In gram-positive bacteria, hydrophobic molecules are able to penetrate the cell and act on the cell wall and cytoplasm. This is exemplified by the phenolic compounds in EOs against gram-positive bacteria [66]. In the gram-negative bacteria, a thin layer of peptidoglycans is present with an outer membrane that contains LPS. LPS consists of lipid A, core LPS, and O-side chain, which makes the gram-negative bacteria more resistant to EOs than gram-positive bacteria [66, 71]. Small hydrophilic solutes will make use of the porin proteins in the gram-negative bacteria to pass through the outer membrane; it is this porin selectivity that also makes the gram-negative bacteria less susceptible to hydrophobic antibiotics [66, 72, 73].

EOs via their different components have different targets against microorganisms such as the membrane and the cytoplasm [8]. Scientists have also found that the solubility of EO in water allowed them to decipher how EOs penetrated the cell wall of microbes; in other words EOs, being soluble in the cell membrane phospholipid bilayer, diffuse through the membrane [74]. A study done using the EO of *Melaleuca alternifolia* (tea tree) against MDR gram-negative bacteria (e.g., *Escherichia coli* and carbapenem-resistant *K. pneumoniae*) and methicillin-resistant *S. aureus* (MRSA) showed that there is a bactericidal effect of tea tree EO on these microorganisms [75]. This indicated that the EO can be used to kill

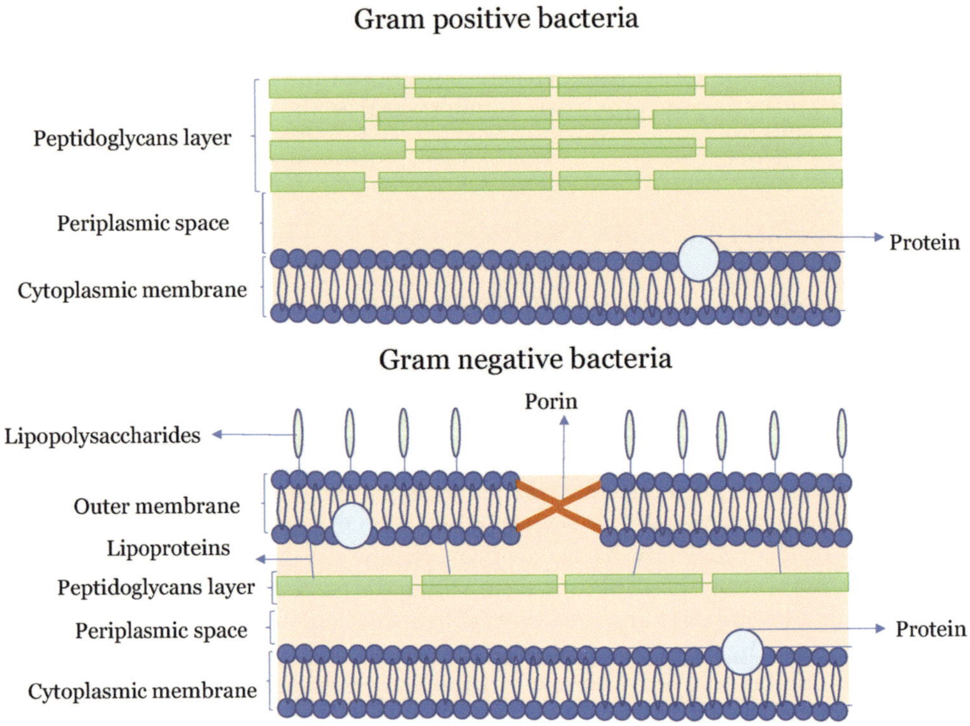

Figure 1.
Schematic of different gram-positive (at the top) and gram-negative (at the bottom) cell walls.

resistant bacteria [74]. Moreover, EO phenolic compounds' effect is concentration-dependent, whereby at low concentrations the phenolic compound will work with enzymes to produce energy, while at high concentration it will denature proteins [66, 76].

5.1 Determination of MIC of EOs for bacteria

Minimal inhibitory concentration (MIC) is the lowest concentration of a specific drug to inhibit the growth of microorganisms such as bacteria [1, 77]. After knowing that a particular EO has bactericidal, viricidal, and antiparasitic effects, the lowest concentration of EO to inhibit microbial growth should be measured [57, 78]. There are many assays to evaluate and screen for antimicrobial activity such as the disk diffusion test, microdilution (resazurin) or broth method, and agar dilution method [79, 80]. The agar disk diffusion test is commonly used to determine the antibacterial activity of the EO, but this method works only for EO with known components. This is because, for the EOs with unknown components, the antimicrobial effect may give rise to false or negative result caused by the unknown components [81]. Previously, in a study performed using the disk diffusion test to examine the antimicrobial activity of *Eucalyptus globulus* leaves, EO showed that there was a bacterial inhibitory effect on *E. coli* and *S. aureus* [82].

The commonly used alternative method to determine antimicrobial activity is the dilution method through a serial dilution of the EO in several tubes, and then determining the MIC after adding the test microorganism, turbidity is measured as a signal for growth [81]. In this method, the EO is first diluted; then it will be added to the medium that contains the broth culture, followed by incubation for 18 h in 37°C [69]. After the incubation period, the tube with the lowest concentration that showed no sign of growth is the MIC of the EO [69, 83]. However, this method requires a large quantity of the plant extract [81]. A study using the redox dye resazurin for the new modified microdilution method has been carried out to determine the MIC for tea tree EO (*Melaleuca alternifolia*) against the gram-positive and gram-negative bacteria. The results showed that the resazurin method is accurate to determine the MIC and is higher in sensitivity than the results obtained from the agar dilution assay [80].

6. Mechanisms of actions with antibiotic

EOs' mechanism of action is poorly understood, but in general it depends on their chemical composition [8, 66, 84]. As antimicrobial resistance to antibiotics is increasing, scientists are currently exploring the ability of the plant extract to modify bacterial resistance against drugs [39]. The three main types of interactions that occur between the combination of antibiotic and EO are synergism, additivity, and antagonism [85]. Synergistic interaction is when the effect of the combined chemicals is greater than the effect of a chemical alone; additive interaction is when the sum of two chemicals is equal to the sum of chemical effect alone, while antagonism is when the whole effect of the two chemicals is less than the sum of effect of a single chemical alone [86]. In a study performed using the tea tree EO against the MDR bacteria, when a combination of tea tree EO with antibiotic (e.g., oxacillin) was tested on the bacteria, in particular the MRSA, a high synergistic index in the sub-inhibitory concentration was recorded [75]. This indicates that the EO can be used to overcome bacterial resistance to antibiotic. The synergism level increases when the combined effect is higher than the individual effect in the combination therapy [39].

Combination therapy is a new method that combines antibiotics and EO to kill resistant bacteria, via enhancement of the antimicrobial activity [39, 87]. Moreover, EOs have more components possessing different mechanisms of actions for many targets than antibiotics that have only one target. Combination therapy would be useful and able to provide a new treatment option for resistance bacteria [39].

7. Application of essential oils in therapy

Daily, the human body comes into contact with EOs through various sources such as herbs, spices, orange, spearmint, lemongrass, etc., but only limited information about the amount of EO uptake is known [4, 88]. Effects of EO begin to appear after it penetrates the human body in several ways such as by ingestion, by absorbing the EO or diffusion, and by inhalation [4, 89]. EOs can be taken by inhalation through the lungs and distributed into the blood because of their volatility [90–92]. Moreover, consumption of EO by ingestion should be taken with care because EOs may cause probable toxicity [4]. EOs are used in folk medicine to treat many health problems and can also be used as food preservatives by giving antimicrobial, antioxidant, and anti-inflammation properties [93, 94].

Many studies investigated the efficiency of EOs in combination with antibiotics to combat bacterial resistance; EOs with its compounds and secondary metabolites have shown promising synergistic interaction as an indication that they would be helpful to treat and decrease bacterial resistance to antibiotics [39, 95]. The advantages that make the EOs preferable are that they will decrease adverse reactions, besides being comparatively more cost-effective, with more public acceptance due to traditional usage, and being renewable with better biodegradability properties [39, 96].

8. Synergistic activity of essential oil

The synergistic effects between the EOs and antibiotics against the MDR bacteria have been investigated [97]. The synergistic effect can be defined as the ability of EO components to act together with the antibiotic component to increase the activity of the EO against MDR bacteria [98]. This is important because it will help to reduce the use of antibiotics and decrease the rates of antibiotic resistance [97]. Some studies have been done to assess the combinatory activities of lavender, cinnamon bark, peppermint, and other EOs against bacteria, and the results show there is a synergistic effect [97]. Some of these EOs will be discussed in the following sections.

8.1 Lavender essential oil

The lavender EO is used in traditional medicine as well as in cosmetic products; this oil is believed to have sedative, anti-inflammatory, and antimicrobial effects [99]. Lavender EO shows a synergistic effect when combined with piperacillin antibiotic against beta-lactamase-producing *Escherichia coli* under study with fractional inhibitory concentration (FIC) index between 0.26 and 0.5 [97]. This finding shows that it's possible to use the lavender EO as an agent in modifying the antibiotic resistance [97]. Another study which aimed to compare the antimicrobial efficacy of four types of lavender oil on MSSA and MSRA shows that by direct contact the oil inhibits the growth of these microbes [100]. Fusidic acid is one of the compounds within this oil which gives it the antimicrobial activity, the mechanism of which is to cause bacterial cell damage by reducing synthesis of proteins [101].

8.2 Cinnamon bark essential oil

The cinnamon bark EO can be obtained from different parts of the tropical evergreen tree, which is important for human health and agriculture uses [102]. Previously, a study reported that a combination of cinnamon bark EO with piperacillin resulted in a synergistic relationship with FIC $\leq$ 0.5, and this result indicates the possibility of using cinnamon bark EO as a resistance-modifying agent against MDR bacteria [97, 103]. Cinnamon bark oil contains cinnamaldehyde which is one of the compounds that inhibit the activity of amino acid decarboxylase; this compound with others within the oil gives this oil the ability to inhibit some pathogenic bacteria [104].

8.3 Peppermint essential oil

Peppermint EO is significant in inhibiting the microbial growth and increasing the shelf-life of food by preventing food spoilage [105]. Combination of piperacillin and peppermint EOs with FIC in the range 0.26–0.5 was found showing a synergistic effect that is absent in 31 other combination pairs that were studied, indicating a promising alternative to reduce the use of antibiotic and achieve the reverse beta-lactam antibiotic resistance [91]. The antibacterial activity for this oil is associated with menthol and ethyl acetate in high concentrations [106].

9. Future perspectives

Research about the reversal antibiotic resistance is important to preserve the healthy microbial ecosystem in the human host. It is imperative to understand the cause of antimicrobial resistance and to find solutions to alleviate the present situation. As discussed above, combination therapy between EOs and antibiotic provides a promising alternative to mitigate MDR bacteria, possibly by disrupting the bacterial cell wall. Although EOs have been proven to be useful for mitigating MDR bacteria spread, there is still much to be done in terms of the combination stability, selectivity, definite mechanism of action, chemical nature, availability of these products in human body, optimal dose, and adverse reactions as a treatment. These gaps need to be taken into consideration before applying EOs for clinical usage. In addition, there is also a need for animal study and human trials in the future, if one intends to employ EOs as a therapeutic option in medical settings.

Acknowledgements

The authors would like to thank the HCT Research Grants from the Higher Colleges of Technology, UAE for supporting this work.

Conflict of interest

The authors declare they have no competing interests.

Author details

Mariam Aljaafari[1], Maryam Sultan Alhosani[1], Aisha Abushelaibi[1], Kok-Song Lai[1,2] and Swee-Hua Erin Lim[1*]

1 Health Science Division, Abu Dhabi Women's College, Higher Colleges of Technology, Abu Dhabi, United Arab Emirates

2 Department of Cell and Molecular Biology, Faculty of Biotechnology and Biomolecular Sciences, University Putra Malaysia, Serdang, Selangor, Malaysia

*Address all correspondence to: lerin@hct.ac.ae

References

[1] Sartoratto A, Machado ALM, Delarmelina C, Figueira GM, Duarte MCT, Rehder VLG. Composition and antimicrobial activity of essential oils from aromatic plants used in Brazil. Brazilian Journal of Microbiology. 2004;**35**(4):275-280

[2] Maia MF, Moore SJ. Plant-based insect repellents: A review of their efficacy, development and testing. Malaria Journal. 2011;**10**(1):S11

[3] Sharifi-Rad J, Sureda A, Tenore GC, Daglia M, Sharifi-Rad M, Valussi M, et al. Biological activities of essential oils: From plant chemoecology to traditional healing systems. Molecules. 2017;**22**(1):70

[4] Jilani A, Dicko A. The therapeutic benefits of essential oils. In: Nutrition, Well-Being and Health [Internet]. Croatia: InTech; 2012. Available from: https://www.researchgate.net/publication/221925405_The_Therapeutic_Benefits_of_Essential_Oils

[5] Eslahi H, Fahimi N, Sardarian AR. Chemical composition of essential oils. In: Essential Oils in Food Processing [Internet]. United States: John Wiley & Sons, Ltd; 2017. pp. 119-171. Available from: https://onlinelibrary.wiley.com/doi/abs/10.1002/9781119149392.ch4 [Cited: January 4, 2019]

[6] Both Methylerythritol Phosphate and Mevalonate Pathways Contribute to Biosynthesis of each of the Major Isoprenoid Classes in Young Cotton Seedlings | Request PDF. ResearchGate [Internet]. Available from: https://www.researchgate.net/publication/259446369_Both_methylerythritol_phosphate_and_mevalonate_pathways_contribute_to_biosynthesis_of_each_of_the_major_isoprenoid_classes_in_young_cotton_seedlings [Cited: January 26, 2019]

[7] Secondary Metabolites [Internet]. Appendix 4. Available from: http://6e.plantphys.net/PlantPhys6e-appendix04.pdf

[8] Burt S. Essential oils: their antibacterial properties and potential applications in foods: A review. International Journal of Food Microbiology. 2004;**94**(3):223-253

[9] Asbahani AE, Miladi K, Badri W, Sala M, Addi EHA, Casabianca H, et al. Essential oils: From extraction to encapsulation. International Journal of Pharmaceutics. 2015;**483**(1):220-243

[10] Isman MB. Plant essential oils for pest and disease management. Crop Protection. 2000;**19**:603-608. [Internet]. Available from: http://citeseerx.ist.psu.edu/viewdoc/download?doi=10.1.1.456.3876&rep=rep1&type=pdf

[11] Dupler D, Odle TG, Newton DE. Essential oils. In: Fundukian LJ, editor. The Gale Encyclopedia of Alternative Medicine [Internet]. 4th ed. Farmington Hills, MI: Gale; 2014. pp. 859-862. Available from: http://link.galegroup.com/apps/doc/CX3189900316/HWRC?u=971hctnet&sid=HWRC&xid=1cc4939e [Cited: December 23, 2018]

[12] Hammer KA, Carson CF, Riley TV. Antimicrobial activity of essential oils and other plant extracts. Journal of Applied Microbiology. 1999;**86**(6):985-990

[13] Cruz O. Biological screening of Brazilian medicinal plants. Memórias do Instituto Oswaldo Cruz. 2000;**95**(3):367-373

[14] Muturi EJ, Ramirez JL, Doll KM, Bowman MJ. Combined toxicity of three essential oils against *Aedes aegypti* (Diptera: Culicidae) larvae.

Journal of Medical Entomology. 2017;**54**(6):1684-1691

[15] Do T, Hadji-Minaglou F, Antoniotti S, Fernandez X. Authenticity of essential oils. TrAC - Trends in Analytical Chemistry [Internet]. 2014;**66**:146-157. Available from: https://www.researchgate.net/publication/269724960_Authenticity_of_essential_oils

[16] Arriaza P. Topic 7 essential oil. In: Industrial Utilization of Medicinal and Aromatic Plants [Internet]. Madrid, Spain: OpenCourseWare of the Polytechnic University of Madrid; 2010. Available from: http://ocw.upm.es/ingenieria-agroforestal/industrial-utilization-of-medicinal-and-aromatic-plants/contenidos/temario/Unit-4/tema_7.pdf

[17] Moghaddam M, Mehdizadeh L. Chemistry of essential oils and factors influencing their constituents. 2017. pp. 379-419. Available from: https://www.researchgate.net/publication/319007448_Chemistry_of_Essential_Oils_and_Factors_Influencing_Their_Constituents?enrichId=rgreq-7a1ddd9a13778839005186ed2cb997e7-XXX&enrichSource=Y292ZXJQYWdlOzMxOTAwNzQ0ODtBUzo1NTAxODA5MDM3NTU3NzZAMTUwODE4NDc1Nzc5NA%3D%3D&el=1_x_3&_esc=publicationCoverPdf [Cited: December 27, 2018]

[18] Extraction Methods of Natural Essential Oils [Internet]. Available from: http://agritech.tnau.ac.in/horticulture/extraction_methods_natural_essential_oil.pdf [Cited: December 30, 2018]

[19] Carlson L, Machado R, Spricigo C, Krücken Pereira L, Bolzan A. Extraction of Lemongrass Essential Oil. 2013. Available from: https://www.researchgate.net/publication/256537287_extraction_of_lemongrass_essential_oil

[20] Yang S-K, Yap PS-X, Krishnan T, Yusoff K, Chan K-G, Yap W-S, et al. Mode of action: Synergistic interaction of peppermint (*Mentha x piperita* L. Carl) essential oil and meropenem against plasmid-mediated resistant *E. coli*. Records of Natural Products. 2018;**12**(6):582-594

[21] Sandalwood Oils [Internet]. Available from: http://www.intracen.org/uploadedFiles/intracenorg/Content/Exporters/Market_Data_and_Information/Market_information/Market_Insider/Essential_Oils/Sandalwood%20oils.pdf [Cited: December 30, 2018]

[22] Baker BP, Grant JA, Malakar-Kuenen R. Cedarwood Oil Profile. New York: New York State IPM Program; 2018. p. 8

[23] Sethi ML, Subba RG, Chowdhury BK, Morton JF, Kapadia GJ. Identification of volatile constituents of *Sassafras albidum* root oil. Phytochemistry. 1976;**15**:1773-1775

[24] Wharf C. Herbal Medicine: Summary for the Public [Internet]. United Kingdom: European Medicines Agency; 2016. Available from: https://www.ema.europa.eu/documents/herbal-summary/valerian-essential-oil-summary-public_en.pdf

[25] Damayanti A, Setyawan E. Essential oil extraction of fennel seed (*Foeniculum vulgare*) using steam distillation. International Journal of Science and Engineering [Internet]. 2012;**3**:12-14. Available from: https://media.neliti.com/media/publications/93541-EN-essential-oil-extraction-of-fennel-seed.pdf

[26] Nurjanah S, Lanti Putri I, Pretti Sugiarti D. Antibacterial Activity of Nutmeg Oil. 2017. Available from: https://www.researchgate.net/publication/322976449_Antibacterial_Activity_of_Nutmeg_Oil

[27] Pai Jakribettu R, Boloor R, Bhat HP, Thaliath A, Haniadka R, Rai MP, et al. Ginger (*Zingiber officinale* Rosc.) Oils. Netherlands: Elsevier; 2016. pp. 447-454. Available from: https://www.researchgate.net/publication/301257332_Ginger_Zingiber_officinale_Rosc_Oils

[28] Olga M. Composition of volatile oil of *Iris pallida* Lam. from Ukraine. Turkish Journal Of Pharmaceutical Sciences. 2017;**15**:85-90

[29] Haypek E, Silva LH, Batista E, Stefanelli M, Meireles MA, Meirelles A. Recovery of aroma compounds from orange essential oil. Brazilian Journal of Chemical Engineering [Internet]. 2000;**17**:4-7. Available from: http://www.scielo.br/scielo.php?script=sci_arttext&pid=S0104-66322000000400034

[30] Chyi EW. Classification of Essential Oil | Essential Oil | Ester [Internet]. Scribd. Available from: https://www.scribd.com/document/88973088/Classification-of-Essential-Oil [Cited: December 24, 2018]

[31] Hashemi SMB, Mousavi Khaneghah A, A. de Souza Sant'Ana, editors. Essential Oils in Food Processing: Chemistry, Safety and Applications [Internet]. 1st ed. United States: Wiley; 2017. Available from: https://onlinelibrary.wiley.com/doi/book/10.1002/9781119149392 [Cited: December 30, 2018]

[32] What are Essential Oils and Do They Work? [Internet]. Healthline. 2017. Available from: https://www.healthline.com/nutrition/what-are-essential-oils [Cited: December 27, 2018]

[33] Introduction to Essential Oils—Essential Oil Families [Internet]. Escents Aromatherapy, Canada. Available from: https://www.escents.ca/pages/introduction-to-essential-oils-essential-oil-families [Cited: December 27, 2018]

[34] Custódio DL, Veiga-Junior VF. True and common balsams. Revista Brasileira de Farmacognosia. 2012;**22**(6):1372-1383

[35] Morson J. 5 Things to Know About Essential Oils [Internet]. MedShadow. 2018. Available from: https://medshadow.org/features/essential-oils/ [Cited: December 27, 2018]

[36] Natural vs Synthetic Essential Oils [Internet]. Maple Holistics. 2016. Available from: https://www.mapleholistics.com/blog/natural-vs-synthetic-essential-oils/ [Cited: December 27, 2018]

[37] Chemat F, Boutekedjiret C. Extraction//steam distillation. In: Reference Module in Chemistry, Molecular Sciences and Chemical Engineering [Internet]. Netherlands: Elsevier; 2015. Available from: https://www.researchgate.net/publication/285643201_Extraction_Steam_Distillation

[38] Extraction_methods_natural_essential_oil.pdf [Internet]. Available from: http://agritech.tnau.ac.in/horticulture/extraction_methods_natural_essential_oil.pdf [Cited: January 2, 2019]

[39] Yap PSX, Yiap BC, Ping HC, Lim SHE. Essential oils, a new horizon in combating bacterial antibiotic resistance. The Open Microbiology Journal. 2014;**8**:6-14

[40] Aromatherapy with Essential Oils [Internet]. National Cancer Institute. 2005. Available from: https://www.cancer.gov/about-cancer/treatment/cam/hp/aromatherapy-pdq [Cited: December 25, 2018]

[41] Stahl-Biskup E, Venskutonis RP. Thyme. In: Handbook of Herbs and Spices [Internet]. Sawston, Cambridge: Elsevier; 2012. pp. 499-525. Available from: https://linkinghub.elsevier.com/

retrieve/pii/B978085709039350027X [Cited: January 2, 2019]

[42] Hydrodistillation—An Overview | ScienceDirect Topics [Internet]. Available from: https://www.sciencedirect.com/topics/immunology-and-microbiology/hydrodistillation [Cited: January 21, 2019]

[43] Charles DJ, Simon JE. Comparison of extraction methods for the rapid determination of essential oil content and composition of basil. Journal of the American Society for Horticultural Science. 1990;**5**:458-462

[44] Okoh OO, Sadimenko AP, Afolayan AJ. Comparative evaluation of the antibacterial activities of the essential oils of *Rosmarinus officinalis* L. obtained by hydrodistillation and solvent free microwave extraction methods. Food Chemistry. 2010;**120**(1):308-312

[45] Bayramoglu B, Sahin S, Sumnu G. Solvent-free microwave extraction of essential oil from oregano. Journal of Food Engineering. 2008;**88**(4):535-540

[46] Tyśkiewicz K, Gieysztor R, Konkol M, Szałas J, Rój E. Essential oils from *Humulus lupulus* scCO2 extract by hydrodistillation and microwave-assisted hydrodistillation. Molecules [Internet]. 2018;**23**(11):2866. Available from: https://www.ncbi.nlm.nih.gov/pmc/articles/PMC6278360/ [Cited: January 2, 2019]

[47] Kusuma HS, Mahfud M. Kinetic studies on extraction of essential oil from sandalwood (*Santalum album*) by microwave air-hydrodistillation method. Alexandria Engineering Journal. 2018;**57**:1163-1172. Available from: https://www.sciencedirect.com/science/article/pii/S1110016817300698?via%3Dihub [Cited: January 2, 2019]

[48] Ames GR, Matthews WSA. The distillation of essential oils. Tropical Science. 1968;**10**:136-148

[49] Rassem HHA, Nour AH, Yunus RM. Techniques for extraction of essential oils from plants: A review. Australian Journal of Basic and Applied Sciences. 2016;**10**:11

[50] Wells MJM. Principles of extraction and the extraction of semivolatile organics from liquids. In: Mitra S, editor. Sample Preparation Techniques in Analytical Chemistry [Internet]. Hoboken, NJ, USA: John Wiley & Sons, Inc.; 2003. pp. 37-138. Available from: http://doi.wiley.com/10.1002/0471457817.ch2 [Cited: January 21, 2019]

[51] Ferhat MA, Tigrine-Kordjani N, Chemat S, Meklati BY, Chemat F. Rapid extraction of volatile compounds using a new simultaneous microwave distillation: Solvent extraction device. Chromatographia. 2007;**65**(3):217-222

[52] Hughes R. Chemistry of Essential Oils [Internet]. UKassays. 2018. Available from: https://www.ukessays.com/essays/chemistry/chemistry-essential-oils-9280.php [Cited: December 31, 2018]

[53] Zwenger S, Basu C. Plant terpenoids: Applications and future potentials. Biotechnology and Molecular Biology Reviews. 2008;**3**:1-7

[54] Cassel E, Vargas RMF, Martinez N, Lorenzo D, Dellacassa E. Steam distillation modeling for essential oil extraction process. Industrial Crops and Products. 2009;**29**(1):171-176

[55] Isoprene Units are the 5 Carbon Units Making up the Terpenoid Molecules Found in Essential Oils [Internet]. Available from: https://essentialoils.co.za/isoprene.htm [Cited: January 1, 2019]

[56] Essential Oils from Steam Distillation [Internet]. Available from: http://www.engineering.iastate.edu/brl/files/2011/10/brl_essentialoils.pdf [Cited: January 1, 2019]

[57] Bakkali F, Averbeck S, Averbeck D, Idaomar M. Biological effects of essential oils: A review. Food and Chemical Toxicology. 2008;**46**(2):446-475

[58] Chamorro ER. Study of the Chemical Composition of Essential Oils by Gas Chromatography. 2012. Available from: https://www.researchgate.net/publication/221926513_Study_of_the_Chemical_Composition_of_Essential_Oils_by_Gas_Chromatography [Cited: December 31, 2018]

[59] Antibiotic Resistance [Internet]. World Health Organisation. 2018. Available from: https://www.who.int/news-room/fact-sheets/detail/antibiotic-resistance [Cited: January 3, 2019]

[60] Jum'a S, Karaman R. Antibiotics. 2015. pp. 41-73. Available from: https://www.researchgate.net/publication/272820065_ANTIBIOTICS

[61] Singh B. Antibiotics: Introduction to Classification. 2015. Available from: https://www.researchgate.net/publication/281405283_Antibiotics_Introduction_to_Classification

[62] Tenover FC. Mechanisms of antimicrobial resistance in bacteria. The American Journal of Medicine. 2006;**119**(6 Suppl 1):S3-S10; Discussion S62-S70

[63] Yang S-K, Low L-Y, Yap PS-X, Yusoff K, Mai C-W, Lai K-S, et al. Plant-derived antimicrobials: Insights into mitigation of antimicrobial resistance. Records of Natural Products. 2018;**12**(4):295-396

[64] Cantón R, Morosini M-I. Emergence and spread of antibiotic resistance following exposure to antibiotics. FEMS Microbiology Reviews. 2011;**35**(5):977-991

[65] MacKenzie FM, Bruce J, Struelens MJ, Goossens H, Mollison J, Gould IM, et al. Antimicrobial drug use and infection control practices associated with the prevalence of methicillin-resistant *Staphylococcus aureus* in European hospitals. Clinical Microbiology and Infection. 2007;**13**(3):269-276

[66] Nazzaro F, Fratianni F, De Martino L, Coppola R, De Feo V. Effect of essential oils on pathogenic bacteria. Pharmaceuticals. 2013;**6**(12):1451-1474

[67] Mahasneh AM, El-Oqlah AA. Antimicrobial activity of extracts of herbal plants used in the traditional medicine of Jordan. Journal of Ethnopharmacology. 1999;**64**(3):271-276

[68] Beveridge TJ. Structures of gram-negative cell walls and their derived membrane vesicles. Journal of Bacteriology. 1999;**181**(16):4725-4733

[69] Bosnić T, Softić D, Grujić-Vasić J. Antimicrobial activity of some essential oils and major constituents of essential oils. Acta Medica Academica. 2006;**4**:19-22

[70] Inouye S, Takizawa T, Yamaguchi H. Antibacterial activity of essential oils and their major constituents against respiratory tract pathogens by gaseous contact. The Journal of Antimicrobial Chemotherapy. 2001;**47**(5):565-573

[71] Structure—Medical Microbiology—NCBI Bookshelf. Available from: https://www.ncbi.nlm.nih.gov/books/NBK8477/ [Cited: January 7, 2019]

[72] Nikaido H. Prevention of drug access to bacterial targets: Permeability barriers and active efflux | Science. American Association for the Advancement of Science. 1994;**264**(5157):382-388

[73] Vaara M. Agents that increase the permeability of the outer membrane. Microbiological Reviews. 1992;**56**(3):395-411

[74] Knobloch K, Pauli A, Iberl B, Weigand H, Weis N. Antibacterial and antifungal properties of essential oil components. Journal of Essential Oil Research. 1989;**1**(3):119-128

[75] Oliva A, Costantini S, De Angelis M, Garzoli S, Božović M, Mascellino MT, et al. High potency of melaleuca alternifolia essential oil against multi-drug resistant gram-negative bacteria and methicillin-resistant *Staphylococcus aureus*. Molecules [Internet]. 2018;**23**(10):1-14. Available from: https://www.ncbi.nlm.nih.gov/pubmed/30304862

[76] Tiwari BK, Valdramidis VP, Donnell CPO, Muthukumarappan K, Bourke P, Cullen PJ. Application of Natural Antimicrobials for Food Preservation [Internet]. 2009. Available from: https://pubs.acs.org/doi/abs/10.1021/jf900668n [Cited: January 7, 2019]

[77] Wiegand I, Hilpert K, Hancock REW. Agar and broth dilution methods to determine the minimal inhibitory concentration (MIC) of antimicrobial substances. Nature Protocols. 2008;**3**(2):163-175

[78] Kalemba D, Kunicka A. Antibacterial and antifungal properties of essential oils. Current Medicinal Chemistry. 2003;**10**(17):813-829

[79] Balouiri M, Sadiki M, Ibnsouda SK. Methods for in vitro evaluating antimicrobial activity: A review. Journal of Pharmaceutical Analysis. 2016;**6**(2):71-79

[80] Mann CM, Markham JL. A new method for determining the minimum inhibitory concentration of essential oils. Journal of Applied Microbiology. 1998;**84**(4):538-544

[81] Nicolaas Eloff J. A sensitive and quick microplate method to determine the minimal inhibitory concentration of plant extracts for bacteria. Planta Medica. 1999;**64**(8):711-713

[82] Bachir RG, Benali M. Antibacterial activity of the essential oils from the leaves of Eucalyptus globulus against *Escherichia coli* and *Staphylococcus aureus*. Asian Pacific Journal of Tropical Biomedicine. 2012;**2**(9):739-742

[83] Andrews JM. Determination of minimum inhibitory concentrations. The Journal of Antimicrobial Chemotherapy. 2001;**48**(suppl_1):5-16

[84] Calo JR, Crandall PG, O'Bryan CA, Ricke SC. Essential oils as antimicrobials in food systems: A review. Food Control. 2015;**54**:111-119

[85] Danish Veterinary and Food Administration. Combined Actions and Interactions of Chemicals in Mixtures: The Toxicological Effects of Exposure to Mixtures of Industrial and Environmental Chemicals. [Internet]. 1st ed. Søborg: Danish State Information Centre; 2003. Available from: https://mst.dk/media/94195/Rapport_kombinationseffekter_2003.pdf

[86] Combined Toxic Effects of Multiple Chemical Exposures [Internet]. Available from: https://vkm.no/download/18.d44969415d027c43cf1e869/1509708687404/Combined%20toxic%20effects%20of%20multiple%20chemical%20exposures.pdf [Cited: January 21, 2019]

[87] Gibbons S, Oluwatuyi M, Veitch NC, Gray AI. Bacterial resistance modifying agents from *Lycopus europaeus*. Phytochemistry. 2003;**62**(1):83-87

[88] Vankar PS. Essential oils and fragrances from natural sources. Resonance. 2004;**9**(4):30-41

[89] Manion CR, Widder RM. Essentials of essential oils. American Journal of Health-System Pharmacy. 2017;**74**(9):e153-e162

[90] Kallithea GIS on AP (1981). Aromatic plants: Basic and applied aspects. In: Proceedings of an International Symposium on Aromatic Plants [Internet]. Springer Science & Business Media; 1982. 298p. Available from: https://books.google.ae/books?id=_z3qCAAAQBAJ&pg=PR7&lpg=PR7&dq=

[91] Moss M, Cook J, Wesnes K, Duckett P. Aromas of rosemary and lavender essential oils differentially affect cognition and mood in healthy adults. The International Journal of Neuroscience. 2003;**113**(1):15-38

[92] Guidelines for Safe and Effective Use of Essential Oils [Internet]. Available from: https://www.aromatherapynaturesway.com/wp-content/uploads/2018/02/Safe-and-Effective-Use-of-Essentuial-Oils.pdf [Cited: January 8, 2019]

[93] Dagli N, Dagli R, Mahmoud RS, Baroudi K. Essential oils, their therapeutic properties, and implication in dentistry: A review. Journal of International Society of Preventive and Community Dentistry. 2015;**5**(5):335-340

[94] Baratta MT, Dorman HJD, Dcans SG, Figueiredo AC, Barroso JG, Ruberto G. Antimicrobial and antioxidant properties of some commercial essential oils. Flavour and Fragrance Journal. 1998;**13**(4):235-244

[95] Swamy MK, Akhtar MS, Sinniah UR. Antimicrobial properties of plant essential oils against human pathogens and their mode of action: An updated review. Evidence-Based Complementary and Alternative Medicine [Internet]. 2016;**2016**:1-21. Available from: https://www.ncbi.nlm.nih.gov/pmc/articles/PMC5206475/ [Cited: January 22, 2019]

[96] Yap PSX, Yang SK, Lai KS, ErinLim SH. Essential oils: The ultimate solution to antimicrobial resistance in *Escherichia coli*? 2017. Available from: https://www.intechopen.com/books/-i-escherichia-coli-i-recent-advances-on-physiology-pathogenesis-and-biotechnological-applications/essential-oils-the-ultimate-solution-to-antimicrobial-resistance-in-i-escherichia-coli-i- [Cited: January 26, 2019]

[97] Yap PSX, Lim SHE, Hu CP, Yiap BC. Combination of essential oils and antibiotics reduce antibiotic resistance in plasmid-conferred multidrug resistant bacteria. Phytomedicine. 2013;**20**(8):710-713

[98] Wendy TL, Edwin JAV, Burt S. Synergy between essential oil components and antibiotics: A review. Journal Critical Reviews in Microbiology 2013;**40**:76-94

[99] Cavanagh HMA, Wilkinson JM. Lavender essential oil: A review. Australian Infection Control. 2005;**10**(1):35-37

[100] Roller S, Ernest N, Buckle J. The antimicrobial activity of high-necrodane and other lavender oils on methicillin-sensitive and -resistant *Staphylococcus aureus* (MSSA and MRSA). Journal of Alternative and Complementary Medicine. 2009;**15**(3):275-279

[101] de Rapper S, Viljoen A, van Vuuren S. The in vitro antimicrobial effects of *Lavandula angustifolia* essential oil in combination with conventional antimicrobial agents. Evidence-Based Complementary and Alternative Medicine [Internet]. 2016;**2016**:1-9. Available from: https://www.hindawi.com/journals/ecam/2016/2752739/ [Cited: February 7, 2019]

[102] Haddi K, Faroni LRA, Eugenio O. Cinnamon oil. In: ResearchGate [Internet]. 1st ed. United States: CRC Press/Taylor & Francis Group; 2017. pp. 118-150. Available from: https://www.researchgate.net/

publication/317752931_Cinnamon_Oil [Cited: January 19, 2019]

[103] Yang S-K, Yusoff K, Mai C-W, Lim W-M, Yap W-S, Lim S-HE, et al. Additivity vs synergism: Investigation of the additive interaction of cinnamon bark oil and meropenem in combinatory therapy. Molecules. 2017;**22**(11):1733

[104] Magetsari R. Effectiveness of cinnamon oil coating on K-wire as an antimicrobial agent against *Staphylococcus epidermidis*. Malaysian Orthopaedic Journal. 2013;7(3):10-14

[105] Kang J, Jin W, Wang J, Sun Y, Wu X, Liu L. Antibacterial and anti-biofilm activities of peppermint essential oil against *Staphylococcus aureus*. LWT. 2019;**101**:639-645

[106] Singh R, Shushni MAM, Belkheir A. Antibacterial and antioxidant activities of *Mentha piperita* L. Arabian Journal of Chemistry. 2015;**8**(3):322-328

Chapter 5

Essential Oils in the Development of New Medicinal Products

Jason Jerry Atoche Medrano

Abstract

The essential oils present a complex composition of different chemical compounds, where they present synergic or complementary action among each other, modifying their activity. Among its main components we can find the terpenoids and phenylpropanoids, which are responsible for giving the medicinal properties. Essential oils generally have a pleasant and intense odor, mostly in liquid form, found in different plant organs and are soluble in polar solvents. Essential oils are volatile, and are widely used in the perfume industry, cosmetics, food and beverage aroma, as well as use in aromatherapy to treat some diseases. The traditional knowledge of some plant species with phytotherapeutic properties is currently a source for research in the search for new biologically active compounds and as effective therapeutics that contemplate current health care.

Keywords: kinds, chemical compounds, productions, medicinal properties, health care

1. Introduction

The essential oils (EOs) have a wide variety in nature with which they turn out to be an important base in the agricultural activity. They can be used as alternative medicine [1], in food products [2], perfume fixatives [3], pharmaceuticals and cosmetics [4], among others. Many types of essential oil oils are obtained from different plant species which means an important production from a commercial point of view, in this way we can conclude that the production and consumption of essential oils are increasing all over the world [5].

The EOs are susceptible to a series of factors that determine the quantity and quality of the same, as well as: the genetic variation, the type or the variety of plants, the geographic location of the plants, the surroundings, the weather, the seasonal variations, stress during growth, the process of obtaining used; affecting its chemical properties, such as composition and phytotherapeutic properties [6, 7].

Essential oils are mixtures of several compounds with low molecular weight; the obtaining can be done through different techniques: steam distillation, hydro distillation or extraction with solvents [8]. Considering the performance for each species should be considered the most convenient technique to be used. Considering commercial scale production, steam distillation is a technique widely used to obtain essential oils [9]. The main components found in the EOs are: monoterpenes, sesquiterpenes and oxygenated derivatives of these. In this way the terpenoids and phenylpropanoids are the components that make up the EOs [10]. Generally, the particular bioactivities of the OEs are due to one or two of the main components, in this way the characteristics of the EOs is the result of a synergic effect between the main components [11].

We can find from natural products, as well as from their derivatives, new therapeutic sources for various treatments of diseases [12, 13]. Humanity has used natural products since antiquity for the treatment of various diseases, so it is not surprising that this knowledge is sought from a new scientific perspective. Within the natural products we can find the EOs, which have diverse applications mainly in health, agriculture, cosmetics and food.

Currently there is a worldwide effort to study and understand the phytotherapeutic, antimicrobial, antimutagenic, anticancer, antioxidant properties, among others, of the EOs [14].

2. Properties and characterization

The characterization of an essential oil starts with the designation of the vegetable source, i.e., plant, from which it was isolated and the part of the used plant (flowers, leaves, fruits, rhizomes, roots, etc.). It is important to supply, together with the vulgar (vernacular) name of the plant, for example, rosemary, its botanical name (taxonomic identification), which consists of the names of the genus and the species, in this case, *Rosmarinus* (genus) *officinalis* (species) of the Labiatae family and, if the subspecies exists or the variety of the plant, it is important to add it.

It is also necessary to specify, if there is, the chemist of the plant, which receives often its name for the compound majority or distinctive, present in the essential oil. Botanical identification, through the scientific name of the plant, it allows to avoid confusion.

For example, under the vulgar name of "chamomile" they can figure different species, with oils essential composition and properties well different, i.e., German chamomile is Matricaria recutita (*Matricaria chamomilla*) and the Roman chamomile is *Anthemis nobilis*; both belong to the family Asteraceae (Compositae) and both are commonly called "chamomile." Along with identification botany, the provenance of the plant, that is, where it was cultivated (country, region) and what was the extraction method of its essential oil (steam drag or hydrodilatation).

Many factors affect the composition and yield of essential oil in the plant. Among the main ones are: geoclimatic localization, type of soil, stage of development of the plant (e.g., before, during or after flowering) and even the time of day when it is harvested, among others. Geoclimatic factors and soil type can give rise to different chemotypes of the plant, from which essential oils with chemical composition, sensory properties and different biological activity are distilled. For example, in thyme, *Thymus vulgaris* (Fam. Labiatae), at least four chemists are distinguished, according to their major compounds in the essential oil: (I) thymol and p-cymene; (II) carvacrol, timol and borneol; (III) linalool, terpinen-4-ol and linalyl acetate and (IV) geraniol and geranyl acetate. Each oil isolated from these chemotypes, smells different and has different biological properties.

While thymus chemotypes I and II have a strong antibacterial activity, they are irritant, chemotypes III and IV are not and have a moderate antibacterial activity. Thyme chemotype III oil has a sedative effect due to the presence of linalool, monoterpene alcohol, and its acetate [15].

Another example, is the essential oils of geranium plants (*Pelargonium graveolens*), cultivated in the Reunion Islands (Indian Ocean, northern Madagascar) and in China. In the international market the first essential oil is known as "Bourbon" and the second is called "Chinese geranium oil"; its chemical compositions vary widely [15], which can be seen in **Table 1**.

Compounds	Geranium (China), %	Geranium (reunion), %
Citronellol	40	22
Citronellyl formate	11	8
Geraniol	6	17
Geranyl formate	2	7.5
Linalool	4	13

Table 1.
Comparison of essential oils, according to their major compounds isolated from geraniums cultivated in China and Reunion Islands.

Another important parameter is the time of harvest of the plant; from this depend both the yield and the composition of the extracted oil [16]. For example, sage essential oil (*Salvia officinalis*, Fam. Labiatae) contains a neurotoxic monoterpene ketone, α-thujone [15], in different amounts, depending on the time when the plant is harvested. The content of the ketone varies as follows: it is high, when the plant is harvested after flowering, and it is low before its flowering. This is precisely the time when the collection of *Salvia officinalis* is done.

The jasmine flowers collected in the morning hours contain in their oil a preferred combination of linalool, benzyl alcohol, cis-jasmone and indole, but when the flowers are collected in the afternoon hours, their oils have high levels of benzyl benzoate, eugenol and methyl salicylate; the last two introduce some unpleasant and undesirable scent notes, which can generate a rejection in the perfume industry or in aromatherapy. For ylang-ylang (*Cananga odorata*, Fam. Anonácea), the state of its flowers, i.e., fresh vs. wilted, or ripe, yellow vs. green and underdeveloped, notably affects the composition of the oil obtained: the Extra quality of the oil is reached, among other factors, when the yellow flowers are exclusively distilled, completely developed, freshly collected during the first hours of the morning [17].

3. Essential oils used against bacteria and microbes

It is a fact that many of the bacterial infections have increased even after the discovery of many antibiotics, among other factors due to the appearance of strains resistant to antibiotics and the increase of the population with less immunity. This being one of the main causes of deaths due to infectious diseases caused by bacteria [18]. Additionally, the effects of toxicity due to side effects restrict the prolonged use of high concentrations of available antibacterial drugs. In this way it is evident the need to explore new molecules and alternative treatments against pathogenic bacteria, obtained from these natural products [19].

Many plant species contain molecules with antimicrobial properties. It has been shown that especially plant OEs exhibit broad-spectrum inhibitory activities against various bacterial pathogens [20]. In the case of the family of grasses, Poaceae, which includes the producer of lemongrass oil (of *Cymbopogon citratus*), citronella oil (of *C. nardus*) and palmarosa oils (*C. martinii*). The medicinally active components of these EOs are citral, geraniol and geranyl acetate. They have demonstrated antimicrobial and anticancer properties.

In the case of citrus oils that constitute limonene and linalool are derived from the fruit peel of the plants belonging to the Rutaceae family, it has been shown that these components exhibit antimicrobial potential. In the case of the plants *Pelargonium graveolens* and *Santalum* spp. of the family Geraniaceae and Santalaceae, respectively, it has been determined that they possess two important oils, namely, geranium oil and sandalwood, with similar properties [21].

4. Antifungal activities of essential oils

Considering the case of fungi (eukaryotes), they have similarities with their guests both at a cellular and molecular level. Therefore, fungi are a difficult target to attack [22]. Currently there is evidence of the emergence of drug-resistant strains, infections associated with biofilms and the side effects of prescription drugs present difficulties for the prevention and treatment of fungal infections.

Therefore, invasive fungal infections are associated with very high morbidity and mortality rates [23]. Studies have been reported where various fungal pathogens of plants and humans, including yeasts, have been found to be susceptible to EOs [24]. There is evidence that pathogenic yeasts sensitive to drugs, as well as resistant ones, including the main pathogen of humans, *C. albicans*, were inhibited using terpenoid-rich EOs [25].

The efficiency using EOs and their components against the biofilms of *C. albicans* resistant to drugs is important. These activities can be mediated through the inhibition of the ergosterol membrane and the signaling pathways involved in the morphogenesis of the hyphae [26].

5. Cancer preventive properties using essential oils

The treatment in malignant cells represents a challenge for current medicine; In this sense, many plants with phytotherapeutic properties (such as taxol) have shown their efficiency as an alternative method in combating and proliferating malignant cells that can lead to cancer such as: colon cancer, gastric cancer, human liver tumor, lung tumors, breast cancer and leukemia, which have reported a decrease after treatment with OEs [27].

For example, there is evidence that *Cymbopogon martini* geraniol (i.e., palmarosa oil) manages to interfere with membrane functions, ionic homeostasis, and cell signaling events in cancer cell lines, which inhibits synthesis of DNA and a subsequent reduction in the size of the colon tumor [28].

In the case of terpenoids and constituent polyphenols, obtained from plant EOs, they can prevent the proliferation of tumor cells by necrosis or induction of apoptosis [29].

Citral present in lemongrass oil is found useful against the early phase hepatocarcinogenesis [30]. Another example, well known for its anticancer properties is the use of EO from *Allium sativum* (garlic). The preventive activity of chemotherapy is limited to the ability of garlic to suppress detoxifying enzymes of drugs [31]. Additionally the use of EO of lemon balm (*M. officinalis*) inhibits the growth of a series of human cancer cell lines [32].

In this way, it is well established that OEs exhibit a capacity to act as antioxidants and interfere with the mitochondrial functions of cells, decreasing metabolic events (for example, increased cellular metabolism, mitochondrial overproduction and permanent oxidative stress) characteristic of the development of malignant tumors [33].

6. Antiviral efficacy of essential oils

In addition to the aforementioned antimicrobial activities, there are plants that have significant antiviral properties, for example: *Origanum vulgare* anti-viral inactivation of enteric virus [34]; *Eucalyptus globulus* (Eucalyptus oil) activity against respiratory viruses [35]; *Salvia fruticosa*, antiviral activities [36], among other. It is believed that the inhibition of viral replication is attributed to the presence of the monoterpene, sesquiterpene and phenylpropanoid components of the EOs [37].

EOs of eucalyptus and thyme possess inhibitory activity against the herpes virus [38]. Evidence has been found that OEs of *Melaleuca alternifolia* showed significant efficacy in the treatment of herpes virus [39]. The way in which the adsorption or entry of virus into host cells is prevented is associated with the ability to interfere with the viral envelope structures. For example, oregano oil causes dissolution of the HSV envelope to attenuate its infectious capacity [40]. There is evidence that the components of EOs specifically inhibit the early expression of the gene in CMV (cytomegalovirus) and thus prevent viral activation [41].

7. Essential oils as antioxidants

A cause of damage suffered by macromolecules is due to the oxidative stress that is associated with the generation of free radicals and reactive oxygen species (ROS) [42]. Published works show that oxidative damage is related to several health problems such as aging, arteriosclerosis, cancer, Alzheimer's disease, Parkinson's disease, diabetes and asthma [43].

Cellular balance of free radicals is maintained by different antioxidants. Flavonoids, terpenoids and phenolic constituents of EOs exhibit significant antioxidant effects [44]. There is evidence that species, such as: *Origanum majorana*, *Tagetes filifolia*, *Bacopa monnieri* and *C. longa* oils have pronounced antioxidant capacities [45]. In general, efficiency among the essential oils with good radical and antioxidant removal properties are made to order. We can mention species such as: cloves > cinnamon > nutmeg > basil > oregano > thyme [46].

8. Obtaining essential oils

Essential oils can be obtained from plant material by three main methods [47] view **Figure 1**.

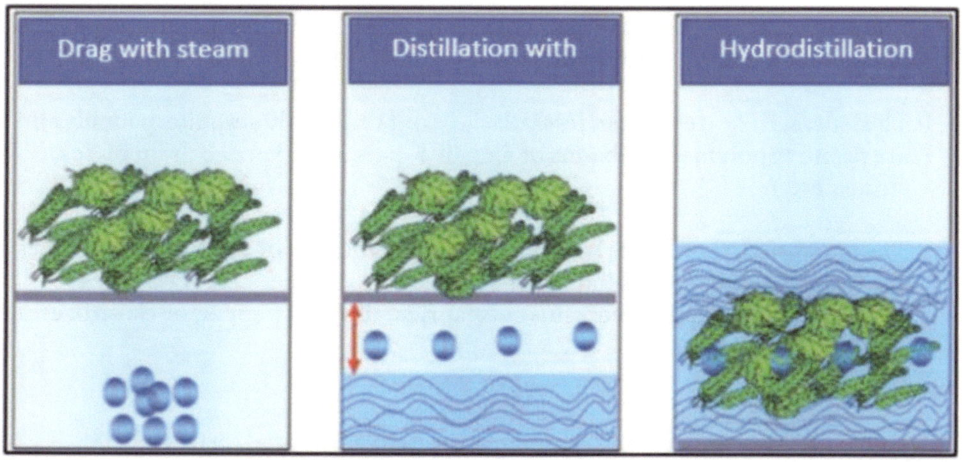

Figure 1.
Obtaining essential oils by entrainment with steam (external source of steam, e.g., boiler); distillation with water vapor (wet steam obtained by heating the water to its vigorous boiling); and hydrodistillation (heating of the water with the submerged plant material, generally very delicate, for example, flowers).

8.1 Drag with steam

This process is carried out with an overheated dry steam, usually generated by a boiler or boiler, which penetrates the plant material under pressure higher than atmospheric, the vapor stream breaks the oil cells or channels in the plant and entrains the volatile mixture, which condenses after passing through a refrigerant. Generally the oils are lighter than the water and very little soluble in it; therefore, can be separated by decantation. The exception is nail oil, which is heavier than water and is collected underneath it. The steam entrainment method is used to extract rhizome oils, roots, seeds (vetiver, valerian, ginger, anise, cardamom, etc.) and dried or fermented leaves of some plants (e.g., patchouli).

8.2 Distillation with water vapor

In this system of extraction uses a humid steam, coming from the water in boiling, that transpasses the vegetal material suspended above and supported on a mesh. Most herbaceous plants are distilled by this method.

8.3 Hydrodistillation

Hydrodistillation is a process when the plant material is submerged directly into the water, which is heated to boil. This method is used for the distillation of the plant material delicate, for example, flowers (e.g., ylang-ylang, roses).

9. Classification of essential oils

We can classify essential oils based on the following criteria: consistency, origin, and chemical nature of the major components.

9.1 Consistency

For its consistency the essences are divided into: fluid essences, balms and oleoresins.

a. The fluid essences, they are very volatile liquids at room temperature (essences of rosemary, mint, sage, lemon, basil).

b. The balms, they are thicker, less volatile, contain mainly sesquiterpenoids and are prone to polymerize (balms of Copaiba, balsam of Peru, balsam of Tolu, storax, etc.).

c. The oleoresins, they have the aroma of the plants in a concentrated form, they are typically very viscous liquids or semi-solid substances (rubber, gutta-percha, chewing gum, oleoresins) of paprica, of black pepper, of clavero, etc.)

9.2 Origin

Regarding the origin, essential oils are classified as: natural, artificial and synthetic.

The natural EOs are obtained directly from the plant and are not subsequently subjected to any physicochemical or chemical modification, they are expensive and varied in composition. The artificial essences are obtained by enrichment of natural essences with one of its components; they are also prepared by mixtures of several natural essences extracted from different plants as a mixture of essences of rose, geranium and jasmine.

The essence of anise enriched with anethole is an artificial essence. The synthetic essences are mixtures of various products obtained by chemical processes. They are more economical and therefore are widely used in the preparation of substances flavors and flavorings, such as essences of vanilla, lemon, and strawberry.

9.3 Chemical composition

Essential oils differ in composition and properties of fatty acids or fixed oils, which are composed of glycerides; and of mineral oils that are composed of hydrocarbons.

Because essential oils are a part of the metabolism of plants, their chemical composition varies permanently, modifying the proportions of their constituents or transforming one another, according to the part of the plant the moment of its development or the time of day when collect the plant.

The proportion of the components of the mixture varies from one oil to another, that is, each essential oil has its own characteristic mixture of compounds with defined qualitative and quantitative variations. Some may be so simple as cinnamon oil formed in 85% of cinnamaldehyde only, or as complex as jasmine, or chamomile oil with about 130 compounds. A discrimination is made between the compounds contained in an essence, and then we speak of major compounds when the compounds are present in the essence in a proportion >1 or 0.5%.

An essence is in permanent change, not only while it is part of the metabolism of the plant, also after it is extracted; this speaks of a reduced stability and a process of continuous transformation, which generates three stages in the life of an essence: that of maturation or aging, that of stability or useful life and that of decomposition or rancidity, each essence has different times for each stage, including depending on the case, the intermediate stage, where it is considered that the changes do not significantly change the quality of the same, may have a positive or negative trend.

It must be taken into account that given its complex composition, essences have a high probability of undergoing physicochemical modifications by reactions between its own constituents or between these and the medium that includes factors such as: light, temperature, presence of enzymes, the components of the part of the plant where the essence is stored, etc.

10. Conclusions

At present, there is a growing demand to explore the vast variety of plant species with biological activities, and their EOs with therapeutic potentials that can help in the alternative treatment of different diseases. But there is also a need to increase awareness of the risks and benefits associated with the medicinal uses of EOs, which is still a matter of research.

Use of plant molecules to treat, for example, infectious diseases is a complementary alternative. The use of EOs in the treatment of malignancy, artemisinin (isolated from *Artemisia annua*) and the anticancer drug, taxol (from *Taxus brevifolia*) are successful examples of the use of these popular medicines, obtained from these plant species at a cost much smaller than conventional medications.

Acknowledgements

Thanks to the professors of the PGGNANO program of the University of Brasilia (UnB), and the professors of the BIONORTE program of the University of Acre (UFAC), for the support in the development of the research that has been carried out as a whole.

Conflict of interest

We mentioned that we have no conflict of interests.

Author details

Jason Jerry Atoche Medrano
University of Brasilia (UnB), Brasilia, DF, Brazil

*Address all correspondence to: jajeatme@gmail.com

References

[1] Ngom S, Perez RC, Mbow MA, Fall R, Niassy S, Cosoveanu A, et al. Larvicidal activity of neem oil and three plant essential oils from Senegal against *Chrysodeixis chalcites* (Esper, 1789). Asian Pacific Journal of Tropical Biomedicine. 2018;**8**(1):67

[2] Ju J, Xu X, Xie Y, Guo Y, Cheng Y, Qian H, et al. Inhibitory effects of cinnamon and clove essential oils on mold growth on baked foods. Food Chemistry. 2018;**240**:850-855

[3] Nour AH. Formulation of solid/liquid perfumes of essential oils from different medicinal. Plants. 2018;**5**:95-115

[4] Sarkic A, Stappen I. Essential oils and their single compounds in cosmetics—A critical review. Cosmetics. 2018;**5**(1):11

[5] Djilani A, Dicko A. The Therapeutic Benefits of Essential Oils. Nutrition, Well-Being and Health. InTech; 2012. DOI: 10.5772/25344

[6] Rguez S, Msaada K, Daami-Remadi M, Chayeb I, Bettaieb Rebey I, Hammami M, et al. Chemical composition and biological activities of essential oils of *Salvia officinalis* aerial parts as affected by diurnal variations. Plant Biosystems-An International Journal Dealing with all Aspects of Plant Biology. 2018;**153**:1-9

[7] Askary M, Behdani MA, Parsa S, Mahmoodi S, Jamialahmadi M. Water stress and manure application affect the quantity and quality of essential oil of *Thymus daenensis* and *Thymus vulgaris*. Industrial Crops and Products. 2018;**111**:336-344

[8] Giacometti J, Kovačević DB, Putnik P, Gabrić D, Bilušić T, Krešić G, et al. Extraction of bioactive compounds and essential oils from mediterranean herbs by conventional and green innovative techniques: A review. Foodservice Research International. 2018:**113**:245-262

[9] Pateiro M, Barba FJ, Domínguez R, Sant'Ana AS, Khaneghah AM, Gavahian M, et al. Essential oils as natural additives to prevent oxidation reactions in meat and meat products: A review. Foodservice Research International. 2018;**113**:156-166

[10] Carson C, Hammer K, Riley T. *Melaleuca alternifolia* (tea tree) oil: A review of antimicrobial and other medicinal properties. Clinical Microbiology Reviews. 2006;**19**(1):50-62

[11] Bakkali F, Averbeck S, Averbeck D, Idaomar M. Biological effects of essential oils—A review. Food and Chemical Toxicology. 2008;**46**(2):446-475

[12] Parvez MK. Natural or plant products for the treatment of neurological disorders: Current knowledge. Current Drug Metabolism. 2018;**19**(5):424-428

[13] Raghunath A, Sundarraj K, Kanagaraj VV, Perumal E. 9 plant sources as potential therapeutics for Alzheimer's disease. In: Medicinal Plants: Promising Future for Health and New Drugs. Parimelazhagan Thangaraj Editorial; 2018. p. 161

[14] Sathya B, Balamurugan K, Anbazhagan S. Bioactive compounds with neuropharmacological properties from medicinal plants—A brief review. World Journal of Pharmacy and Pharmaceutical Sciences. 2018;7:368-388

[15] Czepak MP. Os Recursos Vegetais aromáticos no Brasil: Seu Aproveitamento Industrial Para a produção de Aromas e Sabores. Editorial

da Universidade Federal do Espírito Santo; 2008

[16] Maia J. Plantas aromáticas da região Amazônica. Os recursos vegetais aromáticos no Brasil: Seu aproveitamento industrial para a produção de aromas e sabores. 2008

[17] Pino B, Ramírez G. Traditional knowledge of vegetable species used with magic-religious purpose in communities of the northern Colombian Pacific. Boletin Latinoamericano y del Caribe de Plantas Medicinales y Aromaticas. 2009;**8**(3):180-183

[18] Felden B, Cattoir V. Bacterial adaptation to antibiotics through regulatory RNAs. Antimicrobial Agents and Chemotherapy. 2018;**62**:02503-02517

[19] Iseppi R, Sabia C, de Niederhäusern S, Pellati F, Benvenuti S, Tardugno R, et al. Antibacterial activity of *Rosmarinus officinalis* L. and *Thymus vulgaris* L. essential oils and their combination against food-borne pathogens and spoilage bacteria in ready-to-eat vegetables. Natural Product Research. 2018;**33**:1-5

[20] Lang G, Buchbauer G. A review on recent research results (2008-2010) on essential oils as antimicrobials and antifungals. A Review Flavour and Fragrance Journal. 2012;**27**(1):13-39

[21] Bedi S, Vyas S. A Handbook of Aromatic and Essential Oil Plants: Cultivation, Chemistry, Processing and Uses. Jodhpur: Agrobios Editorial; 2008

[22] Routh MM, Raut JS, Karuppayil SM. Dual properties of anticancer agents: An exploratory study on the in vitro anti-Candida properties of thirty drugs. Chemotherapy. 2011;**57**(5):372-380

[23] Sardi J, Scorzoni L, Bernardi T, Fusco-Almeida A, Giannini MM. Candida species: Current epidemiology, pathogenicity, biofilm formation, natural antifungal products and new therapeutic options. Journal of Medical Microbiology. 2013;**62**(1):10-24

[24] Galvão LCC, Furletti VF, Bersan SMF, da Cunha MG, Ruiz ALTG, Carvalho JE, et al. Antimicrobial activity of essential oils against *Streptococcus mutans* and their antiproliferative effects. Evidence-based Complementary and Alternative Medicine. 2012;**12**:1-12

[25] Zore GB, Thakre AD, Rathod V, Karuppayil SM. Evaluation of anti-Candida potential of geranium oil constituents against clinical isolates of *Candida albicans* differentially sensitive to fluconazole: Inhibition of growth, dimorphism and sensitization. Mycoses. 2011;**54**(4):e99-e109

[26] Raut JS, Shinde RB, Chauhan NM, Mohan Karuppayil S. Terpenoids of plant origin inhibit morphogenesis, adhesion, and biofilm formation by *Candida albicans*. Biofouling. 2013;**29**(1):87-96

[27] Hamid A, Aiyelaagbe O, Usman L. Essential oils: Its medicinal and pharmacological uses. International Journal of Current Research. 2011;**33**(2):86-98

[28] Qi F, Yan Q, Zheng Z, Liu J, Chen Y, Zhang G. Geraniol and geranyl acetate induce potent anticancer effects in colon cancer Colo-205 cells by inducing apoptosis, DNA damage and cell cycle arrest. Journal of BU ON: Official Journal of the Balkan Union of Oncology. 2018;**23**(2):346-352

[29] Raut JS, Karuppayil SM. Bioprospecting of plant essential oils for medicinal uses. In: Fulekar MH, Bhawana Pathak, Kale RK, editors. Environment and Sustainable Development. Springer; 2014. pp. 59-76

[30] MacHraoui M, Kthiri Z, JABEUR M, Hamada W. Ethnobotanical and phytopharmacological notes on *Cymbopogon citratus* (DC.) Stapf. Journal of New Sciences. 2018;**55**:3642-3652

[31] Alam MK, Hoq MO, Uddin MS. Medicinal plant Allium sativum. A review. Journal of Medicinal Plant Studies. 2016;**4**(6):72-79

[32] Magalhães DB, Castro I, Pereira JM, Lopes-Rodrigues V, Ferreira IC, Xavier C, et al. Study of the mechanism of action of *Melissa officinalis* L. extracts in inhibiting the growth of human tumor cellular lines. IUJUP. 2018;**9**:3134-3142

[33] Ali B, Al-Wabel NA, Shams S, Ahamad A, Khan SA, Anwar F. Essential oils used in aromatherapy: A systemic review. Asian Pacific Journal of Tropical Biomedicine. 2015;**5**(8):601-611

[34] Sánchez G, Aznar R. Evaluation of natural compounds of plant origin for inactivation of enteric viruses. Food and Environmental Virology. 2015;**7**(2):183-187

[35] Horváth G, Ács K. Essential oils in the treatment of respiratory tract diseases highlighting their role in bacterial infections and their anti-inflammatory action: A review. Flavour and Fragrance Journal. 2015;**30**(5):331-341

[36] Miraj S, Kiani S. A review study of therapeutic effects of *Salvia officinalis* L. Der Pharmacia Lettre. 2016;**8**(6):299-303

[37] Astani A, Reichling J, Schnitzler P. Screening for antiviral activities of isolated compounds from essential oils. Evidence-based Complementary and Alternative Medicine. 2011;**11**:1-8

[38] Koch C, Reichling J, Schneele J, Schnitzler P. Inhibitory effect of essential oils against herpes simplex virus type 2. Phytomedicine. 2008;**15**(1-2):71-78

[39] Astani A, Schnitzler P. Antiviral activity of monoterpenes beta-pinene and limonene against herpes simplex virus in vitro. Iranian Journal of Microbiology. 2014;**6**(3):149

[40] Akthar MS, Degaga B, Azam T. Antimicrobial activity of essential oils extracted from medicinal plants against the pathogenic microorganisms: A review. Issues in Biological Sciences and Pharmaceutical Research. 2014;**2**(1):1-7

[41] Pusztai R, Abrantes M, Sherly J, Duarte N, Molnar J, Ferreira M-JU. Antitumor-promoting activity of lignans: Inhibition of human cytomegalovirus IE gene expression. Anticancer Research. 2010;**30**(2):451-454

[42] Bhattacharya S. Reactive oxygen species and cellular defense system. In: Rani V, Chand U, Yadav S, editors. Free Radicals in Human Health and Disease. Springer. 2015. pp. 17-29

[43] Kehrer JP, Klotz L-O. Free radicals and related reactive species as mediators of tissue injury and disease: Implications for health. Critical Reviews in Toxicology. 2015;**45**(9):765-798

[44] Sánchez-Vioque R, Polissiou M, Astraka K, De Los Mozos-Pascual M, Tarantilis P, Herraiz-Peñalver D, et al. Polyphenol composition and antioxidant and metal chelating activities of the solid residues from the essential oil industry. Industrial Crops and Products. 2013;**49**:150-159

[45] Erenler R, Sen O, Aksit H, Demirtas I, Yaglioglu AS, Elmastas M, et al. Isolation and identification of chemical constituents from *Origanum majorana* and investigation of antiproliferative and antioxidant activities. Journal of

the Science of Food and Agriculture. 2016;**96**(3):822-836

[46] Asprea M, Leto I, Bergonzi MC, Bilia AR. Thyme essential oil loaded in nanocochleates: Encapsulation efficiency, in vitro release study and antioxidant activity. LWT-Food Science and Technology. 2017;77:497-502

[47] GuzmÃ JD. Fragmentaciã n diastereoespecã ficaennci de dos lignanos dibencilbutã nicos de la corteza de *Dugandiodendron argyrotrichum* (Magnoliaceae). Revista Productos Naturales. 2008;**2**(1):6-12

Section 2

Resources and Structures of Essential Oils

Chapter 6

Essential Oil and Glandular Hairs: Diversity and Roles

Zakaria Hazzoumi, Youssef Moustakime and Khalid Amrani Joutei

Abstract

The accumulation of essential oils in plants is generally limited to specialized secretory structures, namely, glandular trichomes (hairs) which are multicellular epidermal glands, found in some families such as Lamiaceae, Asteraceae, and Solanaceae, and which secrete terpenes in an extracellular cavity at the apex of the trichome. Storage of terpenoids in these structures can also be used to limit the risk of toxicity to the plant itself. The morphology of these structures varies according to the conditions of irrigation and also according to the toxicity of intracuticular contents and can be changed with the phenology of the plant. The secretory glands of aromatic plants come in different shapes and sizes, in order to ensure a specific function. This function consists mainly in the protection of different plant organs and the attraction of pollinators. Some scientist classified these glands into peltate hairs and capitate hairs, based on morphological criteria; however, others classified them into short-term glands and long-term glands, based on the mode of secretion. Short-term glands are glands that secrete rapidly to protect young organs. The long-term glands are glands in which the secretory substance accumulates gradually in the subcuticular space and play a role in the protection of mature organs such as the flower, as well as in pollination. According to this definition, he inferred that the capitate hairs are the short-term glands, while the peltate hairs are long-term glands. The difference between these two types of glands consists several aspects like structure, mode of secretion, and timing of secretion. In this object, this chapter includes some microscopic observation to glandular hairs and their combination with mode of secretion, nature of contents, and phenology of plant to give a good comprehension and classification.

Keywords: capitate, peltate, essential oils, glandular hairs, long term, short term

1. Introduction

Essential oil presents a complicated and heterogenic composition, with some molecules which can cause an injury to the plant itself, as there is evidence that many terpenoids are potentially toxic to plant tissues, when monoterpenes are released to proximate cells [1]. Injury has also been found when some sesquiterpenes are artificially deposited on leaves during tests of their ability to deter herbivores [2]. Therefore, the sequestration of terpenoid in specific compartments by sensitive metabolic processes may be essential to avoid adverse effects. The morphology of these structures varies according to the conditions of irrigation and also according to the toxicity of intracuticular contents [3].

IntechOpen

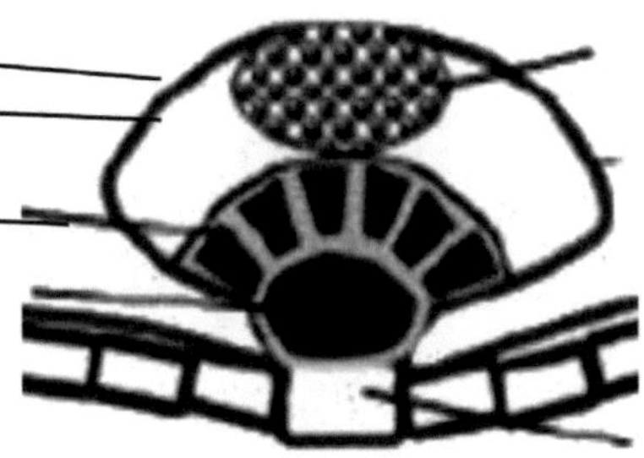

Figure 1.
Glandular hairs structure (Iriti et al. [12]).

These risks implicate the presence of specialized structures for the storage and the secretion of these compounds, those structures changed with the content and function. Largely named glandular trichomes which are multicellular epidermal hairs, found in some families such as Lamiaceae, Asteraceae, and Solanaceae, which secrete terpenes in an extracellular cavity located at the apex of the trichome [4, 5].

Previous studies have been able to observe these structures as well as its different constituents (**Figure 1**). The glandular trichome is composed of several cells performing different functions: the terpenes synthesized by the secretory cells pass into a subcuticular space located at the apex of the trichome for accumulation, and the basal cells ensure the attachment of the structure to the epidermis. These glands vary morphologically, biochemically, and secretionally.

The secretory glands of aromatic plants come in different shapes and sizes, in order to ensure a specific function. This function consists mainly in the protection of the different organs of the plant and the attraction of pollinators. These glands are subdivided into peltate and capitate hairs. Werker et al. [4] classified glands into short-term glands and long-term glands; short-term glands are glands that secrete rapidly to protect young organs. The long-term glands are glands in which the secretory substance accumulates gradually in the subcuticular space and play a role in the protection of mature organs such as the flower, as well as in pollination. According to this definition, he inferred that the capitate hairs are short-term glands, while the peltate hairs are long-term glands. The difference between these two types of glands consists of several aspects like structure, mode of secretion, and timing of secretion.

Capitate usually consists of a single or bicellular head and rarely more than two cells in some species with lipophilic content. This content is ready to be released just after its production via a porous cuticle [3].

2. The glands' location depends on the function

The glands of secretions are localized in all the plant organs, leaves, and stems and even at the root. The location of these structures depends on the organ in which they are located and the function and nature of the substances stored and secreted by these glands.

Due to the enormous diversity of these structures, morphology, origin, size, location, microstructure of the head, secretory capacity, secretion mode, function, etc., their classification is made difficult. The use of one of these criteria renders the classification incomplete which requires a method based on different types of criteria to classify these glands.

The major classification divides the glands into two groups: glandular hair and nonglandular hair. The main characteristic of distinction between these two types is the morphology as well as the nature of the substances to be secreted.

3. Non-glandular trichome

Nonglandular trichomes differ in morphology, anatomy, and microstructure. Essentially, they are classified according to their morphology. They can be unicellular or multicellular and branched or unbranched. The unbranched multicellular trichomes can be uniseriate, biseriate, or multiseriate. They can be distinctly articulated between cells or transverse walls which make the distinction of these structures impossible on the surface. They may differ in length, size, and shape of cells and may be symmetrical or asymmetrical. This type of gland provides a protective function as a mechanical barrier and does not ensure secretion of any substance.

4. Glandular trichome

The term "glandular hair" refers to a wide variety of glands. They differ according to the chemical composition of the substances they secrete, accumulate, or absorb; according to their mode of production, structure, and location; and according to their functions. All of these differences serve a certain level, which overlaps the classification.

In many of the Lamiaceae, two main types of glandular trichomes are encountered, the capitate and the peltate. They differ according to the shape of their secretory head and according to the morphology as well as the nature of the substances to be secreted. The head of the capitate glands consists of 1–4 more or less rounded secretory cells, generally oriented horizontally; a stem, one to several long cells; and a basal cell. The head cell can sometimes be very large, as in some species of *Salvia* [5]. The peltate hairs' head consists of 4–18 more flattened cells on a horizontal plane, a stem cell, and a basal cell. Thus, intermediate shapes can be encountered.

5. Variability and classification

There is great variability in the secreted materials of glandular trichomes: polysaccharides, sugars, salts, lipids, essential oils, resins, proteins, etc. Uphof [6] proposed to classify glandular trichomes according to the nature of their secretory products. Fahn [7] classified secretory substances in plants in general into two groups:

1. Unmodified or slightly modified substances, such as salts secreted by certain glands, as well as nectar.

2. Substances synthesized by secretory cells. This can be hydrophilic (as in trichomes which secrete mucilage and glandular trichomes of digestive carnivorous plants) or lipophilic (such as in the glandular trichomes of Lamiaceae, Asteraceae, Geraniaceae, Solanaceae, and Cannabaceae).

The difficulty with this type of classification is that some glands secrete more than one type of substance. For example, the glands of *Inula viscosa* secrete lipophilic substances, polysaccharides, and proteins at different stages of life and by different organelles [8]. In some carnivorous plants, the same glandular hairs produce both seductive (nectar) and digestive substances (enzymes).

In Lamiaceae, in addition to lipophilic substances, relatively large amounts of polysaccharides, which vary in amount between species, are secreted by certain capitate glands [8]. Many of these secreted materials are phytotoxic.

6. Glandular hair functions

The function of glandular trichomes varies according to their location, the substances it secretes, and the moment of secretion. Structurally, similar trichomes can produce different materials, but even when similar materials are produced, when the trichomes are located differently, they may have different functions. For example, the functions of seed mucilage trichomes differ considerably from those of leaves.

Different types of glandular trichomes may differ according to the stage of development of the organ. In Lamiaceae, from a functional viewpoint and according to their mode of secretion, the secretion materials of certain types of capitate glands are released outside just after production, when the leaves are still young and developing, often with a porous cuticle or mechanical breakage. The lipophilic fraction acts as a repellent on young leaves even before being touched. In the peltate glands, the secretory material is gradually secreted from the glandular cells in a space formed by elevation of the cuticle. The cuticle can break through a predator, releasing secreted material, which can then act as repellents. The glandular hairs can also be classified in "short term" corresponding with the first group of capitate glands and "long term" corresponding with the peltate glands [3].

a. **Protection**: the plant needs protection against various external factors such as herbivores and pathogens, extreme temperatures, excessive loss of water, allelopathy against competing plants, etc. When nonglandular hairs are densely distributed (sometimes forming a carpet thicker than the leaf itself, found in many plants growing in arid conditions), they can serve as a mechanical barrier against most of the factors externally. A correlation between trichome density and pest resistance has been demonstrated by Levin [9]. The glands, which secrete lipophilic substances, can be used in chemical protection against various types of herbivores and pathogens by repulsing or poisoning them. Secreted mucilage can trap insects by sticking.

b. **Absorption of water**: ensure a permanent function of water absorption and soil minerals. They are usually short-lived, unicellular, unbranched, and well elongated or with one or more swellings.

The epiphytic plants, which have no contact with the soil, as well as the plants that grow in xeric conditions, sometimes have other types of glands for the collection of fog water such as pineapple (*Ananas comosus*) [10].

c. **Salt secretion**: consist of cells that actively secrete a solution of different minerals.

d. **Seductive trichomes**: secretory glands of certain nectars, such as the unicellular glands of *Lonicera* and *Tropaeolum* and the multicellular glands of *Abutilon*. The most striking examples are carnivorous plants that have special requirements to catch their prey; they need to seduce and trap the various preys (sensory-triggering, digestion, and absorption devices). These functions are mainly fulfilled by trichomes [11].

6.1 Histological variations of the glandular hairs of three aromatic and medicinal plants: *Ocimum gratissimum*, *Salvia officinalis*, and *Pelargonium* sp.

6.1.1 Histological study of glandular structures of Salvia officinalis

The microscopic observations made on the *Salvia* plants (**Figure 2**) show the existence of two types of structures (secretion glands and protective hairs) which play various roles during the growth of the plants.

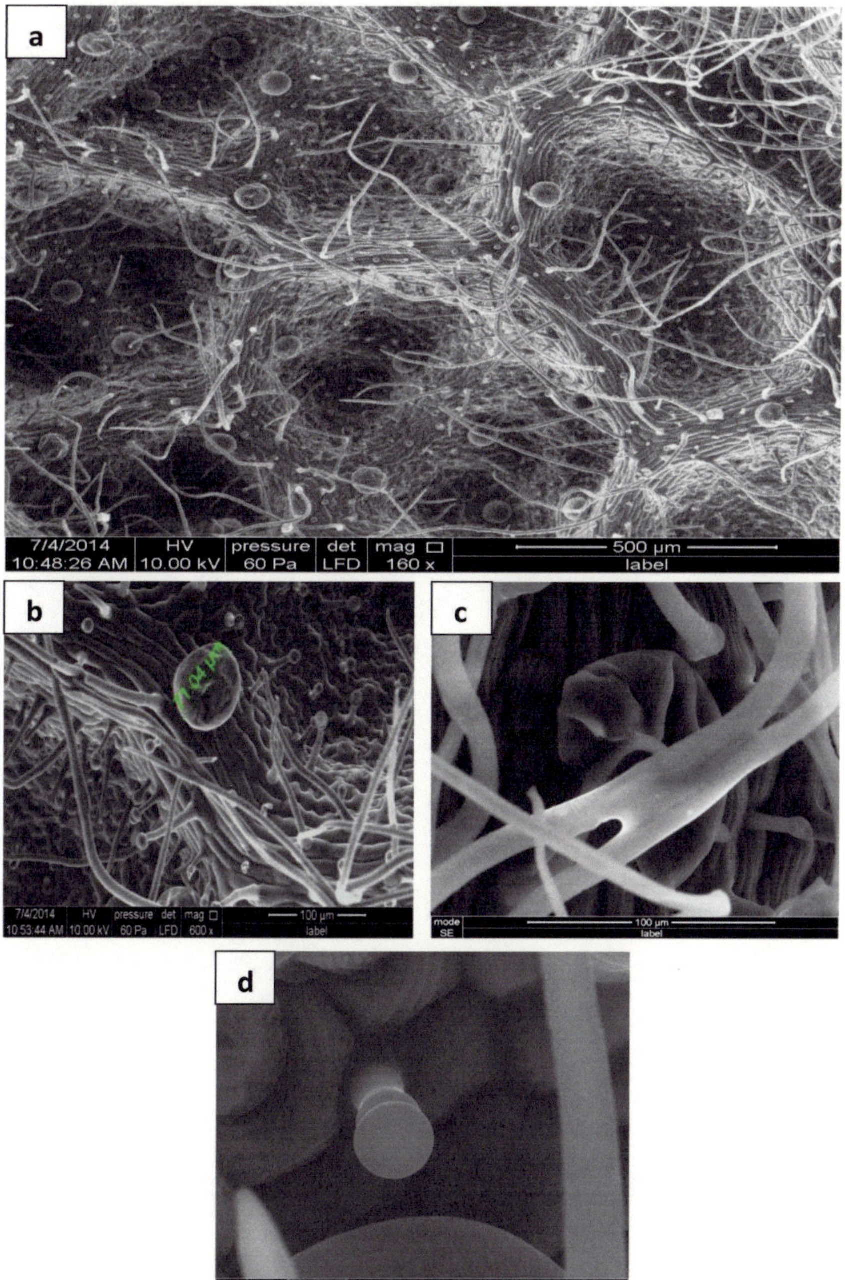

Figure 2.
Scanning electron microscope observations of the central region of Salvia officinalis leaf (a), showing the presence of two types of glands: peltate glands (b, c) and capitate glands (d) with GP (peltate glands) and P (protective hairs).

Glandular hairs consist of a base cell, a short unicellular stem, and a secretion head generally composed of 12 cells (sometimes 16) arranged in the form of a shield of 4 central cells surrounded by 8 peripheral cells (**Figure 2a, b**). Moreover, **Figure 2c** shows a sessile gland with multiple cells of the head. These structures are called peltate glands.

The second category of glands or capitate glands have a variable morphology since one distinguishes two types of glands: glands consisting of a short stem with one or two cells (**Figures 3a** and **4a–c**) and a head that can be unicellular or bicellular. In this type of gland, the cuticle is thin, with the presence or absence of a small subcuticular space. The second type consists of glands observed in old leaves at the end of flowering (**Figure 5**) and has a thin cuticle with the presence of a large subcuticular space directly to the epidermis. This structure is characterized by the presence of an accumulation pocket which is used to store the secretory material suggesting the presence of secretory cells below this pocket and whose synthesis products would arrive at the accumulation pocket through a structure (channel or pore) located under this pocket. The absence of this structure in young leaves and their appearance in old leaves suggests that these structures have a specific function related to the age of the leaf and are therefore specialized in the accumulation of a different substance.

Capitate glands are usually characterized by a single-cell head (**Figure 4a, b**). However, the light microscopic observations of the *Salvia officinalis* leaves show the presence of capitate glands that can present a two-celled head (**Figure 4c**).

According to the classification proposed by Werker et al. [4], the capitate glands are short-term glands specialized in the synthesis of non-terpenic substances frequently found in young leaves to provide a protective function against predators.

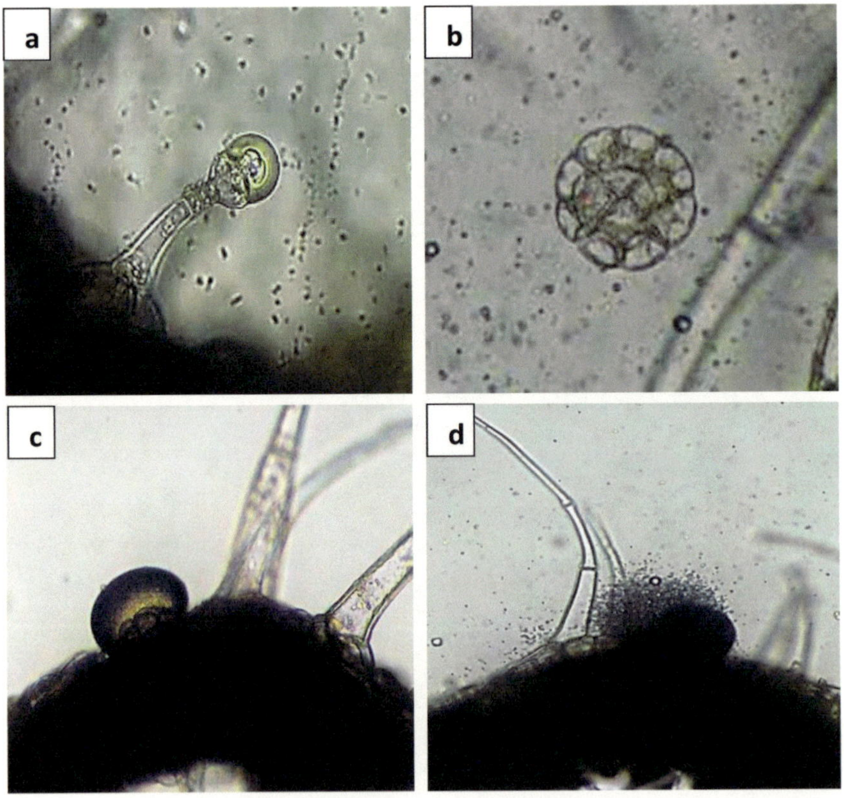

Figure 3.
Light microscopic observations of the different glands in Salvia officinalis. (a) Capitate gland, (b) detached peltate gland, (c) peltate gland in secretion phase, and (d) post-secretion peltate gland.

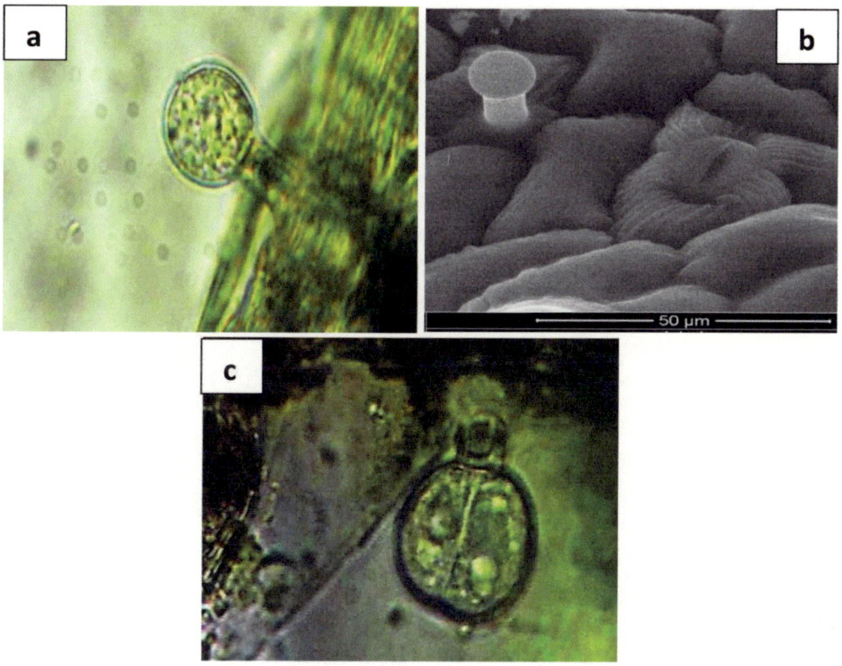

Figure 4.
Observations in light and scanning electron microscopy of capitate glands in Salvia officinalis. (a & b) Glands with single-cell head; (c) Gland with two-cell head.

The peltate glands are called long-term glands that are present in the leaves throughout the growth phases and whose number increases at flowering and are specialized in the synthesis of terpene substances.

Observations by the LM as well as SEM of the peltate glands (**Figure 3c, d**) do not reveal the presence of pores through which the secretory material can exude, which makes the material not released until the break of the cuticle, whether due to mechanical events or at the end of gland life.

6.1.2 Histological study of Pelargonium sp. glandular hairs

As in the case of sage, light microscopy observations made on *Pelargonium* sp. leaves show that this plant is characterized by the presence of two types of secretory glands, the peltate glands or long-term glands and the capitate glands or short-term glands, as well as protective hairs or non-secretory glands (**Figure 6**).

Capitate glands (short-term glands) are divided into two categories according to their forms: capitate glands with a stem consisting of four cells (**Figure 7b**) and capitate glands with one head and unicellular stem (**Figure 7c**). At the level of these glands, the synthesized secretion material is ready to be released just after its production via a porous cuticle (**Figure 7b**).

Moreover, we could only meet one type of peltate glands (long term), with a multicellular stem and a large head in which one finds the cells of synthesis as well as a storage pocket of essential oils. The appearance of these glands varies with their stage of maturity. **Figure 8a** shows a gland that has an empty aspect of any content and confirms the juvenile condition of the gland. These glands are gradually filled by the EO as they mature (**Figure 8b, c**), and their cuticle eventually breaks to release their contents at the end of their maturation (**Figure 8c**).

These peltate glands have a thick and rigid cuticle, but along the maturity stages of these glands, the thickness of this cuticle decreases to facilitate bursting and

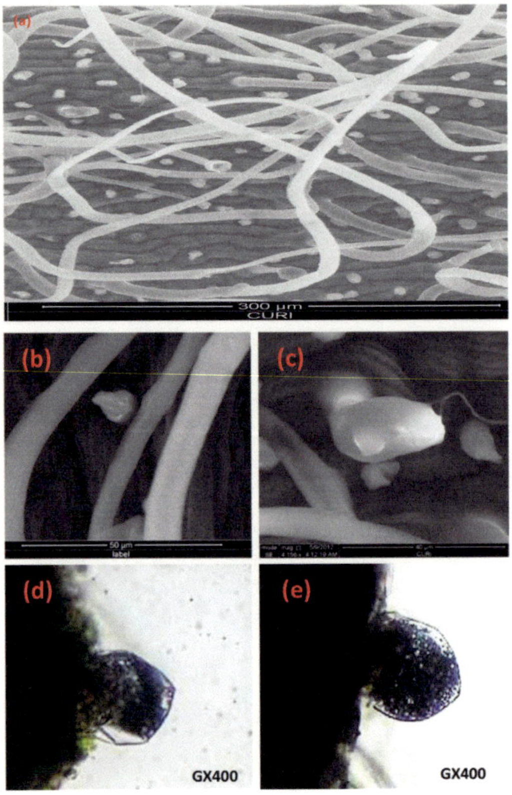

Figure 5.
Observations with light microscopy and scanning electron microscopy of capitate glands in Salvia officinalis. (a) Distribution of the pocket; (b&d) empty pocket; (c&e) pocket fill with secretory material.

release of contents toward the end of the maturity of the gland or by contact with some insects (**Figure 8**).

6.1.3 Histological study of Ocimum gratissimum glandular hairs

Environmental scanning electron microscopy observations of basil leaves show the presence of two types of glands (**Figure 9**), peltate gland and capitate gland. The peltate glands have a round shape and a fairly large diameter that can exceed 70 microns. The capitate glands are small not exceeding 40 microns and come in two forms, glands capitate with one head (CH) and two bilobed heads (C2H).

The observations made by the light microscopy of the capitate glands confirm those made by scanning electron microscopy, with the presence of capitate glands with a single head (**Figure 10a**) and capitate glands with two heads (**Figure 11b**). On the other hand, the glands with a single head can be unicellular (**Figure 10a**) or bicellular (**Figure 10b**).

In basil plants, the peltate glands show a variability in the number of secretory cells. There are large glands with four cells (**Figure 12a**). In addition, we have been able to highlight the presence of glands that have eight secretory cells (**Figure 12b, c**). A thick cuticle covers both types of glands. With the filling of the essential oil structure, this cuticle becomes thinner to facilitate the release of the contents.

Figure 13 shows peltate glands of basil after bursting and release of their terpenic contents; according to this figure, we notice the existence of four secretory cells

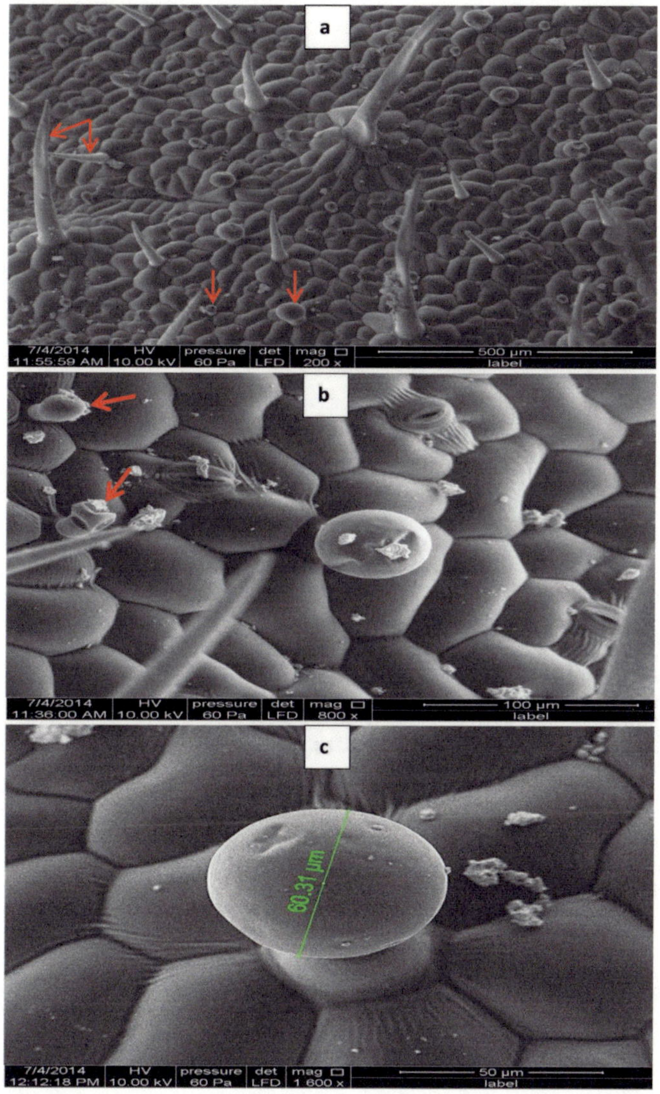

Figure 6.
Scanning electron microscopic observations of the ventro-central part of a Pelargonium *sp. leaf, showing the presence of two types of secretion glands, peltate. (a): observation of non glandular hairs and glandular hairs (c) and capitate (with red arrow).*

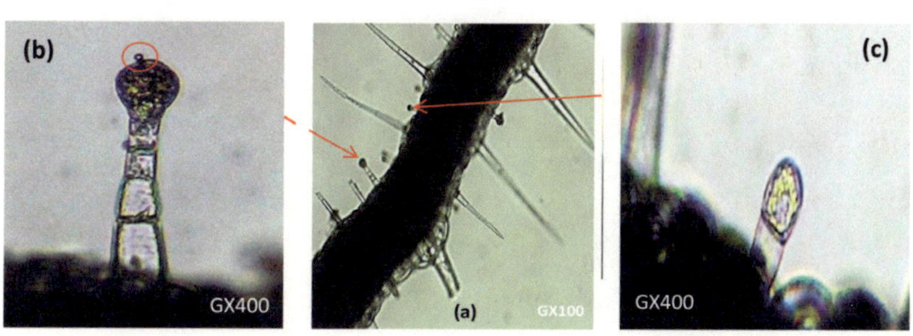

Figure 7.
Observations made by light microscopy on Pelargonium sp. leaf showing the two types of capitate glands. (a) Global vision; (b) capitate glands with a stem consisting of four cells; (c) capitate glands with one head and unicellular stem.

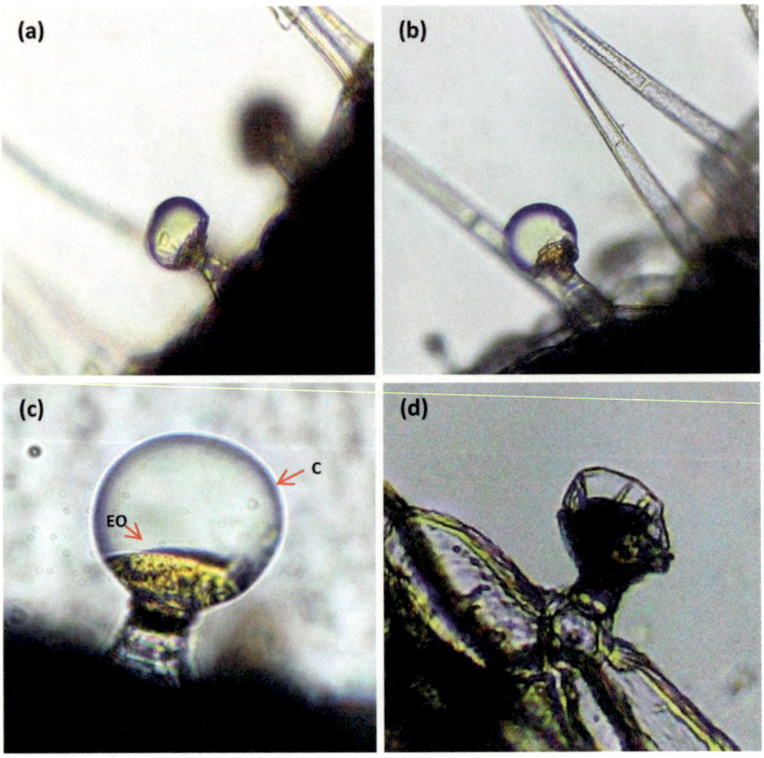

Figure 8.
Observations in light microscopy of peltate glands, secreting essential oils at different stages of maturity in Pelargonium *sp. (a) Beginning of maturation, (b, c) filling with the EO, and (d) bursting of the gland. C, cuticle; EO, essential oil (Gx 400).*

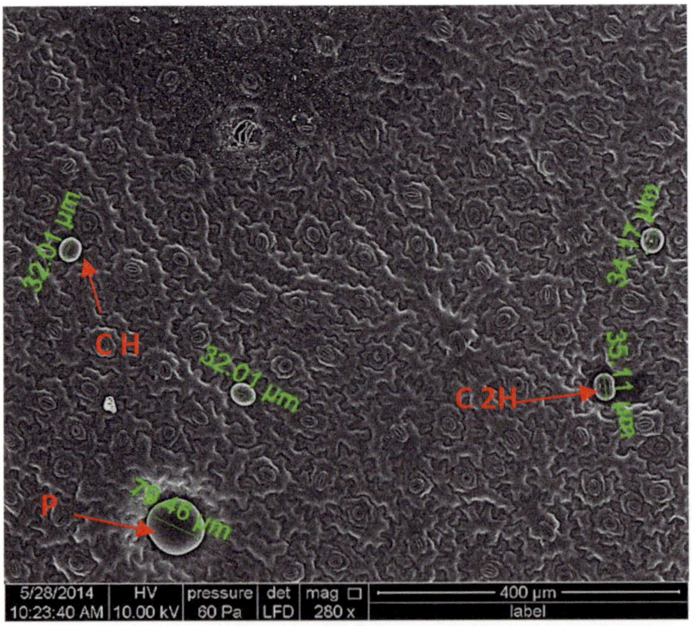

Figure 9.
Observation by scanning electron microscopy of basil leaf showing the presence of pelt glands (P) and capitate glands (CH: capitate gland with one head and capitates gland with head bilobed C2H).

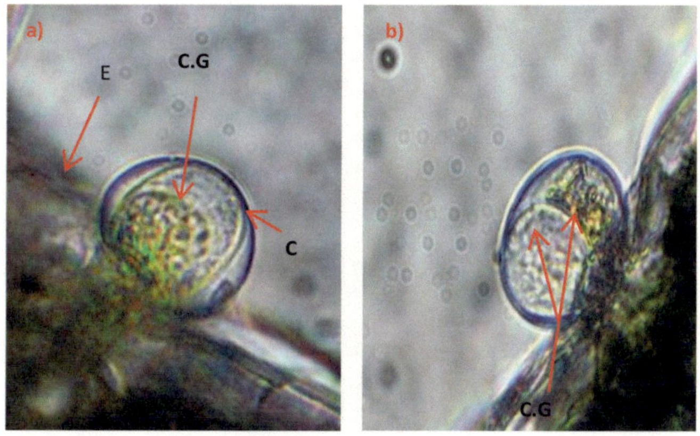

Figure 10.
Light microscopy observations of a capitate gland with a single head and a single cell (a) and with two cells (b). C, cuticle; E, epidermis; CG, glandular cell (GX400).

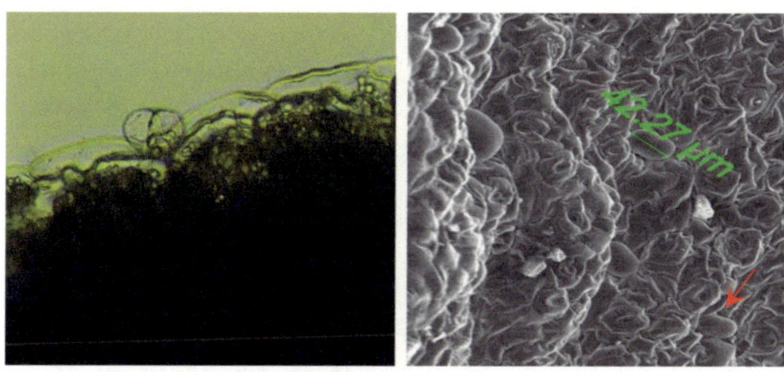

Figure 11.
Observations made by light microscopy and scanning electron microscopy of a double-headed gland in basil plants.

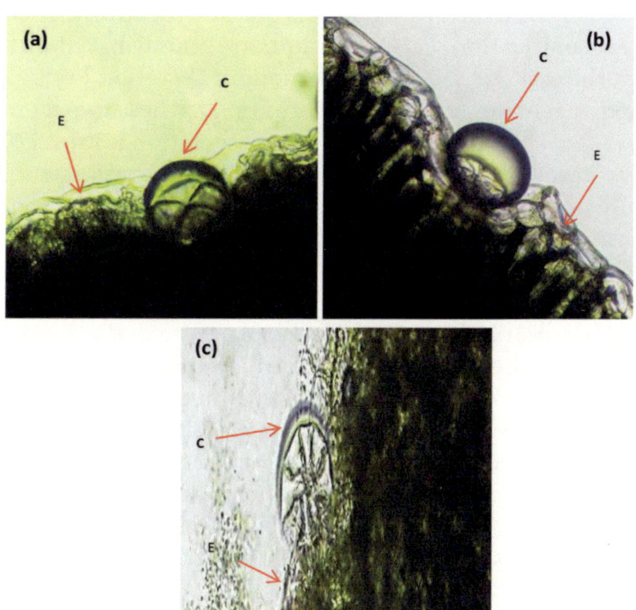

Figure 12.
Light microscopic observations of a cross section (central part of a basil leaf O. gratissimum*) showing two types of peltate glands: glands with four secretory cells (a) and glands with eight secretory cells (b, c).*

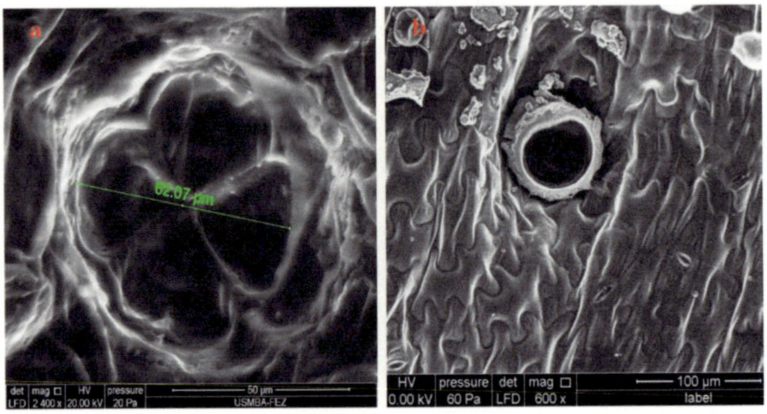

Figure 13.
Observation by scanning electron microscopy of a pelted basil gland after natural bursting (a) and mechanics (b).

(**Figure 13a**) with an internal diameter (without the cuticle which encompasses the gland) that touches 62 µm. Furthermore, **Figure 13b** shows an immature gland, but burst under pressure (exerted by SEM), and shows a thick cuticle which confirms the state of non-maturity that has already been reported.

7. Discussion

The microscopic observations made in this work show the presence of two major groups of secretory glands in the three plants (sage, pelargonium, and basil). The first group is peltate glands or long-term glands, and the second group is capitate glands or short-term glands. Within the same group, we can see a morphological variation from one plant to another, depending on the content, the role, and the phenological stage of the plant, which makes the classification difficult. But to ensure a credible classification, various authors are mainly based on the morphological criterion, the content, as well as the mode of secretion [4–8, 12].

In this context, scientists have given a definition to these two names; they have described a short-term gland as any gland whose secretion is rapid to protect young organs and a long-term gland as any gland in which the secretory substance accumulates gradually in the subcuticular space and serves for the protection of mature organs in the flower, as well as for pollination. The difference between the two types consists of several aspects, namely, the structure, the mode of secretion, and the moment of secretion.

The microscopic observations made on our plants go in the direction of this classification and show the two groups of glands, peltate and capitate. Moreover, two major types of peltate glands could be determined in this study, peltate glands with stem consisting of a single-cell and peltate glands glued directly to the epidermis. In the latter group, there is a major difference in the number of cells of each gland, since there are 4 and 8 cells in *O. gratissimum* and up to 12 in *Salvia*. These results are confirmed by the work of Tissier [13], Werker [8], and Gang et al. [14]. According to these authors, the peltate glands can have a large number of cells that can reach 16 cells.

These peltate glands are generally characterized by a large subcuticular space and a rigid cuticle which thins during the maturation of the gland which facilitates bursting at maturity or after mechanical contact with predators.

The captioned glands are subdivided into several categories according to their forms:

- Capitate glands possessing a basal cell and a united or multicellular stem (1–3 cells) and a head
- Capitate glands with a stem cell and a united or bicellular head
- Capitate glands with a basal cell and a head with one or two cells
- Capitate glands in the form of a pocket glued directly to the epidermis

These capitate glands are characterized by a thin cuticle with the synthesized secretory material which is ready to be released just after its production via a porous cuticle and usually consist of a united or bicellular head with lipophilic content.

Similar observations have been found in *Salvia* [15], since according to these authors the capitate glands are morphologically very variable and four types can be distinguished. Type I has a single or double bicellular stem and a united or bicellular head. The cuticle is thin and there is no subcuticular space. The secretory material is released slowly through the cuticle and can be released suddenly if the cuticle is broken. Type II is very small, possessing a unicellular stem and a small subcuticular space. The secretory material is probably secreted by a pore. Type III is a large gland with a long stem made up of several cells (about three or more), a neck cell, and a unicellular head, which probably releases the secretory material often collected as a drop on the head. Type IV is a large gland with a long thin stem, made up by four cells.

According to these authors, each type of gland is responsible for the synthesis of a distinct substance. Venkatachalam et al. [16] found a correlation between the increased number of peltate glands and increased camphor and thujane synthesis.

These observations are in agreement with those of Pedro et al. [17] who deduced the presence of three types of glands, a peltate type and two capitate types. According to these authors, the glands with a unicellular stem represent a distinct capitate gland with a specific content different from that of the peltate structure. According to the same work, these three types of glands ensure the synthesis of three types of different substances, terpenoids, phenols, and flavonoids.

In other works by Ko et al. [18], *Pelargonium*'s peltate and capitate glands all have a single basal cell. These two types of glands differ in the number of stem cells and in the diameter of the head depending on the stage of maturity [3]. According to these authors, the glands defined in our study as glands capitate to a head and a unicellular stem represent only one stage of the formation of the peltate glands. However, observations made in *Salvia* leaves show structures identical to this structure, possessing a small unicellular head and a single-celled stem. This structure has been classified as a capitate gland.

These kinds of peltate and capitate glands are found in several species (*Mentha piperita*, *Origanum dictamnus*, *Monarda fistulosa*, *Pogostemon cablin*, etc.) and have been described in ancient works [16]. In mint, the peltate glands have eight head cells, and the capitate glands have a single-stem cell and a head cell. In *Origanum dictamnus*, the head of the peltate glands has 12 cells, whereas in *Pogostemon cablin*, it has only 4 cells.

According to Werker [8], young basil leaves (*O. basilicum*) that are highly vulnerable to predators are heavily covered by capitate glands that are characterized by rapid secretion to repel herbivores. When the plants become older, the rate of the peltate glands increases; these terpenoid-rich glands will be used for defense against pathogens in case of injury and pollinator attraction.

8. Conclusion

The microscopic observations made on the three plants (basil, pelargonium, and sage) show the presence of two major types of secretory glands. The first type is peltate glands or long-term glands, and the second types are capitate glands or short-term glands. Within the same type of gland, we can see a morphological variation from one plant to another, depending on the content, the role, and the phenological stage of the plant.

In basil, we have demonstrated for the first time the existence of eight secretory cell glands in the peltate glands.

In *Salvia*, we noted the presence of two types of glands, capitate and peltate, the latter being constituted by 12–16 cells arranged in the form of a shield. On late flowering, we noted the presence of pocket secretions in older leaves. In this plant, we have mainly highlighted the presence of sessile glands, and this observation has never been confirmed in the literature.

Conflict of interest

The authors of this work (Zakaria Hazzoumi, Youssef Moustakime, Khalid Amrani Joutei) declare that they have no conflict of interest.

Notes/thanks/other declarations

"The scientist is not a person who gives the right answers, he is one who asks the right questions. Claude Levi-Strauss".

Author details

Zakaria Hazzoumi*, Youssef Moustakime and Khalid Amrani Joutei
Laboratory of Bioactive Molecules, Faculty of Science and Technology, Sidi Mohamed Ben Abdellah University, Fez, Morocco

*Address all correspondence to: zakaria.hazzoumi@yahoo.fr

References

[1] Levy Y, Krikun J. Effect of vesicular-arbuscular mycorrhiza on *Citrus jambhiri* water relations. The New Phytologist. 1980;**85**:25-31

[2] Polonsky J, Bhatnagar SC, Griffiths DC, Pickett JA, Woodcock CM. Activity of quassinoids as antifeedants against aphids. Journal of Chemical Ecology. 1989;**15**:993-998

[3] Zakaria H, Moustakime Y, Elharchli EH, Joutei KA. Effect of arbuscular mycorrhizal fungi (AMF) and water stress on growth, phenolic compounds, glandular hairs, and yield of essential oil in basil (*Ocimum gratissimum* L.). Chemical and Biological Technologies in Agriculture. 2015;**2**:10. DOI: 10.1186/s40538-015-0035-3

[4] Werker E, Putievsky E, Ravid U, Dudai N, Katzir I. Glandular hairs and essential oil in developing leaves of *Ocimum basilicum* L. (Lamiaceae). Annals of Botany. 1993;**71**:43-50

[5] Werker E, Ravid U, Putievsky E. Structure of glandular hairs and identification of the main components of their secreted material in some species of the Labiatae. Israel Journal of Botany. 1985;**34**:31-45

[6] Uphof JCT. Plant hairs. Encyclopedia of Plant Anatomy IV. 1962;**5**:1-206

[7] Fahn A. Secretory Tissues in Plants. New York: Academic Press; 1979. pp. 162-164

[8] Werker E. Trichome diversity and development. Advances in Botanical Research. 2000;**31**:1-35

[9] Levin DA. The role of trichomes in plant defence. Quarterly Reviews of Biology. 1973;**48**:3-15

[10] Sakai WS, Sanford WG. Ultrastructure of the water-absorbing trichomes of pineapple (*Ananas comosus*, Bromeliaceae). Annals of Botany. 1980;**46**:7-11

[11] Thompson JD. The biology of an invasive plant: What makes *Spartina anglica* so successful. BioScience. 1991;**41**:393-401

[12] Iriti MGC, Chemat F, Smadja J, Faoro F and Franco A. Visinoni Histo-cytochemistry and scanning electron microscopy of lavender glandular trichomes following conventional and microwave-assisted hydrodistillation of essential oils: A Comparative Study. Flavour and Fragrance Journal. 2006;**21**:704-712

[13] Alain T. Glandular trichomes: What comes after expressed sequence tags. The Plant Journal. 2012;**70**:51-68

[14] Gang DR, Wang J, Dudareva N, Nam KH, Simon JE, Lewinsohn E, et al. An investigation of the storage and biosynthesis of phenylpropenes in sweet basil. Plant Physiology. 2001;**125**:539-555

[15] Corsi G, Bottega S. Glandular hairs of *Salvia officinalis*: New data on morphology, localization and histochemistry in relation to function. Annals of Botany. 1999;**84**:657-664

[16] Venkatachalam kv, Kionaas R, Croteau R. Development and essential oil content of secretory glands of Sage (*Salvia officinalis*). Plant Physiology. 1984;**76**:148-150

[17] Pedro L, Campos P, Pais MS. Morphology, ontogeny and histochemistry of secretory trichomes of *Geranium robertianum* (Geraniaceae). Nordic Journal of Botany - Section of Structural Botany. 1990;**10**(5):501-509

[18] Ko KN, Lee KW, Lee SE, Kim ES. Development and ultrastructure of glandular trichomes in *Pelargonium x fragrans* 'Mabel Grey' (Geraniaceae). Journal of Plant Biology. 2007;**50**(3):362-368

Chapter 7

Effect of Essential Oils on Storability and Preservation of Some Vegetable Crops

Aml Abo El-Fetouh El-Awady

Abstract

Essential oils, as natural sprout inhibitor and safe fungicides, are a promising tool and good alternative compounds otherwise synthetic due to their high efficacy, biodegradability, eco-safety and volatile nature. They are consisting of a number of various components, i.e., terpenes, phenols, alcohols, esters, aldehydes and ketones in different composition or combinations. These effective compounds supply excess to prevent sprouting in potatoes and Jerusalem artichoke (JA) and less chance to development of resistance in fungi in JA, strawberry and broccoli with low concentrations. On contrary, high concentration of these oils induce the germination of seeds like broccoli and carob. This chapter explains the practical application of using essential oils as natural antisprouting, inducing quality, preserving fungal diseases, eco-friendly compounds, alternating synthetic chemicals, giving high benefits and easy to apply. The foliar application with essential oils increases the productivity, quality and marketable yield and storability and reduces weight losses and decay. Moreover, the essential oils increase broccoli seed germination, antioxidant content and other phytochemical parameters. The chapter provides a novel anti-sprouting agent for inhibiting growth of processing potato tubers and identification of terpenoids that use to inhibit tuber sprouting as well as application of Chloropropham (CIPC) isopropyl-N-(3-chlorophenyl) carbamate as a conventional chemical inhibitor.

Keywords: essential oils, constituents, anti-sprouting agent, antifungal, sprout growth, postharvest, vegetables, structure, extraction, application

1. Introduction

Recently, the natural alternatives such as plant essential oils provide a promising control of plant diseases and anti-sprout agent because they virtually constitute a rich source of bioactive chemicals such as phenols, flavonoids, quinones, tannins, alkaloids, saponins, sterols terpenes, aromatic and aldehydes [1]. Moreover, these natural alternatives can also maintain the biochemical constituents of tubers during storage, they are biodegradable to nontoxic products, and are potentially suitable for use in integrated pest management programs.

Jerusalem artichoke JA or sun choke (*Helianthus tuberosus* L.) is a perennial plant which has a high economic value. JA used traditionally for human food and livestock feed due to its high nutritive value. JA tubers used for production of biofuels (ethanol) and some functional food like inulin, fructooligosaccharides and fructose. Moreover, some bioactive metabolites from its leaves and stems have been used in some pharmaceutical industries [1]. Storage JA tuber, controlled rots can be done by various techniques including; cold temperature, removal of diseases in tubers and minimizing mechanical injuries or application of synthetic fungicides. Another simple applied method in developing countries is keeping the tubers in the soil. Unfortunately, many fungi diseases can grow at cold storage temperatures and lead to damage, especially in extending long storage [2], However, storage of the harvested tubers usually results in high losses in quality, caused mainly by desiccation, rotting, sprouting, freezing and inulin degradation. A common solution is the use of synthetic chemical fungicides, however, their use is accompanied by threatening human health and the environment by supporting the emergence of resistant pathogens and by contamination of food with the pesticide deposits [3]. Essential oils, as green fungicides, are emerging as a better alternative of synthetic fungicides due to their high efficacy, biodegradability, eco-safety and volatile nature.

Respiration of potato tubers during storage and breakdown of dormancy during storage result in sprouting and loss of nutritive value of tubers [4]. Sprouting reduces the weight, the nutritional and processing quality of tubers and the number of marketable potatoes, being responsible for important economic losses during potatoes storage [5]. These physiological changes affect the internal composition of the tuber and destruction of edible material and changes in nutritional quality [6]. Various methods are available to control sprouting during storage. The primary method to control sprouting in storage is with postharvest application of isopropyl N-(3-chlorophenyl) carbamate (chlorpropham; CIPC). CIPC inhibits sprout development by interfering with cell division [7]. Therefore, a pressing need exists to find other, more environmentally acceptable sprout inhibitors for tubers. Nowadays it's very important to use natural products compounds such as essential oils.

Broccoli sprouts are considered as a functional food. Essential nutrient content provides diverse secondary metabolites and phytochemicals [8]. The phenolic compounds, especially flavonoids and anthocyanin, show a great ability capture free radical that leading to oxidative stress, to these compounds are attributed a beneficial effect in the prevention of cardiovascular diseases, circulatory problems, neurological disorders and cancer [9]. Broccoli has been identified as a vegetable with potential anti-cancer activity due to high levels of glucosinolates. The use of essential oil treatments rich in antioxidant to stimulate broccoli seed germination should be considered. Application of thyme and basil oil at 4% reduced the pathogenic fungi from seed to seedling and had a positive effect on the seed germination of infected seeds [10]. Aromatic plants especially essential oils are well known for their antioxidant and antimicrobial properties that prevent food degradation and alteration [11], as they are rich in phenolic substances, usually referred to as polyphenols, which are ubiquitous components of plants and herbs.

2. Application of essential oils

2.1 Alternative preservation method against sclerotium tuber rot of Jerusalem artichoke using natural essential oils

2.1.1 Methodology

Two experiments were conducted in Mansoura laboratory for vegetable crop handling and postharvest according to the storage method. In the first experiment, the tubers were kept in perforated polyethylene bags (0.075 mm thickness), and stored at 4°C and 90–95% relative humidity RH. In the second experiment, the tubers were stored in carton boxes (3 m^3) at 25 ± 2°C with moistened peat moss layers at the rate of peat moss: JA tubers (1.5: 1, kg/kg). The treatments applied for each experiment can be summarized as follows: Control (C), infected with fungal pathogen *S. rolfsii* (P), treated with caraway essential oil (O) and treated with caraway oil and infected with pathogen *S. rolfsii* (O + P). About 30 kg of tubers was used for each treatment.

2.1.2 Important results

2.1.2.1 Antifungal activity of the essential oils

Assessment of antifungal activity in vitro of caraway and spearmint essential oils was evaluated against *S. rolfsii* (**Figure 1**). Caraway essential oil completely inhibited the growth of the fungal pathogen even at the lowest concentration (2%). On the other hand, spearmint essential oil showed slight reduction in the fungal pathogen growth. The antifungal activity of caraway essential oil may be attributed to some antifungal phytochemicals that constitute a large fraction of the oil like carvone, limonene, carveol, pinene and thujone [12].

2.1.2.2 Evaluation of the caraway essential oil and peat moss application under storage conditions

2.1.2.2.1 Severity of disease

Data presented in **Table 1** show the rot fugal disease severity of JA tubers exposed to caraway oil and infected with fungal pathogen *S. rolfsii* under the two storage methods. The disease severity increased with the increasing the storage period over to the storage methods. JA tubers infected with *S. rolfsii* and exposed to emulsion of caraway essential oil (O + P) in peat moss layer at 25°C significantly reduced the disease severity compared to the cold storage method after 4 months storage. Infected-control JA tubers (P) and storage in peat moss layer at 25°C significantly reduced the disease severity for 2 months of storage compared with cold storage method, after which, the tubers were fully decadent. On the other hand, control-uninfected JA tubers (C) and storage in peat moss layer at 25°C significantly reduced the disease severity (caused by reasons other than *S. rolfsii*) compared with the storage under cold storage method. Caraway essential oil had the antimicrobial effects due to its content of basic constituents of monoterpenes, carvone and limonene. The basic constituents had a permeability effect on fungal

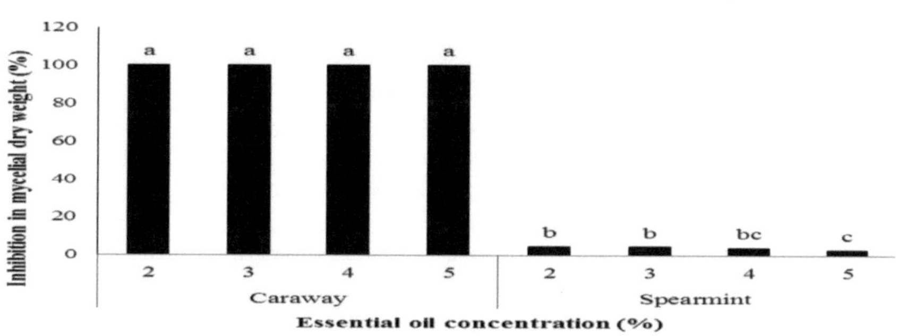

Figure 1.
Inhibition in mycelial dry weight of S. rolfsii *as a function of the tested oils. Columns superscripted with the same letter are not significantly different at P ≤ 0.05 (Duncan test).*

Storage method*	Treatment**	Storage period (day)			
		30	60	90	120
1	C	28.5[b,***]	73.3[a]	100.0[a]	NA
	P	100.0[a]	NA	NA	NA
	O	0.0[e]	0.0[e]	0.0[d]	0.0[d]
	O + P	8.9[d]	24.4[d]	37.5[b]	53.6[b]
2	C	20.1[c]	33.6[c]	40.0[b]	66.5[a]
	P	20.0[c]	47.2[b]	100.0[a]	NA
	O	0.0[e]	0.0[e]	0.0[d]	0.0[d]
	O + P	0.0[e]	0.0[e]	8.9[c]	13.3[c]

1 = storage JA tubers in polyethylene bags at 4°C and 2 = storage JA tubers in peat moss layer at 25°C.
**C = untreated control, P = infected tubers with pathogen, O = treated JA tubers with caraway essential oil and O + P = infected tubers with pathogen and treated with caraway essential oil. NA = not applicable due to full decay.*
***Means in each column followed by the same letter(s) were not significant at P= 0.05; where, (a) refer to the highest mean values, and (e) refer to the lowest mean values according to Duncan multiple range test.*
Tubers rot in the untreated control (C) treatment was due to many reasons rather than the fungal pathogens.

Table 1.
Mean rot severity (%) of JA tubers treated with caraway essential oil and infected with S. rolfsii using two storage methods.

cell membrane, inactivation of some organic compounds and enzymes and/or the inhibition of conidial germination, eventually, the death of fungal pathogen [13]. Moreover, the use of essential oils in storage of JA may have many benefits, including, they are natural-ecofriendly products, biodegradable and multifunctional purposes.

Moreover, the activity of *essential oils against* may tend to reduce pathogenic fungi resistance reinforcement against chemical fungicide because they contain two or more stereo-isomers that may be located on multi-sites on the pathogen's plasma membrane. One of the valuable applications for peat moss is the traditional use in food preservation [14]. The antifungal effect of the use of peat mosses has been reported by many investigators against *Aspergillus niger*, *A. flavus*, *Candida albicans*, *Cryptococcus albidus* and *Trichophyton rubrum* [15]. The antifungal effect of peat moss may be related to some of its contains of extranutritional constituents or bioactive components like a pectin-like polymer and sphagnan, that inhibit fungal mycelium growth via electrostatic immobilization of extracellular enzymes and/or nitrogen deprivation, phenolics that inhibit the activity of extracellular enzymes of microbes or other constituents like sterols and polyacetylenes [16].

2.1.2.2.2 Sprouting, weight loss and dry matter percentages of JA tubers

Table 2 show the mean data of weight loss and dry matter percentages of JA tubers exposed to emulsion of caraway essential oil and infected with fungal pathogenic *S. rolfsii* under using the two different storage methods. Results indicated that, the treatment of healthy JA tubers with emulsion of caraway essential oil completely inhibited the tubers sprouting and weight loss, but recorded the highest dry mater weight percentage along the storage period compared with the untreated-uninfected control treatment over the use of the two different methods. Even after 120 days of storage period, the treatment of the use of JA tubers with caraway oil and infected with pathogenic fungi significantly decreased sprouting and weight loss percentages and increased dry matter content for JA tubers that stored in peat moss layers at 25°C than those stored in polyethylene bags at 4°C when compared

Criterion	Storage method*	Treatment**	Storage period (day)			
			30	60	90	120
Sprouting	1	C	40.0 ± 10.0	70.0 ± 4.0	90.0 ± 4.0	NA
		P	NA***	NA	NA	NA
		O	0.0	0.0	0.0	0.0
		O + P	3.8 ± 0.1	3.9 ± 0.1	4.0 ± 0.4	4.3 ± 0.2
	2	C	50.0 ± 10.0	80.0 ± 14.0	95.0 ± 6.0	98.0 ± 2.0
		P	3.6 ± 0.4	4.8 ± 0.4	NA	NA
		O	0.0	0.0	0.0	0.0
		O + P	0.0	0.0	0.0	0.0
Weight loss	1	C	20.5 ± 2.9	35.7 ± 5.3	59.9 ± 8.0	NA
		P	NA	NA	NA	NA
		O	0.0	0.0	0.0	0.0
		O + P	2.9 ± 0.2	3.0 ± 0.2	3.8 ± 0.6	4.6 ± 0.5
	2	C	7.1 ± 0.2	21.7 ± 3.3	50.9 ± 5.8	70.9 ± 8.2
		P	30.7 ± 4.0	39.5 ± 9.9	NA	NA
		O	0.0	0.0	0.0	0.0
		O + P	0.0	1.0 ± 0.1	1.0 ± 0.2	1.7 ± 0.4
Dry matter weight	1	C	17.2 ± 0.5	18.0 ± 0.7	19.2 ± 1.5	NA
		P	NA	NA	NA	NA
		O	22.5 ± 0.4	22.9 ± 0.6	23.6 ± 0.3	24.9 ± 0.6
		O + P	22.0 ± 0.2	22.3 ± 0.4	23.3 ± 0.9	23.5 ± 0.8
	2	C	22.6 ± 0.4	23.0 ± 0.3	21.0 ± 0.5	21.5 ± 1.1
		P	18.6 ± 0.6	17.0 ± 0.5	NA	NA
		O	23.6 ± 0.5	24.6 ± 0.8	24.9 ± 0.5	25.8 ± 0.5
		O + P	23.5 ± 0.4	23.7 ± 0.6	24.2 ± 0.4	25.5 ± 0.3

**1 = storage JA tubers in polyethylene bags at 4°C and 2 = storage JA tubers in peat moss layer at 25°C.*
***C = untreated control, P = infected tubers with pathogen, O = treated JA tubers with caraway essential oil and O + P = infected tubers with pathogen and treated with caraway essential oil.*
****NA = not applicable due to full decay.*

Table 2.
Mean sprouting, weight loss and dry matter weight (% ± 2SD) of JA tubers treated with caraway essential oil and infected with S. rolfsii using two storage methods.

with the control (infected-untreated) tubers. On the other hand, storage of the untreated-uninfected JA tubers in peat moss layers at 25°C increased the sprouting, and dry matter content and decreased the weight of the tubers compared to the storage of tubers in polyethylene bags at 4°C. The bioactive components like limonene and carvone, in caraway essential oil are known to inhabit sprouting percentage of JA tubers by the suppressing of mitochondrial respiration and reducing carbohydrate deterioration sugar content. Carvone had a specific tool for inhibition of sprout growth of potato tubers, such as the repression of key enzyme in the mevalonate acid pathway, which is the main precursor of gibberellin biosynthesis [17]. On the other hand, peat moss has a relatively high water retention capacity; their cells can hold 16–25 times their dry weight of water [14]. This in turns encourages such amendment for its use in the preservation of JA tubers by increasing a relative humidity around the tubers and preventing heat transfer within the peat moss layer leading to the decrease of the water loss from fresh *tubers* depends on the difference between the *water* vapor pressure within the *tubers* and the *water* vapor pressure of the surrounding air, with *moisture* passing from the higher pressure to the lower even at 25°C. Cabezas et al. [18] reported that dry matter content in JA tubers decreased significantly depends on many factors, such as storage conditions, storage periods and keeping tubers for 30 days at 18°C, this leads to loosing water above 20%.

Criterion	Storage method*	Treatment**	Storage period (day)			
			30	60	90	120
Carbohydrates	1	C	42.5 ± 1.4	41.7 ± 1.5	38.4 ± 4.3	NA
		P	NA***	NA	NA	NA
		O	44.6 ± 1.3	44.2 ± 1.3	43.3 ± 1.3	43.0 ± 1.0
		O + P	42.7 ± 1.7	42.0 ± 0.7	41.6 ± 1.1	41.3 ± 0.3
	2	C	42.0 ± 0.3	40.2 ± 1.4	37.7 ± 1.0	36.0 ± 0.5
		P	36.7 ± 1.4	36.0 ± 0.3	NA	NA
		O	46.0 ± 0.3	46.9 ± 0.5	47.0 ± 0.8	47.7 ± 0.9
		O + P	42.0 ± 0.8	41.9 ± 0.7	41.5 ± 1.2	41.0 ± 1.4
Inulin	1	C	14.2 ± 0.6	13.6 ± 0.7	12.8 ± 1.2	NA
		P	NA	NA	NA	NA
		O	15.6 ± 0.8	15.0 ± 0.3	16.7 ± 0.6	15.0 ± 0.2
		O + P	14.9 ± 0.6	14.0 ± 0.1	14.0 ± 0.0	13.9 ± 0.4
	2	C	14.3 ± 0.5	14.0 ± 0.4	13.8 ± 0.6	13.0 ± 0.2
		P	13.0 ± 0.4	12.0 ± 0.3	NA	NA
		O	18.9 ± 0.5	18.0 ± 0.4	17.9 ± 0.2	17.6 ± 0.6
		O + P	17.9 ± 0.5	17.6 ± 0.8	17.0 ± 0.2	17.0 ± 0.0
Protein	1	C	12.2 ± 0.4	12.0 ± 0.2	11.9 ± 0.4	NA
		P	NA	NA	NA	NA
		O	12.8 ± 0.3	12.7 ± 0.3	12.7 ± 0.4	12.6 ± 0.4
		O + P	12.6 ± 0.5	12.6 ± 0.2	12.4 ± 0.1	12.0 ± 0.1
	2	C	12.5 ± 0.3	12.3 ± 0.4	12.0 ± 0.4	11.9 ± 0.2
		P	9.9 ± 0.2	9.0 ± 0.4	NA	NA
		O	13.0 ± 0.3	13.0 ± 0.6	12.7 ± 0.5	12.7 ± 0.2
		O + P	12.6 ± 0.5	12.4 ± 0.5	12.3 ± 0.4	12.0 ± 0.4

**1 = storage JA tubers in polyethylene bags at 4°C and 2 = storage JA tubers in peat moss layer at 25°C.*
***C = untreated control, P = infected tubers with pathogen, O = treated JA tubers with caraway essential oil and O + P = infected tubers with pathogen and treated with caraway essential oil.*
****NA = not applicable due to full decay.*

Table 3.
Mean contents of carbohydrates, inulin (mg/g ± 2SD) and protein (% ± 2SD) of JA tubers treated with caraway essential oil and infected with S. rolfsii using two storage methods.

2.1.2.2.3 Biochemical constituents of JA tubers

Table 3 show the data of carbohydrates content, inulin and protein in JA tubers exposed to emulsion of caraway essential oil and then infected with fungal pathogenic *S. rolfsii* over the use of two storage methods. The application of caraway essential oil and uninfected JA tubers had significant effects on total carbohydrates, inulin and protein contents compared with the untreated-uninfected control in both storage methods. Along 4 months of storage, the treatment of infected JA tubers with pathogen and treated with caraway essential oil effectively decreased the carbohydrate, inulin and protein contents compared with the infected-untreated control JA tubers in both methods of storage. A fresh JA tuber contains 80% water, 15% carbohydrates, mainly in the form of inulin and about 2% protein in dry matter [19]. There are many changes in fresh JA tubers with long term storage, i.e., physical, biochemical, microbiological and enzymatic and which may lead to tuber decay. To inhibit these biochemical activities, natural or artificial drying products are widely used [20]. Davies [21] reported that the basic constituents of caraway oil (monoterpenes) tend to delay and the deterioration of carbohydrates and protein contents associated with the enzymatic system as well as respiration and energy metabolism enzyme keeping the internal biochemical enzymatic activities in minimum level.

2.1.2.2.4 Peroxidase, polyphenoloxidase enzymes and phenol content in JA tubers

The mean activities data of peroxidase, polyphenoloxidase and phenol contents of JA fresh tubers treated with caraway essential oils and infected with pathogenic fungi over the use of two different two storage methods are presented on **Table 4**. Results revealed that infection with *S. rolfsii* had significant effects on total phenol and the activity of peroxidase and polyhenoloxidase enzymes in JA tubers than those of the uninfected JA tubers control in the two different storage methods. On the contrary, the application of caraway essential oil to infected/uninfected JA tubers increased peroxidase and polyphenoloxidase and phenol content compared with the untreated-uninfected JA tubers in both methods. These results are in a line with those obtained by [22] who reported an increasing in peroxidase and polyhenoloxidase enzymes in potato fresh tubers when treated with caraway essential oil. Although regulatory mechanisms of plant *enzyme* complexes and the most enzymatic reactions are reduced at low temperature degree, JA tubers metabolism could continue at a slow rate even at minimum temperature (2°C) during cold storage. The enzymatic activation due to the exogenous application of caraway essential oil treatment could be directly related to its content of bioconstituents like carvone.

2.2 Inhibition of sprout growth and increase storability of processing potato by antisprouting agent

2.2.1 Methodology

2.2.1.1 Tuber material

Fresh local potato cv. Fridor and uniformly size of 60–80 mm in diameter (weighing 180–250 g) were selected without any sprouting in eyes and no antisprouting treatment was used. Each treatment was treated with natural and safe

Criterion	Storage method*	Treatment**	Storage period (day)			
			30	60	90	120
Peroxidase	1	C	0.40 ± 0.02	0.30 ± 0.0	0.28 ± 0.05	NA
		P	NA***	NA	NA	NA
		O	0.40 ± 0.03	0.38 ± 0.01	0.38 ± 0.01	0.37 ± 0.02
		O + P	2.67 ± 0.16	2.69 ± 0.26	2.73 ± 0.04	2.74 ± 0.03
	2	C	0.23 ± 0.01	0.22 ± 0.02	0.21 ± 0.01	0.20 ± 0.06
		P	1.90 ± 0.02	1.92 ± 0.03	NA	NA
		O	0.33 ± 0.02	0.34 ± 0.02	0.34 ± 0.02	0.34 ± 0.03
		O + P	1.77 ± 0.03	1.86 ± 0.04	1.87 ± 0.02	1.87 ± 0.26
Polyphenoloxidase	1	C	0.39 ± 0.01	0.36 ± 0.01	0.35 ± 0.02	NA
		P	NA	NA	NA	NA
		O	0.47 ± 0.03	0.47 ± 0.02	0.45 ± 0.03	0.42 ± 0.06
		O + P	1.46 ± 0.02	1.57 ± 0.05	1.47 ± 0.02	1.57 ± 0.03
	2	C	0.40 ± 0.04	0.40 ± 0.03	0.37 ± 0.03	0.35 ± 0.03
		P	1.49 ± 0.03	1.50 ± 0.04	NA	NA
		O	0.46 ± 0.02	0.43 ± 0.01	0.41 ± 0.06	0.40 ± 0.01
		O + P	1.46 ± 0.02	1.57 ± 0.02	1.57 ± 0.03	1.67 ± 0.02
Total phenol	1	C	0.29 ± 0.02	0.28 ± 0.01	0.27 ± 0.01	NA
		P	NA	NA	NA	NA
		O	0.32 ± 0.03	0.32 ± 0.03	0.32 ± 0.02	0.31 ± 0.02
		O + P	0.52 ± 0.02	0.52 ± 0.02	0.51 ± 0.02	0.52 ± 0.01
	2	C	0.35 ± 0.02	0.35 ± 0.02	0.35 ± 0.02	0.34 ± 0.03
		P	0.57 ± 0.04	0.57 ± 0.04	NA	NA
		O	0.26 ± 0.01	0.26 ± 0.01	0.26 ± 0.02	0.26 ± 0.02
		O + P	0.55 ± 0.02	0.55 ± 0.02	0.55 ± 0.01	0.56 ± 0.01

**1 = storage JA tubers in polyethylene bags at 4°C and 2 = storage JA tubers in peat moss layer at 25°C.*
***C = untreated control, P = infected tubers with pathogen, O = treated JA tubers with caraway essential oil and O + P = infected tubers with pathogen and treated with caraway essential oil.*
****NA = not applicable due to full decay.*

Table 4.
Mean activities of peroxidase, polyphenoloxidase enzymes and phenol content (% ± 2SD) of JA tubers treated with caraway essential oil and infected with S. rolfsii using two storage methods.

antisprouting agent and stored at ambient temperature (average: 35/15°C day/night and 70% RH) in Laboratory for 4 months.

2.2.1.2 Treatments

The experiment included seven treatments, which were as follows: *Cymbopogon martini* (rich in geraniol and geranyl acetate), *C. flexuosus* (rich in citral), *C. winterianus* (rich in rich in citronellal and citronellol), *Ocimum sanctum* (rich in rich in ketone and camphor), *Carum carvi* (rich in rich in carvone), *Artemisia annua* (rich in ketone camphor) and *Lavendula officinalis* (rich in linalool). The isolated terpenoids were purified by HPLC. Essential oils were purified by column chromatography and substantially pure compounds were used. Tubers dipped in emulsions of 8 mm concentration of each compound in distilled water and Tween 20 (6%) for 30 min after 1 month of harvest or at such time that the tubers begin to sprout.

2.2.2 Results and discussion

2.2.2.1 Sprouting, weight loss and dry matter content

All control tubers had significant values of sprouting and weight loss percentages at the end of storage period (**Table 5**). Geraniol and citral completely inhibited sprouting by 100%, decreased weight loss and increase tuber dry matter content in both seasons. Application of geranyl acetate inhibited sprouting by 95%. On the other hand, linalool and L-carvone had no significant effect on tuber sprouting. It has been reported that L-carvone and D-carvone displayed little or no inhibition of sprouting in potatoes [17]. Geraniol and citral have a high content in monoterpenes such as benzaldehyde, eugenol and thymol [23]. CIPC inhibited sprouting over 98.5%.

Under this study condition, the beneficial effect of the applied anti-sprouting agent (geraniol and citral) on controlling tubers sprouting and increasing dry matter content could be associated with their similar advantages effect in preservation of their tubers starch, carbohydrates, sugars and amino acid content (**Table 6**). Suppression of sprouting and weight loss logically associated with maintenance of dry matter. Furthermore, monoterpenes acts as antioxidant and had a protective role against oxidative stress under normal conditions of storage.

2.2.2.2 Reducing sugars, amino acids and peroxidase POD activity

All storage treatments gave significant lower values on reducing sugars and amino acids content during two seasons of study as compared to the control (**Table 6**). In the ambient temperature, the lowest significant values of reducing

Treatments	Sprouting (%)		Weight loss (%)		Dry matter (%)	
	2012	2013	2012	2013	2012	2013
1. Control	100.0^{a}	96.00^{a}	25.12^{a}	26.18^{a}	21.65^{f}	22.80^{e}
2. CIPC	2.49^{e}	1.20^{c}	4.33^{e}	2.80^{ef}	23.60$^{a–d}$	23.66^{d}
3. Geranyl acetate	4.68^{d}	4.33^{c}	3.41^{f}	4.65^{d}	22.50^{ef}	24.55^{ab}
4. Geraniol	0.00^{f}	0.00^{c}	2.19^{h}	1.45^{g}	24.56^{a}	25.30^{a}
5. Camphor	6.92^{c}	5.98^{c}	2.88^{g}	2.95^{ef}	23.33$^{b–e}$	24.38^{bc}
6. Citral	0.00^{f}	0.00^{c}	1.51^{i}	1.26^{g}	24.00^{ab}	24.95^{ab}
7. Linalool	100.00^{a}	72.00^{b}	9.50^{b}	8.00^{b}	22.66^{de}	23.60^{d}
8. L-Carvone	70.58^{b}	62.00^{b}	9.50^{b}	6.25^{c}	22.80$^{c–e}$	23.70^{cd}
9. D-Carvone	72.00^{b}	76.98^{b}	8.03^{c}	3.45^{e}	22.90$^{c–e}$	24.89 ab
10. D-Citronellol	2.89^{e}	2.00^{c}	6.75^{d}	5.73^{c}	23.60$^{a–d}$	24.68^{ab}
11. L-Citronellol	0.00^{f}	0.00^{c}	2.25^{gh}	2.10^{fg}	23.80$^{a–c}$	24.55^{ab}

Means followed by the same letter(s) within each column do not significantly differ using Duncan's multiple range test at the level of 5%; where, (a) refer to the highest mean values, and (h) refer to the lowest mean values according to Duncan Multiple Range Test.

Table 5.
Sprouting behavior characters and dry matter of potato tubers as affected by anti-sprouting agent during 2012 and 2013 seasons (after 4 months of storage period).

Treatments	Reducing sugars (%)		Total free amino acids (%)		Peroxidase activity POD (%)	
	2012	2013	2012	2013	2012	2013
1. Control	4.29^{a}	4.52^{a}	0.352^{a}	0.348^{a}	56.77^{g}	55.51^{g}
2. CIPC	2.05^{c}	3.18^{d}	0.307^{ab}	0.301^{ab}	95.81^{b}	94.63^{b}
3. Geranyl acetate	1.39^{cd}	3.93^{b}	0.084^{bc}	0.047^{c}	79.75^{e}	79.33^{e}
4. Geraniol	1.24^{d}	1.51^{f}	0.030^{c}	0.028^{c}	97.33^{a}	96.29^{a}
5. Camphor	3.41^{b}	3.48^{c}	0.152^{a-c}	0.153^{a-c}	80.68^{e}	80.26^{e}
6. Citral	1.25^{d}	1.52^{f}	0.045^{c}	0.045^{c}	97.68^{a}	96.46^{a}
7. Linalool	4.07^{ab}	4.13^{b}	0.106^{bc}	0.108^{bc}	80.67^{e}	79.06^{c}
8. L-Carvone	3.81^{ab}	1.83^{e}	0.084^{bc}	0.151^{a-c}	81.67^{e}	80.50^{e}
9. D-Carvone	1.45 cd	1.68^{ef}	0.146^{a-c}	0.157^{a-c}	77.55^{f}	76.77^{f}
10. D-Citronellol	1.76^{cd}	1.54^{f}	0.186^{a-c}	0.187^{a-c}	84.50^{d}	83.62^{d}
11. L-Citronellol	1.29^{d}	1.58^{f}	0.147^{a-c}	0.059^{c}	87.67^{c}	86.65^{c}

Means followed by the same letter(s) within each column do not significantly differ using Duncan's multiple range test at the level of 5%; where, (a) refer to the highest mean values, and (g) refer to the lowest mean values according to Duncan Multiple Range Test.

Table 6.
Reducing sugars, amino acids and peroxidase enzyme of potato tubers as affected by anti-sprouting agent during 2012 and 2013 seasons (after 4 months of storage period).

sugars and amino acids content were found in tubers exposed to emulsion of geraniol and citral, without significant difference between the two treatments.

The monoterpenes rich in compounds had a potential role in preservation and maintenance of the stored tubers reserves, keeping the enzymatic activities in a minimal level and in more stable case thereby prolonged their dormancy period. Also, application of these treatments were highly effective in tuber protection against the degradable effects of oxidative stressful during high temperature storage conditions and accordance to the findings of [20] who indicated that monoterpenes and antioxidants tended to slow down the activity of carbohydrates, breakdown of protein and enzymatic activity as well as reduce respiration rate and metabolism enzyme. The role of POD in sprouting of potatoes was widely reported, particularly its degrading activity of IAA, and cytokinin which is considered an effective promote oxidative stress is of great importance and depending on the activation degree of peroxidase as affected by storage treatments.

2.2.2.3 Processing quality of potato fries and chips

All storage treatments and CIPC treatment at ambient temperature had significant differences on quality characters of potato chips and French fries, i.e., color, crispiness and taste in comparison with the control treatment (**Table 7**).

The same treatments prevented and blocked the accumulation of total sugars, and kept the reducing sugars and amino acids in optimize levels in the stored tubers at ambient temperature. This is true in the end of storage (4 months). Thus, we noticed the worst processing quality (dark potato chips and crispness with bad taste) of storage treatments due to the appearance of Millard reaction during frying process and the accumulation of reducing sugars and amino acids [23]. The same processing quality parameters were correlated with dry matter content (**Table 8**) and with amino acids content (**Table 9**) in both seasons. These results are in harmony with those previously obtained by [24]. Meanwhile, we also noticed that

Treatments	Chips						French fries					
	Color		Taste		Crispness		Color		Taste		Crispness	
	2012	2013	2012	2013	2012	2013	2012	2013	2012	2013	2012	2013
1. Control	3.00^{e}	3.33^{c}	3.00^{d}	3.33^{bc}	4.33^{a-c}	4.33^{a-c}	3.33^{de}	3.00^{d}	3.33^{cd}	4.00^{b-d}	4.67^{ab}	4.67^{ab}
2. CIPC	3.33^{de}	3.33^{c}	4.33^{a-c}	4.33^{ab}	4.33^{a-c}	4.67^{ab}	3.67^{cde}	3.33^{cd}	4.00^{a-c}	4.33^{a-c}	4.67^{ab}	4.67^{ab}
3. Geranyl acetate	4.67^{ab}	4.67^{ab}	5.00^{a}	4.67^{a}	5.00^{a}	5.00^{a}	4.67^{ab}	4.67^{ab}	5.00^{a}	4.67^{ab}	5.00^{a}	4.33^{abc}
4. Geraniol	5.00^{a}	5.00^{a}	5.00^{a}	4.67^{a}	5.00^{a}	5.00^{a}	4.67^{ab}	4.67^{ab}	5.00^{a}	4.67^{ab}	5.00^{a}	4.33^{a-c}
5. Camphor	4.67^{ab}	4.67^{ab}	5.00^{a}	4.67^{a}	5.00^{a}	5.00^{a}	4.67^{ab}	4.67^{ab}	4.67^{ab}	4.67^{ab}	5.00^{a}	5.00^{a}
6. Citral	5.00^{a}	5.00^{a}	5.00^{a}	4.67^{a}	5.00^{a}	5.00^{a}	5.00^{a}	5.00^{a}	4.67^{ab}	5.00^{a}	5.00^{a}	5.00^{a}
7. Linalool	4.67^{ab}	4.67 ab	4.64^{ab}	4.67^{a}	5.00^{a}	5.00^{a}	4.67^{ab}	4.67^{ab}	4.67^{ab}	4.67^{ab}	5.00^{a}	5.00^{a}
8. L-Carvone	4.67^{ab}	4.67^{ab}	4.67^{ab}	4.67^{a}	5.00^{a}	5.00^{a}	4.67^{ab}	4.67^{ab}	5.00^{a}	5.00^{a}	5.00^{a}	5.00^{a}
9. D-Carvone	5.00^{a}	5.00^{a}	5.00^{a}	4.67^{a}	5.00^{a}	5.00^{a}	5.00^{a}	5.00^{a}	4.67^{ab}	5.00^{a}	5.00^{a}	5.00^{a}
10. D-Citronellol	4.00^{b-d}	4.67^{ab}	4.67^{ab}	4.67^{a}	4.67^{ab}	4.67^{a}	4.00^{b-d}	4.00^{a-c}	4.33^{a-c}	4.33^{a-c}	4.67^{ab}	5.00^{a}
11. L-Citronellol	4.33^{a-c}	4.67^{ab}	4.67^{ab}	4.67^{a}	4.67^{ab}	4.67^{a}	4.00^{b-d}	4.33^{a-c}	3.67^{b-d}	4.33^{a-c}	4.67^{ab}	4.67^{ab}

Means followed by the same letter(s) within each column do not significantly differ using Duncan's multiple range test at the level of 5%; where, (a) refer to the highest mean values, and (d) refer to the lowest mean values according to Duncan Multiple Range Test.

Table 7.
Quality processing of potato tubers as affected by anti-sprouting agent during 2012 and 2013 seasons (after 4 months of storage period).

Treatment		Seed germination index [%]		Seed germination [%]		Seedling length [cm]		Seedling vigor index [cm]		Yield [g] container/ 242 cm^2	
		2012	2013	2012	2013	2012	2013	2012	2013	2012	2013
1	Water (control)	13.36^{e}	12.96^{d}	86.67^{c}	86.0^{a}	4.67^{c}	4.00^{b}	4.03^{c}	3.44^{c}	36.40^{e}	34.20^{d}
2	Hot water	14.61^{de}	13.02^{d}	93.78^{b}	90.44^{bc}	5.00^{c}	4.80^{b}	4.71^{c}	4.33^{c}	40.88^{de}	37.21^{d}
3	Fennel oil	22.01^{a}	23.01^{a}	97.33^{ab}	97.33^{a}	7.33^{ab}	7.67^{a}	7.13^{ab}	7.47^{ab}	56.90^{b}	49.17^{c}
4	Caraway oil	21.94^{a}	22.88^{a}	97.33^{ab}	99.00^{a}	8.00^{ab}	8.33^{a}	7.79^{a}	8.25^{a}	54.97^{bc}	67.75^{a}
5	Basil oil	20.22^{ab}	21.82^{ab}	94.67^{b}	92.33^{b}	7.00^{b}	7.67^{a}	6.63^{b}	7.07^{b}	64.87^{a}	68.17^{a}
6	Thyme oil	18.81^{bc}	20.14^{b}	100.00^{a}	100.0^{a}	8.20^{a}	8.30^{a}	8.20^{a}	8.30^{a}	66.54^{a}	67.75^{a}
7	Sage oil	16.91^{cd}	17.76^{c}	100.00^{a}	100.0^{a}	7.83^{ab}	7.83^{a}	7.83^{a}	7.83^{ab}	47.83^{cd}	49.17^{c}

Means followed by the same letter(s) within each column do not significantly differ using Duncan's multiple range test at the level of 5%; where, (a) refer to the highest mean values, and (e) refer to the lowest mean values according to Duncan Multiple Range Test.

Table 8.
Vegetative characters of treated broccoli seeds with different essential oils before cold storage.

the best processing quality of basic constituents of essential oils produced chips, the optimization of reducing sugars and amino acids of their tubers thereby, the prevention of Millard reaction occurrence during frying processes and thus it turn reflects on best color, crispiness and taste.

2.3 Increasing antioxidant content of broccoli sprouts using essential oils during cold storage

2.3.1 Methodology

2.3.1.1 Plant material and germination condition

(*Brassica oleracea* L. var. italica and the variety name is F1 Hybrid Sakura) from Tokita Seeds CO., LTD (Saitama, Japan). The seeds (1000 seeds, nearly 5 g) for each treatment were soaked in a sodium hypochlorite solution at 0.5% v/v for 15 min then were dipped in 50 ml of deionized water for 5½ h with shaking every 30 min and washed with deionized and sterilized water. On 15th of September, broadcast the seeds were done over absorbent medical cotton in sprouting plastic containers (220 × 110 mm). The emulsions of various natural essential oil at the concentration of 0.05% were emulsified in tween 80 (1.5 ml/l) in the cotton media and the containers were getting closed immediately. The containers were maintained at 25 ± 2°C with and 16 h light/8 h darkness, 80–90% relative humidity and 7.4 lmol/m^2/s light intensity to give the best germination conditions. All sprouts in containers were cut above their root mats after 3 days from sowing. The sprouts were weighed for 20 g for each placed container and stored at 4°C in the dark to simulate a domestic refrigerator for 15 days. The sprouts of best treatment with control were stored only.

2.3.1.2 Application and extraction of essential oils

The essential oils of fennel seeds (*Foeniculum vulgare*), caraway seeds (*Carum carvi*), thyme herbs (*Thymus vulgaris*), basil herbs (*Ocimum basilicum*) and sage leaves (*Salvia officinalis*) (200 g from each one) were used for oil extraction by

	Treatment	Total phenolic acid (mg/100 g F.W.)		Total flavonoids (mg/100 g F.W.)		Anthocyanin (mg/100 g F.W.)		Ascorbic acid (mg/100 g F.W.)		DPPH (Mmol TE/g F.W.)	
		2012	2013	2012	2013	2012	2013	2012	2013	2012	2013
1	Water (control)	83.33^{d}	84.11^{e}	91.99^{d}	95.18^{e}	7.13^{d}	7.70^{d}	70.58^{e}	81.23^{d}	23.66^{a}	24.66^{a}
2	Hot water	88.71^{c}	88.56^{c}	100.95^{c}	101.03^{d}	8.62^{c}	8.77^{c}	86.81^{c}	86.81^{c}	23.54^{b}	23.66^{a}
3	Fennel oil	88.46^{c}	88.90^{c}	107.66^{b}	107.72^{c}	8.86^{c}	8.87^{bc}	87.66^{c}	88.00^{c}	21.98^{d}	21.98^{c}
4	Caraway oil	87.90^{c}	88.13^{cd}	104.66^{b}	104.73^{c}	9.84^{bc}	9.84^{bc}	77.33^{d}	85.80^{c}	21.96^{de}	21.96^{c}
5	Basil oil	122.06^{b}	122.29^{b}	113.00^{a}	113.00^{b}	11.71^{a}	12.05^{a}	94.67^{b}	94.67^{b}	21.94^{de}	21.94^{c}
6	Thyme oil	131.66^{a}	131.60^{a}	115.66^{a}	116.24^{a}	12.09^{a}	12.14^{a}	102.33^{a}	103.33^{a}	21.86^{e}	20.03^{d}
7	Sage oil	87.9^{c}	84.74^{de}	104.33^{bc}	104.59^{c}	10.38^{b}	10.38^{b}	82.33^{cd}	86.69^{c}	22.79^{c}	22.79^{bc}

Means followed by the same letter(s) within each column do not significantly differ using Duncan's multiple range test at the level of 5%; where, (a) refer to the highest mean values, and (e) refer to the lowest mean values according to Duncan Multiple Range Test.

Table 9.
Phytochemical screening by GLC for 3-days-old broccoli sprouts produced from treated seeds with essential oils before cold storage.

hydro-distillation for 2–3 h. After extraction, essential oils were analyzed by Gas Liquid Chromatography (GLC) to separate and identify their basic constituents.

2.3.2 Results and discussion

2.3.2.1 Vegetative characters of broccoli sprout

All essential oil treatments rich in antioxidant stimulate the germination of broccoli seeds. All essential oils treatments significantly increased germination, germination index, seedling length, seedling vigor index and container yield compared with the control (tap water) during the two seasons (**Table 8**). The essential oils of fennel, caraway and thyme increased the seed germination index by 171.43, 170.29 and 148.02%, respectively, compared to the control 100%. The increases of seed germination % over the control reached to 12.73, 13.74 and 15.82% for the effective treatments, respectively. The essential oils of thyme, caraway and fennel had significant increases in seedling vigor and yield container over the control to 50.25, 73.82 and 90.22%, respectively.

The allelochemical effects of essential oils for induce stimulatory or inhibitory of seed germination and other physiological process varied depending on the dose, tested species, concentration and basic components. Under our study, the lower doses of essential oils had a stimulatory effect [25]. The obtained results reveal that the applications of essential oils at a low level improve seed germination of broccoli. However, application of thyme oil reaches 100% of sprouts after seed germination (**Table 8**). Impact of essential oils on seed germination of other plant species was reported as 24 out of 47 tested terpenoids enhanced the seed germination of *Lactuca sativa* [26]. Also, the positive impact of thyme essential oil on broccoli seeds could be because of its active ingredients.

2.3.2.2 Phytochemical characters

All treatments significantly surpassed over the control in Broccoli sprout bioconstituents, i.e., total phenolic acid, total flavonoids, anthocyanin and ascorbic acid, while the control treatment gave the highest DPPH radical scavenging capacity (**Table 9**). Application of thyme oil treatment produced significant increases of total phenol, total flavonoids, anthocyanin and ascorbic acid content. Moreover, thyme and basil essential oils decreased significantly the DPPH free radical scavenging capacity. Accordingly, it has been chosen to study the storage behavior characters, in addition to control treatment. The majority of the antioxidant activity attributes to phenolic compounds, flavonoids and ascorbic acidin essential oils [27]. Moreover, the effect of antioxidant on DPPH free radicle was due to the presence of hydroxyl groups in their chemical structure. In this respect, [28] found that the oregano essential oil inhibited hydro-peroxide formation and that the CHO fraction showed the highest antioxidants activity.

The thyme oil showed significant lowest radical scavenging capacity compared to the control and other treatments (**Table 9**). All other antioxidants/essential oils showed high and almost the same antioxidant capacity effect. It was known that the free radical scavenging DPPH intensity of some compounds can be influenced by their different kinetic behavior [29]. For slow reacting compounds the influence was attributed to the complex reacting mechanism. In our study, probably, the constituents from thyme essential oil involved one or more secondary reactions, which result the slower reduction of DPPH solutions [29].

2.3.2.3 Antioxidant activity during cold storage

2.3.2.3.1 Total phenolic compounds and DPPH radical scavenging capacity

Figure 2 illustrate that there was a gradual increase in the total phenolic acid content, and reaching a maximum value at day 5 and 10 (132.67 and 135.04 mg GAE/100 g F.W.) compared to the initial time. This concentration decreased in to 129.03 mg at day 15 due to thyme oil application (**Figure 2**). Keeping in view that the control treatment decreased to 73.84 GAE/100 g FW at day 5. On the 15th day, the old-sprout from storage, the control was reduced by 28.57% compared to thyme oil (1.98%). The control treatment of antioxidant capacity increased significantly until day 10 (29.43 mg/100 g F.W.), and finally decrease (28.46% mg 100/g F.W.) at day 15 increased from initial period (20.28%). While, application of thyme oil the change was not clear at the end of storage (1.98%) (**Figure 3**). During cold storage (**Figure 3**), the control was reduced DPPH by 28.57% compared to thyme oil at 15 day old-sprout (1.98%). Nath et al. [30] observed a constant decrease in the antioxidant capacity for 144 h of storage of broccoli inflorescences. This behavior in DPPH may be due to the steady changes in plant metabolism during storage period as a result of oxidative stress, which may include structural and chemical changes in synthesis or antioxidant content [31].

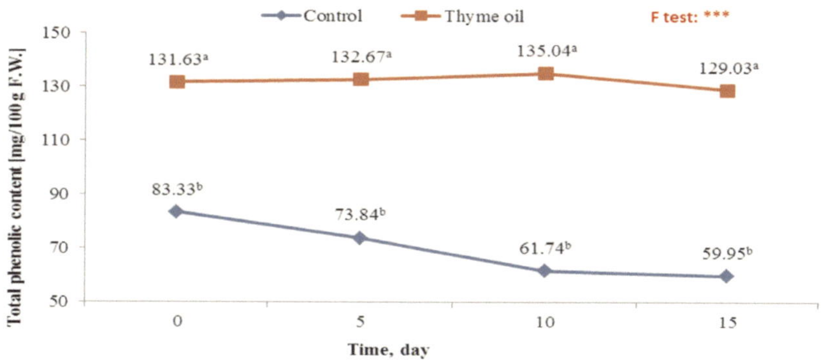

Figure 2.
Total phenolic content as affected by thyme oil compared to control treatment at different storage period.

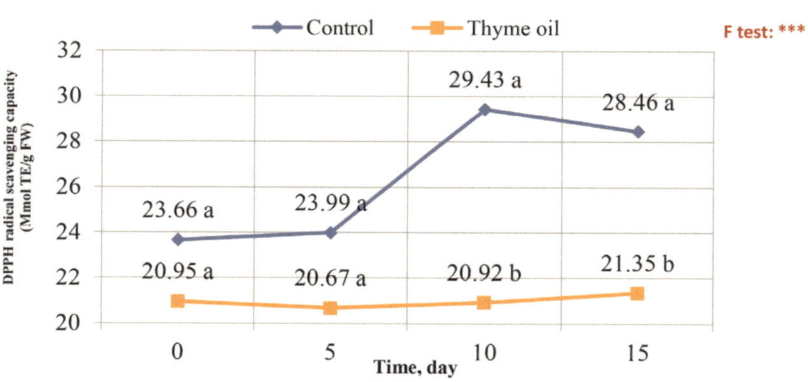

Figure 3.
DPPH radical scavenging capacity as affected by thyme oil compared to control treatment at different storage period.

2.3.2.3.2 Total flavonoids

Total flavonoids (**Figure 4**) were found in a higher concentration in 3-day-old sprouts of thyme treatment, with values of 115.95 mg/100 g F.W., after 5 and

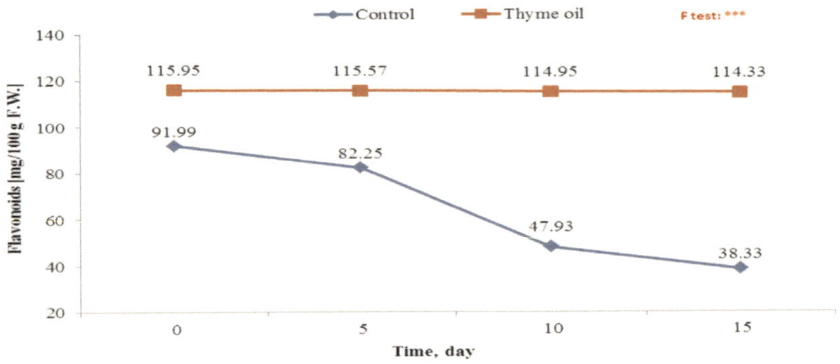

Figure 4.
Total flavonoids content as affected by thyme oil compared to control treatment at different storage period.

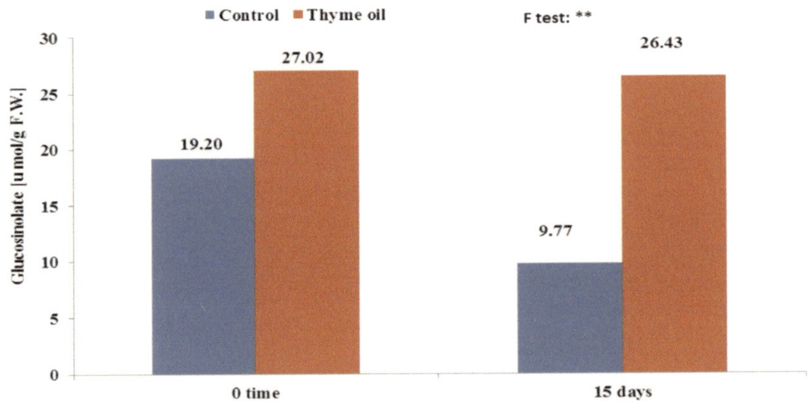

Figure 5.
Total glucosinolates content as affected by thyme oil compared to control treatment at 0 time and 15 DAS.

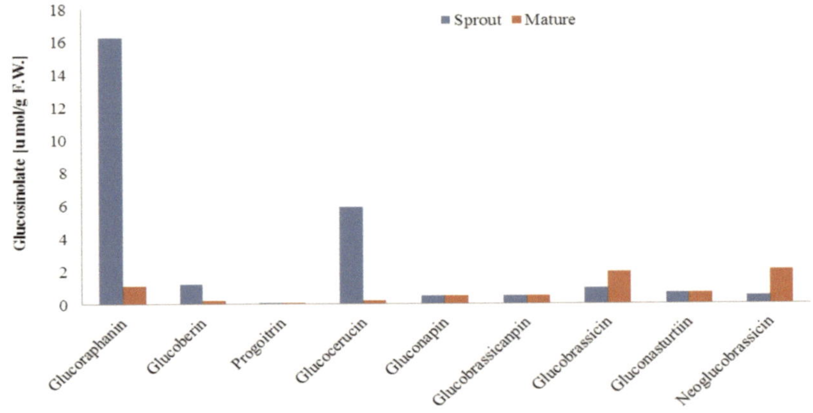

Figure 6.
Total and individual aliphatic, aromatic/indole glucosinolates levels in broccoli in 3-days-old sprout and mature at harvest.

10 days of storage slight decrease to 0.021 and 0.086%, respectively, when compared with the initial value, and finally reduced by 1.39%. The high loss of flavonoids reached to 10.59 and 47.89%, after 5 and 10 days, respectively, and at 15 days the loss increased to 58.33% for control treatment (average two seasons).

2.3.2.3.3 Glucosinolates content

Storage time had significant differences in glucosinolates content of the samples analyzed. **Figure 5** illustrate that the thyme oil increased significantly glucosinolates content in 3-day-old sprouts, compared to control treatments. Moreover, thyme oil had a high value of total glucosinolates (27.02 μg/g F.W.) and slightly decreased up to 26.43 μg/g F.W. on day 15. At the end of storage, the decreasing changes percent was about 2.18%. In the control treatment, the highest decrease in total glucosinolates content was observed, where reached about 49.12% at the end of storage.

2.3.2.3.4 Glucosinolates content of mature head versus sprout broccoli

In sprout, the total glucosinolates level (27.02 μg/g F.W.) is higher than in florets or heads (7.37) (**Figure 6**). Glucoraphanin is the powerful of antioxidant and the most abundant aliphatic glucosinolates present in sprout. The glucoraphanin reached the highest 16.24 followed by glucoerucin 5.9 and glucoiberm 1.2 μg/g F.W. On the other hand, the florets/heads contain the highest level of aromatic/indolylglucosinolates, neoglucobrassicin (2.11) followed by glucobrassicin (1.67). Our results are in agreement with those obtained by [32].

Author details

Aml Abo El-Fetouh El-Awady
Horticulture Research Institute, Agriculture Research Center, Giza, Egypt

*Address all correspondence to: aml.elawady@yahoo.com

References

[1] Tesio F, Weston LA, Ferrero A. Allelochemicals identified from Jerusalem artichoke (*Helianthus tuberosus* L.) residues and their potential inhibitory activity in the field and laboratory. Scientia Horticulturae Amsterdam. 2011;**129**:361-368. DOI: 10.1016/j.scienta.2011.04.003

[2] Yang L, He QS, Corscadden K, Udenigwe CC. The prospects of Jerusalem artichoke in functional food ingredients and bioenergy production. Biotechnology Reports. 2015;**5**:77-88. DOI: 10.1016/j.btre.2014.12.004

[3] Zhang Y, Li S, Jiang D, Kong L, Zhang P, Xu J. Antifungal activities of metabolites produced by a termite-associated *Streptomyces canus* BYB02. Journal of Agricultural and Food Chemistry. 2013;**61**:1521-1524. DOI: 10.1021/jf305210u

[4] Suhag M, Nehra BK, Singh N, Khurana SC. Storage behavior of potato under ambient condition affected by curing and crop duration. Haryana Journal Of Horticultural Sciences. 2006; **35**:357-360

[5] Delaplace P, Brostaux Y, Fauconnier ML, du Jardin P. Potato (*Solanum tuberosum* L.) tuber physiological age index is a valid reference frame in postharvest ageing studies. Postharvest Biology and Technology. 2008;**50**: 103-106. DOI: 10.1093/jxb/erp008

[6] De Carvalho C, Da Fonseca MMR. Carvone: Why and how should one bother to produce this terpene. Food Chemistry. 2006;**95**:413-422. DOI: 10.1016/j.foodchem.2005.01.003

[7] Pringle B, Bishop C, Clayton R. Potatoes Postharvest. UK: CAB International; 2009. p. 427

[8] Villarreal-Garcia D, Nair V, Gisneros-Zevallos L, Jacobo-Veazquez DA. Plants as biofacttories: Postharvest stress-induced accumulation of phenolic compounds and glucosinolates in broccoli subjected to wounding stress and exogenous phytohormones. In Fronties in Plant Science. 2016;**7**(45): 1-11. DOI: 10.3389/fpls.2016.00045

[9] Baenas N, García-Viguera C, Moreno DA. Biotic elicitors effectively increase the glucosinolates content in brassicaceae sprouts. Journal of Agricultural and Food Chemistry. 2014; **62**:1881-1889. DOI: 10.1021/jf404876z

[10] Nguefack J, Somda I, Mortensen CN, Amvam Zollo PH. Evaluation of five essential oils from aromatic plants of Cameroon for controlling seed-borne bacteria of rice (*Oryza sativa* L.). Seed Science and Technology. 2005;**33**: 397-407. DOI: 10.15258/sst.2005.33.2.12

[11] Justesen U, Knuthsen P. Composition of flavonoids in fresh herbs and calculation of flavonoid intake by use of herbs in traditional Danish dishes. Food Chemistry. 2001; **73**:245-250

[12] Darougheh F, Barzegar M, Sahari M. Antioxidant and anti-fungal effect of caraway (*Carum carvi* L.). Essential oil in real food system. Current Nutrition & Food Science. 2014;**10**(1):70-76

[13] Ma B, Ban X, Huang B, He J, Tian J, Zeng H, et al. Interference and mechanism of dill seed essential oil and contribution of carvone and limonene in preventing Sclerotinia rot of rapeseed. PLoS ONE. 2015;**10**(7):e0131733. DOI: 10.1371/journal.pone.0131733

[14] Taskila S, Särkelä R, Tanskanen J. Valuable applications for peat moss. Biomass Conversion and Biorefinery. 2016;**6**:115-126

[15] Zaitseva N. A polysaccharide extracted from Sphagnum moss as

antifungal agent in archaeological conservation [master's thesis]. Ontario, Canada: Queen's University, Kingston; 2009. p. 282

[16] Borsheim KY, Christensen BE, Painter T. Preservation of fish by embedment in sphagnum moss, peat, or holocellulose: Experimental proof of the oxopolysaccharidic nature of the preservative substance and its antimicrobial and tanning action. Innovative Food Science and Emerging Technologies. 2012;**2**(1):63-74

[17] Oosterhaven K, Hartmans KJ, Scheffer JJC. Inhibition of potato sprouts growth by carvone enantiomers and their bioconversion in sprout. Potato Research. 1995;**38**:219-230

[18] Cabezas MJ, Rabert C, Bravo S, Shene C. Inulin and sugar contents in *Helianthus tuberosus* and *Cichorium intybus* tubers: Effect of post-harvest storage temperature. Journal of Food Science. 2002;**67**:2860-2865

[19] Brkljaca J, Bodroza-Solarov M, Krulj J, Terzic S, Mikic A, Marjanovic-Jeromela A. Quantification of inulin content in selected accessions of Jerusalem artichoke (*Helianthus tuberosus* L.). Helia. 2014;**37**(60):105-112

[20] Norkulova KT, Safarov JE. Research of sorption characteristics of tubers Jerusalem artichoke (*Helianthus tuberosus*). Journal of Food Processing and Technology. 2015;**6**(6):453-454

[21] Davies HV. Carbohydrate metabolism during sprouting. American Potato Journal. 1990;**67**:815-827

[22] Afify AMR, El-Beltagi HS, Aly AA, El-Ansary AE. Antioxidant enzyme activities and lipid peroxidation as biomarker for potato tuber stored by two essential oils from Caraway and Clove and its main component carvone and eugenol. Asian Pacific Journal of Tropical Biomedicine. 2012;**2**:S772-S780

[23] Hartmans KJ, Diepenhorst P, Bakker W, Gorris LGM. The use of carvone in agriculture: Sprout suppression of potatoes and antifungal activity against potato tuber and other plant diseases. Industrial Crops and Products. 1995;**4**: 3-13

[24] El-Awady AA. Studies on storing potato tubers out refrigerator using natural essential oils [Ph.D. thesis]. Fac. Agriculture: Mansoura University; 2006. p. 166

[25] Leth V. Use of essential oils as seed treatment. IPGRI Newsletter. 2002;**9**: 15-16

[26] Vokou D, Douvli P, Blionis GJ, Halley JM. Effects of monoterpenoids, acting alone or in pairs, on seed germination and subsequent seedling growth. Journal of Chemical Ecololgy. 2003;**29**:2281-2301. DOI: 10.1023/A: 1026274430898

[27] Heim KE, Tagliaferro AR, Bobilya DJ. Flavonoid antioxidants: Chemistry, metabolism and structure-activity relationships. The Journal of Nutritional Biochemistry. 2002;**13**:572-584. DOI: 10.1016/S0955-2863(02)00208-5

[28] Milos M, Mastelic J, Jerkovic I. Chemical composition and antioxidant effect of glycosidically bound volatile compound from oregano (*Origanum vulgare* L. ssp. *hirtum*). Food Chemistry. 2000;**71**:79-83

[29] Konczak I, Zhang W. Anthocyanins —More than nature's colours. Journal of Biomedicine Biotechnology. 2004;**5**: 239-240. DOI: 10.1155S11107243044 07013

[30] Nath A, Bagchi B, Misra LK, Deka BC. Changes in post-harvest phytochemical qualities of broccoli florets during ambient and refrigerated storage. Food Chemistry. 2011;**127**: 1510-1514. DOI: 10.1016/j. foodchem.2011.02.007

[31] Xiao Z, Lester GE, Luo Y, Xie Z, Yu L, Wang Q. Effect of light exposure on sensorial quality, concentrations of bioactive compounds and antioxidant capacity of radish microgreens during low temperature storage. Food Chemistry. 2014;**151**:472-479. DOI: 10.1016/j.foodchem.2013.11.086

[32] Fahey JW, Zhang Y, Talalay P. Broccoli sprouts: An exceptionally rich source of inducers of enzymes that protect against chemical carcinogens. Proceedings National Academy of Sciences of the United States of America, USA. 1997;**94**:10367-10372

Chapter 8

Study of Essential Oils Obtained from Tropical Plants Grown in Colombia

Elena Stashenko and Jairo René Martínez

Abstract

Researchers from several Colombian universities have joined efforts for over 15 years to characterize the composition and biological properties of more than a thousand samples of essential oils (EOs) obtained from aromatic plants collected during at least 30 botanical outings in different regions of Colombia. This chapter presents a brief description of essential oil extraction and chemical characterization techniques, followed by a representative list of references to publications on EO composition obtained from tropical aromatic plants that grow in Colombia. Opportunities for the development of interesting products for the pharmaceutical, cosmetics, hygiene, and food industries are illustrated with a few selected works on the evaluation of cytotoxicity, antioxidant, antiviral, antigenotoxic activities, and repellence of these essential oils.

Keywords: *Lippia*, CENIVAM , antioxidant, antigenotoxic, antiviral, repellence

1. Introduction

Colombia, located at South America's northwest, has coasts on the Caribbean and the Pacific Ocean, extensive prairies and mountains with many forests, wild pastures and cultivated land, rivers, and lakes. The country is rich in many natural resources and water. Contrasting landscapes and varied climatic conditions have made it after Brazil, the second most biodiverse country. This biodiversity includes medicinal and aromatic plants; most native aromatic plants remain unexamined. The aromatic herbs and spices commonly used in everyday life were brought to Colombia by the Spanish conquerors five centuries ago (basil, chamomile, mint, parsley, oregano, rosemary, sage, thyme, etc.); some (citronella, lemongrass, clove, ginger, cinnamon) were introduced later, in the last two centuries.

The extension of land cultivated with medicinal and aromatic plants more than doubled between 2007 (1253 ha) and 2015 (2709 ha) [1]. These plantations are located mainly in the Andean region, some in the Eastern Plains of Colombia. The crop of medicinal and aromatic plants amounted to 16,188 tons in 2015. This vegetal material was used for many applications different from essential oil (EO) extraction, in over 100 companies and 2500 commercial establishments [1]. Aromatic plants are used in Colombia's food industry for beverages, and a portion of the crop is exported in fresh (8288 tons in 2017) [2]. Colombia has currently no commercial enterprise dedicated to the cultivation of aromatic plants destined to produce essential oils for export or the national market. Brazil, India, China, Indonesia, and the United States are Colombia's main essential oil suppliers. In 2017, the total cost of the country's essential oil imports was 14.289 million dollars, while the country exported just 298 thousand dollars [2]. Since there is no essential oil production, the EO exported amounts corresponded to commercialization of previously imported oils.

The publication of Colombian scientific articles on EO research started in 1974 and grew slowly during the following 30 years (less than three articles per year). The transition point was marked by the creation in 2005 of a network of research groups that joined their expertise around the development of the EOs agroindustry. The Research Center of Excellence for the Agroindustrialization of Aromatic and Tropical Medicinal Species (CENIVAM), under technical and administrative coordination at the Industrial University of Santander (Bucaramanga), has been a leader in aromatic plants and EO studies in Colombia for more than a decade. Over 250 scientific articles comprise the results of its investigations, which have been focused on the multidisciplinary and systematic search of promising native plants and on introduced species such as ylang-ylang, palmarosa, turmeric, patchouli, mints, basils, citrus, geraniums, and others. Researchers from more than 10 universities have carried out their work in areas of botany and taxonomy, plant physiology, and ecology; on the study of secondary plant metabolites, crop and post-harvest improvement, EO distillation and its optimization, and design of rural stills; on the study of volatile fractions from plants and flowers, obtainment of extracts with solvents and supercritical fluids (SFE-CO2), and catalytic transformation of EOs or their main components; and on the study of their diverse biological properties (antioxidant, antimicrobial, insecticidal, antiviral, and others).

2. Essential oil isolation

The primary metabolites (proteins, lipids, sugars, etc.) in plants are vital for the plant to grow, multiply, and live, while secondary metabolites are required by the plant to survive. For sure and with all the experimental details studied, the role played by secondary metabolites in plants is not completely known, because they fulfill several functions and operate through different mechanisms. Among many secondary metabolites isolated from plants, there are some very special, widely used in various branches of industry, medicine, and in many products of everyday life. This class of substances is called EOs, volatile oils, ethereal oils, or essences. Numerous substances are part of these oils; they are a complex mixture of volatile compounds with very diverse chemical nature. What most characterizes and highlights them is their smell, generally pleasant and intense, that evokes the fragrance of the plant or of the fruit or wood, from which these oils come. The essence can be remembered as the smell, for example, of a freshly cut grass or vanilla, sweet and cloying, among other aromatic tones that an EO has, formed by a complex range of volatile substances with different fragrant notes and different sensory thresholds for their perception.

Isolated from flowers (rose, orange blossom, lily, ylang-ylang), seeds (coriander, celery, carrot, anise, cardamom), leaves and stems (basil, thyme, mint, lavender, oregano), bark (cinnamon), wood (pine, sandalwood), roots (valerian, vetiver), and rhizomes (ginger, turmeric). EOs can be considered as the soul of the plants, their spirit, which characterizes, highlights, evokes, and makes them memorable in time; oils, generally, produce a pleasant sensation, especially when diluted. The EOs in the plants can be found in the different oil cells (ginger, turmeric, vanilla), in the secretory channels (pine, artemisia, anise, angelica), in the glands (citrus, eucalyptus), or in the trichomes (many plants of Labiatae, Asteraceae, Solanaceae, Geraniaceae families). The plant material (aromatic plant), when subjected to water vapor, releases a liquid odoriferous mixture (EO) of various volatile substances; this mixture can have from 50 to more than 300 chemical substances and is composed of terpene hydrocarbons, their oxygenated derivatives, alcohols, aldehydes, and ketones, as well as ethers, esters, phenolic compounds, phenylpropanoids, and other derivatives [3].

EOs can be obtained from plant material by three main methods (**Figure 1**). (1) Steam distillation. This process is carried out with a superheated dry steam, usually generated by a boiler or steam generator, which penetrates the plant material at higher than atmospheric pressure; the steam current breaks the cells or oil channels in the plant and drags the volatile mixture, which condenses after passing through a cooling system (heat exchanger). Generally, the oils are lighter than water and with very little soluble in it; therefore, they can be separated by decantation. The exception is the clove oil, which is heavier than water and is collected under it. The steam distillation method is used to extract oils from rhizomes, roots, seeds (vetiver, valerian, ginger, anise, cardamom, etc.), and dried or fermented leaves of some plants (*e.g.*, patchouli). (2) Distillation with water-steam. In this extraction system, wet steam is used, coming from the boiling water, which passes through the plant material suspended above and supported on a mesh. Most herbaceous plants are distilled by this method. (3) Hydro-distillation is a process in which the plant material is directly immersed in water, heated to a boil. This method is used for the distillation of more delicate plant material, for example, flowers (*e.g.*, ylang-ylang, roses). The citrus peel (orange, tangerine, lime) EOs are also obtained by cold-pressing or by scraping their surfaces. The mixtures obtained by the methods mentioned above are called "essential oils"; other products, isolated by maceration in different solvents or with supercritical fluid (CO_2), are generally called "extracts" and not "oils"; among them are concrete—obtained

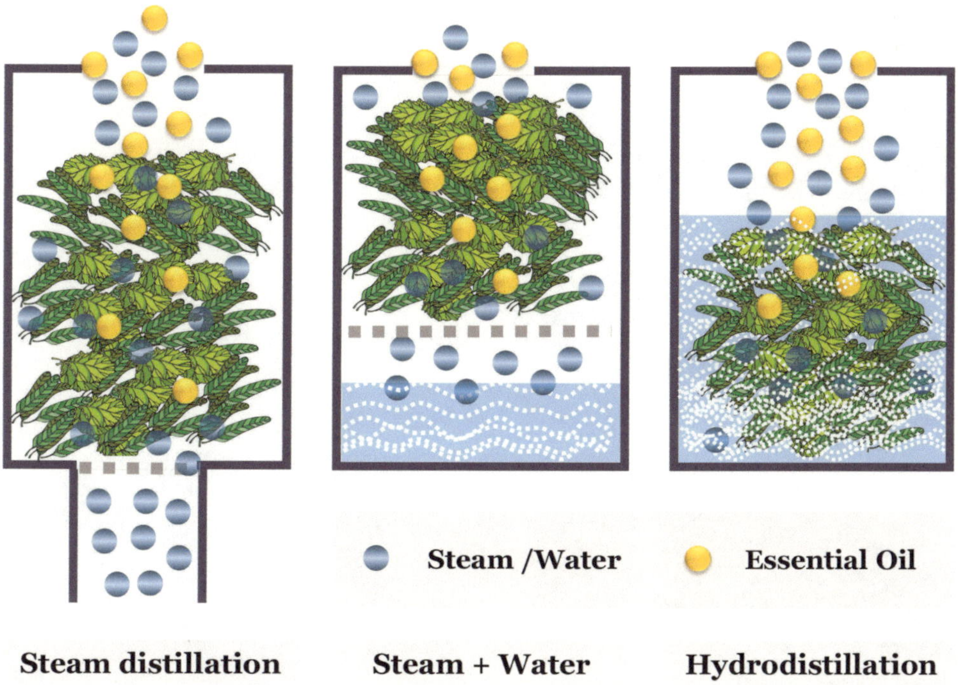

Figure 1.
Main methods of essential oil isolation.

by extraction with hydrocarbons from aromatic plants or, more frequently, flowers —and absolutes, which are separated from concrete or pomade (obtained by enfleurage) by alcohol.

The EO industrial production involves field distillation, in order to avoid the high transportation costs of large vegetal material loads from which only about 1% is going to be obtained as EO. Steam generation is one of the main components of the operation costs. Current trends point toward the use of lignocellulosic waste as biofuel for the furnace. Still capacity is determined by the crop size. The goal is to maintain the still operating for at least 300 days of the year and to schedule the harvests to avoid long storage (more than a week) of the cut vegetal material waiting for its distillation. This is mainly to prevent mold formation. Patchouli and vetiver are two exceptions to this rule, because a curing period of several days or months (vetiver) recommended to enhance oil yield and organoleptic quality.

The reality is that a large part of Colombian small growers have low purchasing power, low economic performance and productivity, and not very sophisticated technology level in rural operations and processes. Traditional agricultural production faces a complex problem that includes low prices, low profitability, and the increasingly acute lack of rural labor, because young people migrate to the cities. The EO industry is a very important rural development alternative in which the harvested vegetal material is no longer the final product, but the start of an added-value product chain. Several pilot projects, financed by the Ministry of Agriculture and Rural Development and Colciencias (Colombia's Science Funding Agency), have been carried out in the past 15 years by CENIVAM with the participation of small rural farmers associations. The common goal of these projects has been the development of the EO value chain. The economic, agronomic, and quality viability of EOs obtained in several productive units have been studied. Each unit has characteristics, as follows: 5–8 ha crop extension, 20–22 families of small growers

Figure 2.
Rural production and distillation of essential oils in Colombia (Barbosa, Santander).

involved in each project, 3 or 4 plant species cultivated per unit (palmarosa, citronella, *Lippia origanoides*, *L. alba*, rosemary, or thyme), plant nurseries, and the facilities for EO extraction in the field. Several rural stills (1 m^3 retort capacity) have been designed and built by CENIVAM. A mobile autonomous version that uses a radiator as condenser received a patent [4]. The farmers are trained on good agricultural practices, post-harvest treatment, and steam distillation. All activities are accompanied by permanent technical assistance (**Figure 2**). These small rural projects constitute an opportunity for a commercializing enterprise that consolidates the various producers around quality control guidelines and provides the technical support to connect them with buyers abroad. The university provides the technical support for chemical characterization with modern instrumentation, production of technical data sheets, and quality assurance.

3. Essential oil characterization

3.1 Gas chromatographic analysis

The analytical technique routinely used for the instrumental chemical analysis of EOs is gas chromatography (GC), because the constituents of oils are volatile (monoterpenoids, esters, etc.) or semi-volatile substances (sesquiterpenoids, phenolic derivatives, etc.), whose molecular masses and boiling points do not exceed 300 a.m.u. and 300 °C, respectively. A chromatographic system comprises four fundamental blocks: (1) sample introduction system (injector), (2) separation system (column), (3) detection system for analytes eluted from the column (detector), and (4) data analysis and operation control system.

The GC can have conventional, *e.g.*, flame ionization detector (FID), or thermal conductivity detector (TCD), and spectral detectors can have an external device attached, for example, a headspace sampler, a pyrolyzer, a purge and trap (P&T) system or a thermal desorption setup, among others. Each block of the chromatographic system has its own function and its "responsibility" for the quality of the analysis and the results obtained; for example, the function of the injection system is to transfer the sample to the column quantitatively, without discrimination by molecular weight or by the volatility of the components and without their chemical alteration (decomposition, isomerization, or polymerization) (**Figure 3**). The "responsibility" of the chromatographic column in the EO analysis is high: the clear, complete (ideally) separation of all the components of the mixture must be accomplished. The separation is based on achieving different distribution constants of the components between the two phases, stationary and mobile. This is obtained by establishing the optimal operational conditions (temperature, type of mobile phase, its velocity, stationary phase polarity, carrier gas pressure, temperature program, etc.) (**Figure 4**) and by correctly choosing the chromatographic column, *i.e.*, its dimensions (length, internal diameter), chemical composition of stationary phase, its polarity and thickness, and among other factors. For the EO analysis, long columns (50 and 60 m) are used, since the oils are complex multicomponent mixtures and, above all, they have structurally very similar compounds (isomers), which require that the column has a very high resolution, which, among other factors, is achieved by increasing its length. The EOs contain compounds of very different polarities, both nonpolar (terpene hydrocarbons) and polar (alcohols, aldehydes). This implies that for their analysis, columns with different stationary phase polarities will be required.

The detection system differentiates the analyte molecules from those of the mobile phase (carrier gas), to which the detector is transparent. The response of the

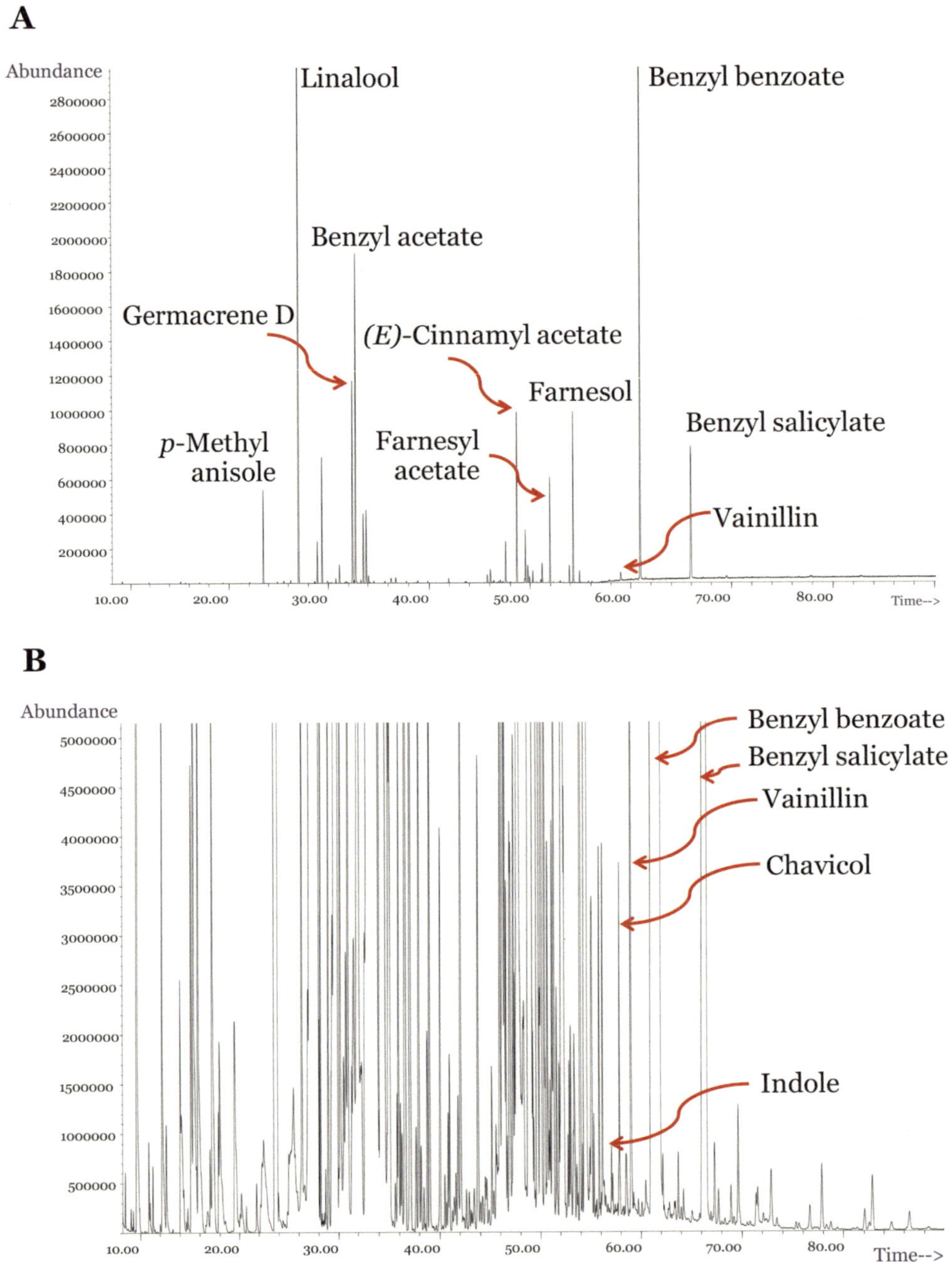

Figure 3.
Ylang-ylang essential oil obtained by hydro-distillation from flowers. GC-MS analysis. DB-WAX column (60 m). Injection modes: ***A.*** *Split 1, 100;* ***B.*** *Splitless. Injection volume—1 μL (250°C). Many "new" compounds appear in the chromatogram obtained in splitless mode.*

detector is based on the measurement of one of the physical properties of the system, *e.g.*, ion current, thermal conductivity, photon emission, etc. The analog signal becomes digital, graphic, *i.e.*, a chromatographic peak, which is characterized by its area (A), which is proportional to the analyte quantity or concentration (C). This permits to establish an interdependent relationship, A = f (C), and to carry out a quantitative analysis, to determine not only how many components there are in a mixture but in what proportion (quantity) they are present. Through a combination

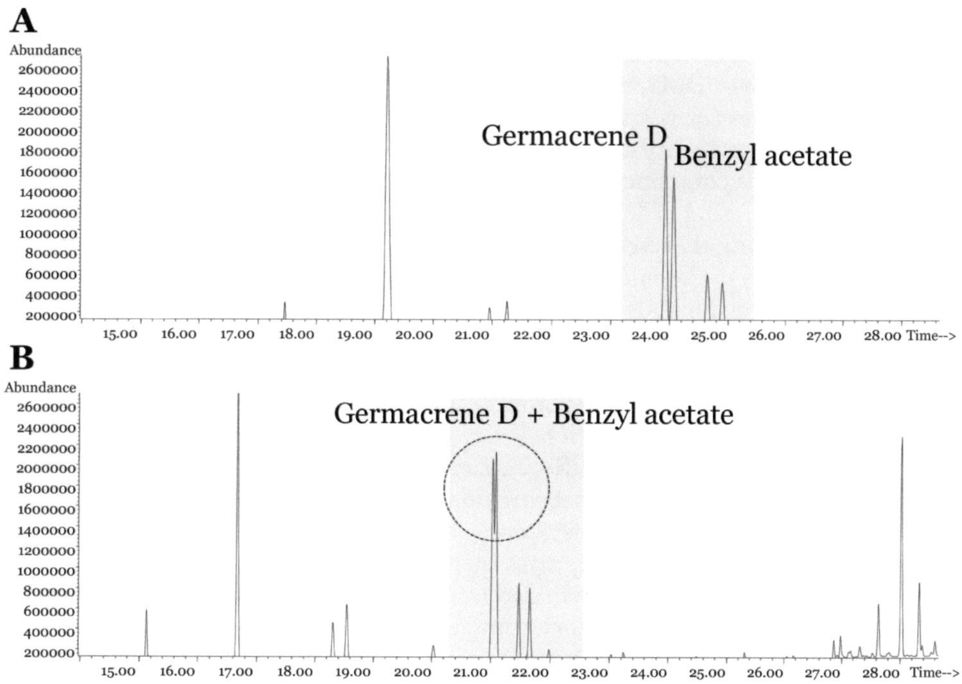

Figure 4.
Ylang-ylang essential oil obtained from flowers by hydro-distillation and analyzed by GC-MS on a polar column (DB-WAX, 60 m), using different temperature programs: ***A.*** *12 °C/min and* ***B.*** *8 °C/min (chromatogram fragment). With higher temperature rate, poorer separation of germacrene D and benzyl acetate is observed.*

of specialized software (data system), its accessories, interfaces, and analog-digital converters, the work of the chromatographic system and all its operational parts (hardware) is harmonized. For the EO analysis, which are very complex mixtures, two GC detectors are mainly used, namely, the flame ionization detector and the mass selective detector (MSD) or the mass spectrometry (MS) detector. The GC-FID is used to quantify the oil components.

3.2 Tentative and confirmatory identification of essential oil components

The preliminary or presumptive (tentative) identification of the EO components may be obtained once the retention indices are determined. The analysis in modern equipment uses a program for the column temperature; in these cases, linear retention indexes are calculated, which are part of many databases and bibliographic references [5, 6]. The confirmatory identification of a compound in a complex mixture analyzed by GC needs to obtain its "fingerprint," which is the mass spectrum (MS) represented by a unique combination of charged fragments (ions) generated during the breakup of the previously ionized molecule. The complementarity of the chromatographic analysis (screening) with confirmatory spectral data (mass spectra) is achieved using the combination of two techniques, GC and MS. The GC-MS coupling complements the quantitative analysis carried out by GC-FID and provides important additional information, *i.e.*, the mass spectra of all components, through which their identity can be established.

EOs contain both nonpolar (monoterpene and sesquiterpene hydrocarbons) and polar compounds (their oxygenated derivatives, aliphatic alcohols, ketones, oxides, phenolic compounds and their derivatives, phenylpropanoids, and rarely acids, among others). Their analysis is performed by GC-FID (quantitative analysis) and by GC-MS (qualitative analysis), in two columns, with polar and nonpolar stationary phases. In columns with the nonpolar stationary phase, poly(dimethylsiloxane), PDMS, or 5% phenyl-PDMS, the elution of components happens depending on their boiling temperatures (or volatilities), that is, the retention times, t_R, increase with the decrease of the volatility and with the increase of the molecular masses and boiling points of the components (**Figure 5**). The compounds reach the end of the column in the increasing order of their boiling points. In the polar column, poly (ethylene glycol), the elution order of the components is more difficult to predict, because it is related to the intermolecular forces between the analyte and the stationary phase and depends both on the dipole moment of the molecule (the polarity) and on the possibility of hydrogen bond formation between the substance and the stationary phase.

The elution order of some compounds in columns with different stationary phases may be reversed. This often helps, together with the mass spectra and the fragmentation pattern study, to differentiate, for example, terpene alcohols from their acetates, since the latter sometimes do not exhibit molecular ions, $M^{+\bullet}$ in their mass spectra. When the chromatographic parameters (t_R, t_{RR}, or retention indices) and spectroscopic parameters, *i.e.*, mass spectra (characteristic fragmentation pattern; see **Figure 6**) of the analyte and reference substance (certified standard) coincide, a complete or confirmatory compound identification is achieved. However, when retention indexes and mass spectra are used, extracted from the specialized literature [6, 7] or from the databases (*e.g.*, spectral libraries, NIST, WILEY, Adams [7], others), and compared with the spectroscopic and chromatographic parameters of the EO component, their coincidence leads only to a recognition of the chemical structure, but not to its unambiguous, absolute identification, which requires the use of a standard compound, a pure substance with certified chemical structure. Frequently, it is necessary to isolate the compound from the mixture and purify it for further characterization through the UV, IR, MS, X-ray diffraction, NMR, elemental analysis, or high-resolution mass spectrometry (HRMS). Each one of the mentioned spectroscopic techniques contributes with some structural information, but the combined results allow to assemble the puzzle and elucidate the chemical structure unequivocally.

The biggest challenge in EO analysis is the complete separation of its components (**Figure 7**) because their frequent coelution occurs due to their very close or equal distribution constants. Some conventional strategies, *e.g.*, change of the column (polarity), temperature program, use of selective detectors, etc., can often fail or be insufficient to determine all the compounds present in the oil. Multidimensional chromatography makes it possible to separate the peaks of partially or totally co-eluted substances. For this, it uses a second column, usually orthogonal, through the "heart-cutting" operation by means of pneumatic switching valves—today with the micro-fluidic technology, between the two columns and diverting part of the eluent from the first to the second column. This method has played an important role in the development of separation techniques for complex mixtures, including EOs [8, 9]. Multidimensional chromatography requires at least two detectors and can have configurations of up to three columns in the same oven or in separate chromatographic ovens. Along with this, today one of the most modern, complete solutions for the separation of multicomponent mixtures—although not very affordable for most laboratories because of its high price—is comprehensive or total gas chromatography (Comprehensive GC × GC), whose applications and developments have grown day after day for more than two decades [10, 11].

In comprehensive gas chromatography (GC x GC), two columns are used, linked together by means of a modulator. In contrast to conventional multidimensional gas

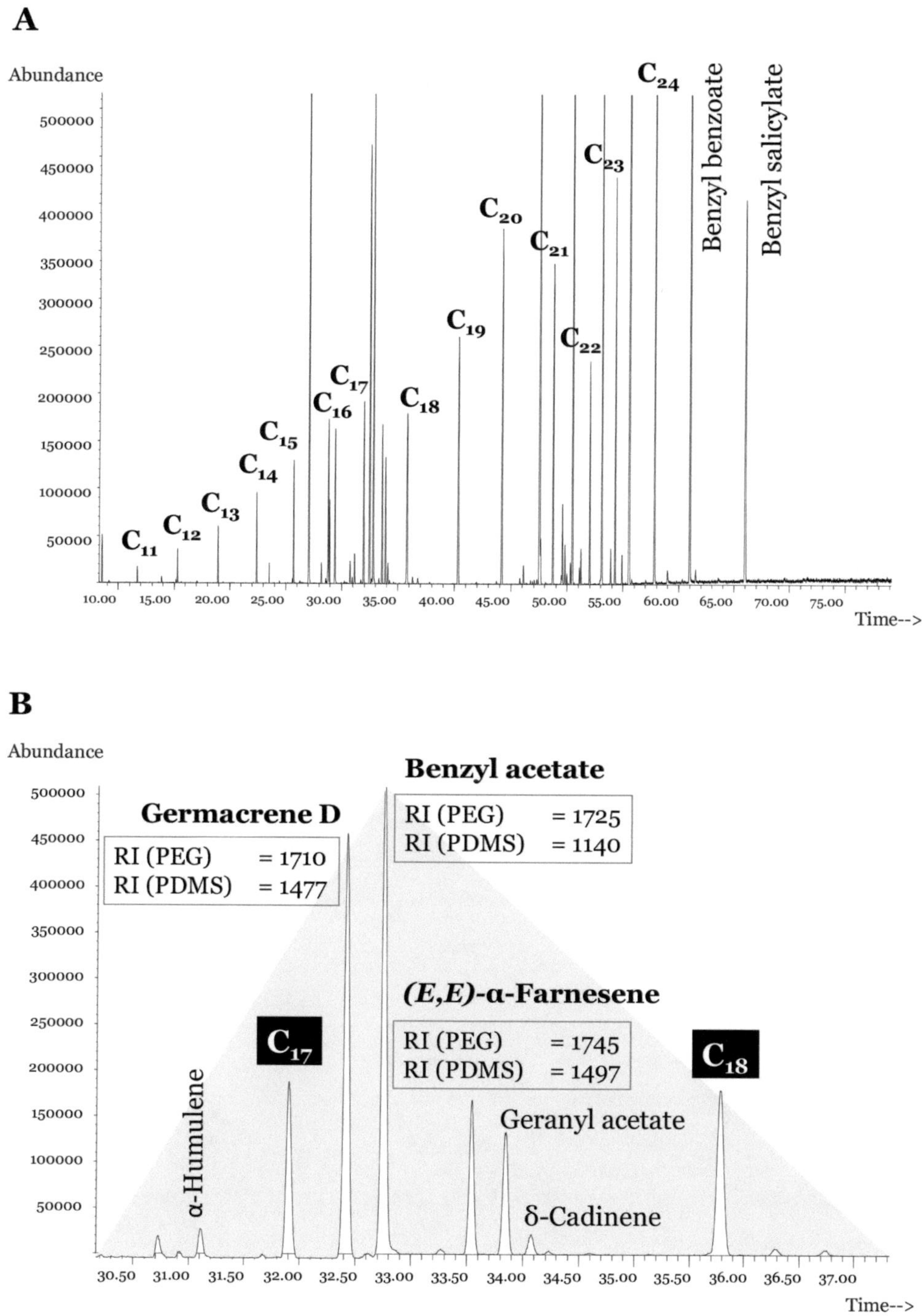

Figure 5.
GC-MS chromatogram of ylang-ylang essential oil isolated from flowers by hydro-distillation and analyzed by GC-MS on a polar column (DB-WAX, 60 m). ***A.*** *Co-injection of the essential oil and* n*-paraffin mixture to calculate retention indices (RI).* ***B.*** *RIs of germacrene D, benzyl acetate, and (E,E)-α-farnesene measured in nonpolar [poly(dimethylsiloxane)] and polar [poly(ethylenglycol), PEG] columns.*

chromatography, the GC x GC requires a single detector with high processing frequency; both columns can be in the same oven or in two separate ovens. There are different types of modulators, *e.g.*, rotary thermal modulator ("sweeper"), cryogenic "jet" system, modulators of valves, or longitudinal cryogenic modulator, among others, which also vary in the cryogenic agent employed; more modern modulators are not cryogenic in nature. The eluent of the first column, by means of the modulator, is "split" into very small "slices," which, one after the other, enter the second column from the first column. The first column (1D) is a conventional column, with length of 25 or 30 m, and the second column (2D) is of rapid chromatography, that is, short and with a very thin internal diameter (0.1 mm or less). The stationary phases in both columns are "orthogonal," *i.e.*, if the first is nonpolar,

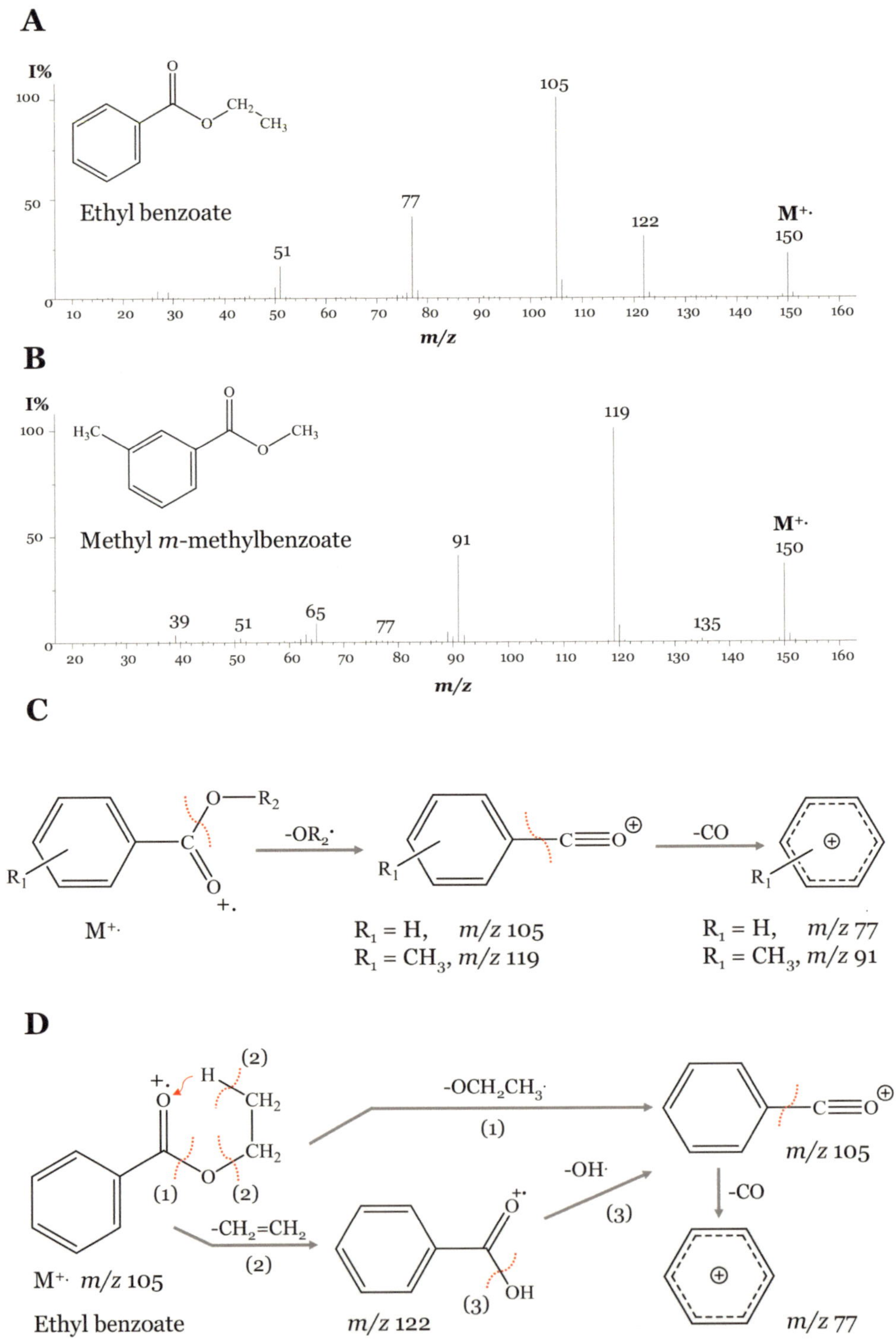

E

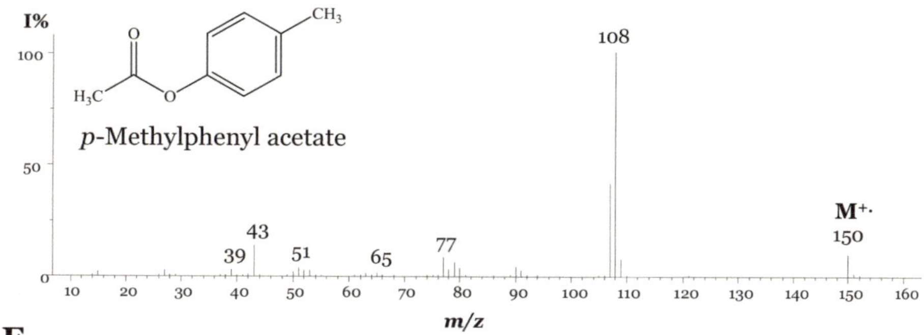

F

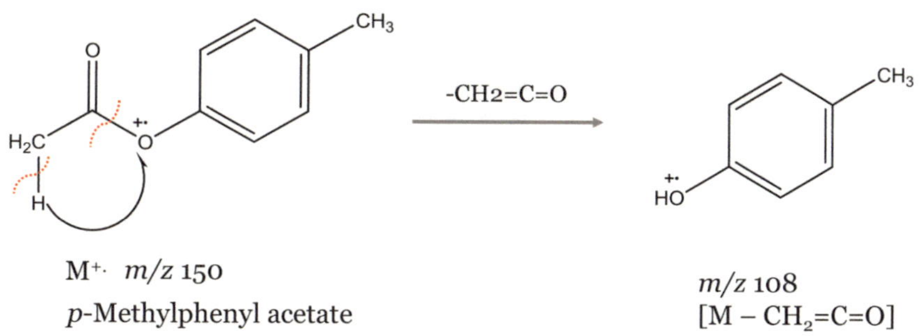

G

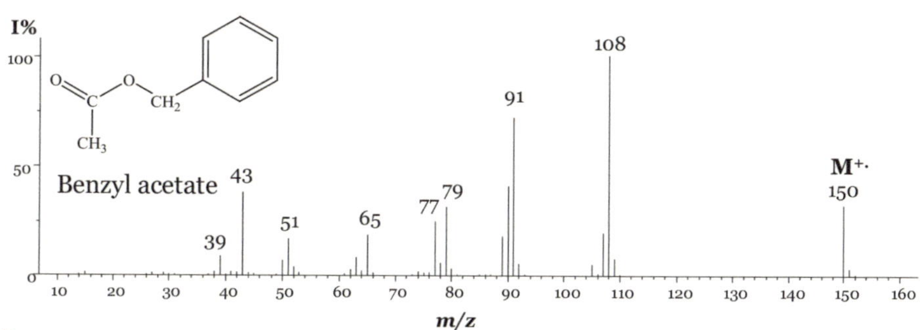

H

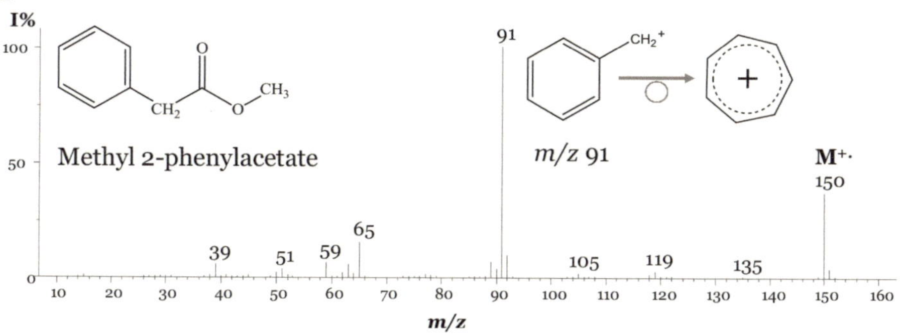

Figure 6.

Fragmentation pattern in mass spectra (electron impact, 70 eV) of some essential oil components. ***A.*** *Ethyl benzoate mass spectrum.* ***B.*** *Methyl* m*-methyl benzoate mass spectrum.* ***C.*** *α-Rupture and typical benzoyl ion (m/z 105, 119) formation.* ***D.*** *McLafferty transposition of ethyl benzoate molecular ion* $M^{+\bullet}$ *and formation of* $[M–CH_2=CH_2]^{+\bullet}$ *(m/z 122) fragment.* ***E.*** p*-Methylphenyl acetate mass spectrum.* ***F.*** *Elimination of* $CH_2=C=O$ *neutral fragment and formation of* $[M—42]^{+\bullet}$*, diagnostic ion for acetates.* ***G.*** *Benzyl acetate mass spectrum.* ***H.*** *Methyl 2-phenylacetate mass spectrum and the formation of tropylium ion (m/z 91) generated through benzylic excision.*

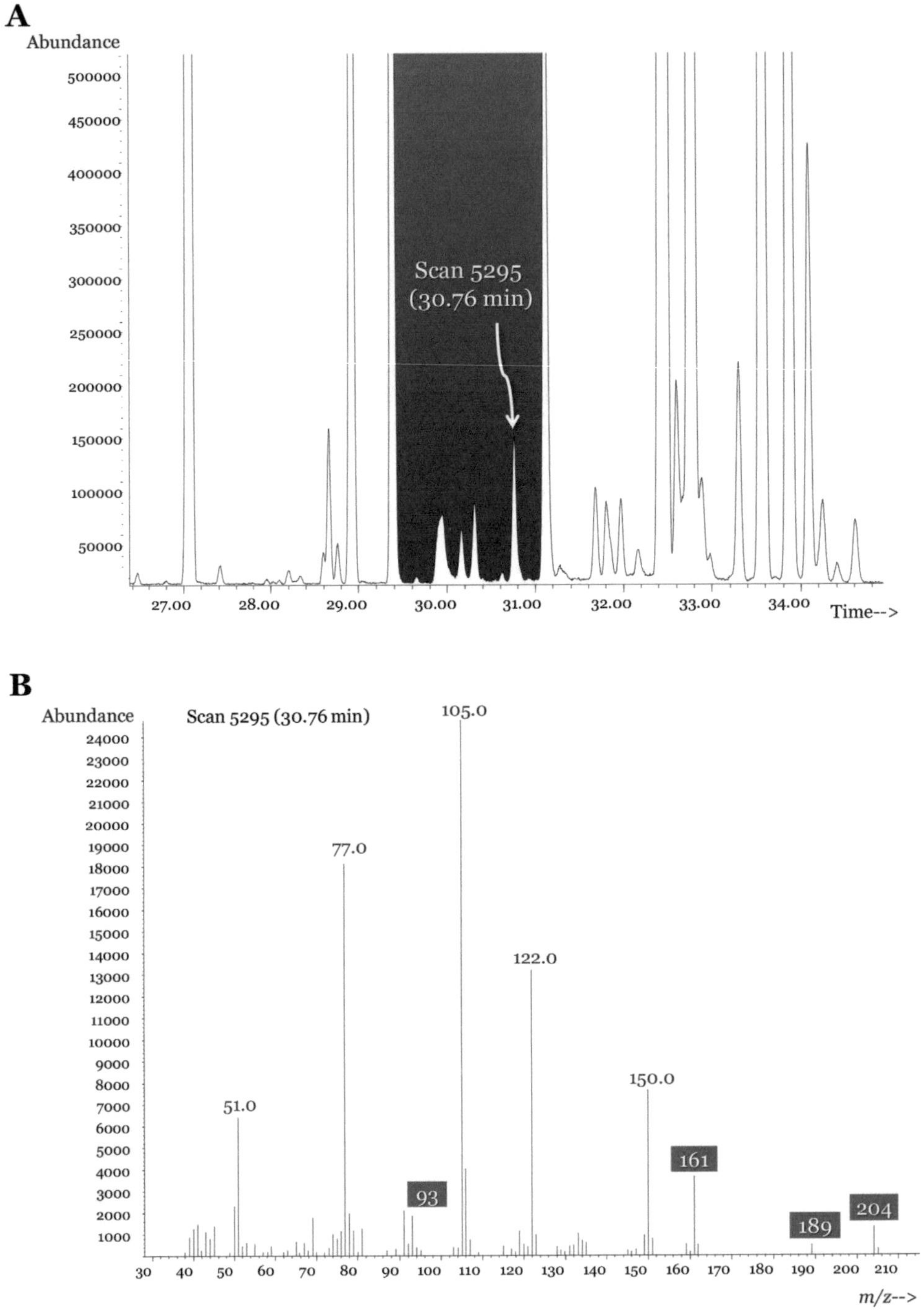

Figure 7.
Ylang-ylang essential oil chromatogram (GC-MS) fragment. ***A.*** *Chromatographic peak at* t_R *= 30.76 min is highlighted.* ***B.*** *Mass spectrum corresponded to this peak, identified by the software as ethyl benzoate (****Figure 6A****); nevertheless, the peak is "contaminated" with other compounds; the presence of typical sesquiterpene hydrocarbon ions (m/z 204, 189, 161,) is observed.*

the second column is polar, and *vice versa*. The modulation time, required for the transfer of a very small portion of eluent from the first column to the second, must be very short and similar but never longer than the elution time of the "slowest" component in the second column. The second column, therefore, must be short and very thin and separate the components in just a few seconds. Since the second column is connected to the detector (MSD, FID, or ECD), it must have a very-high-reading and signal-processing frequency. In most cases, a time-of-flight (TOF) mass detector is used, which is the best option—though expensive—to make a quantitative analysis and identification of compounds in such complex mixtures, as are EOs (**Figure 8**).

Further technical details for EO chemical characterization and that of their components can be found elsewhere [12, 13]. In summary, EO characterization necessary for its quality control and the determination of authenticity, as part of a technical data sheet necessary for its commercialization, can be divided into four main stages or areas: (1) organoleptic properties, (2) physicochemical determinations, (3) qualitative and quantitative analysis of the components present in the oil (chemical composition), and, finally, (4) some other determinations, *e.g.*, pesticide residues, traces of heavy metals, etc.

3.3 Chemical compositions of essential oils obtained from tropical plants grown in Colombia

CENIVAM has studied Colombian plants widely used in popular medicine or in culinary, for example, anise [14], oregano [15], rue [16, 17], and other species introduced from Asia, such as lemongrass, citronella, ginger, citrics [18–20], vetiver, and ylang-ylang [21–23], as well as several native species, among others, *Copaifera officinalis* [24], *Spilanthes americana* [25], *Lepechinia schiedeana* [26], *Lippia alba* [27], *Xylopia americana* [28], *Hyptis umbrosa* [29], *Callistemon speciosus* (sims) DC. [30], *Swinglea glutinosa* [31], *Satureja viminea* [32], and *Lippia origanoides* [33], with emphasis on the comparative study of extraction methods [34–40]. **Table 1** summarizes the composition of several *Lippia* EOs, according to compound families. The knowledge of the chemical composition has been the basis for the interpretation of the results of bioactivity assays such as genotoxicity

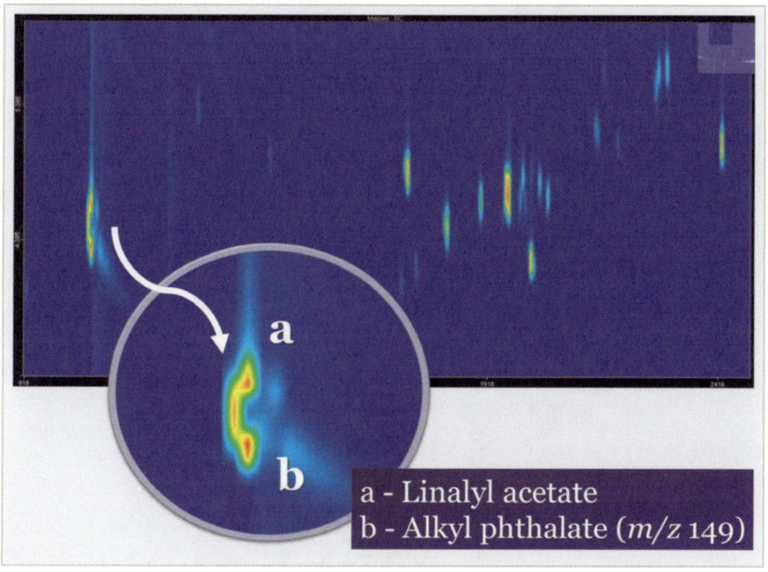

Figure 8.
Fragment of GCxGC-HRMS-TOF chromatogram of ylang-ylang essential oil contaminated with plasticizer (phthalate) traces. m/z 149 is a base peak in alkyl phthalates' mass spectra.

Compound family	Relative amount, %											
	L. alba, chemotypes				*L. origanoides*, chemotypes			*L. citriodora*	*L. micromera*	*L. americana*	*L. graveolens*	*L. dulcis*
	Carvone	Citral	Citral + Carvone	Myrcenone	Thymol	Carvacrol	Phellandrene					
Monoterpene hydrocarbons	31.5	4.1	24.6	12.6	14.7	31.0	45.7	15.4	44.9	19.1	12.2	0.3
Oxygenated monoterpenes	61.4	62.4	52.7	71.6	2.1	2.0	7.4	49.7	2.8	1.6	2.0	5.6
Oxygenated compounds (phenylpropanoids)	—	6.9	2.7	0.0	63.3	57.7	2.0	4.4	47.7	0.2	78.1	7..9
Sesquiterpene hydrocarbons	4.9	11.9	16.2	6.6	5.3	3.9	29.1	11.0	3.2	52.3	5.3	54.0
Oxygenated sesquiterpenes	0.3	5.1	1.1	—	0.4	—	5.2	12.8	0.5	10.7	0.2	19.1

Table 1.
Relative amounts of compound families in essential oils of various Lippia *species grown in Colombia.*

[41–43], antiviral [44] and antifungal [45–52] activities, insect repellence [53–59], antioxidant capacity [60–65], cytotoxic [30, 66–68], antituberculosis [69], and antiprotozoarial activities [70, 71]. A few examples are highlighted in the following section.

4. Some biological activities of essential oils obtained in Colombia

EOs have been used in phytotherapy and folk medicine for their good odor and antibacterial, antifungal or insecticidal activities. Phenols, alcohols and aldehydes are EO components capable of crossing the cell wall, and in doing so, they alter its permeability and may cause leakage of macromolecules, loss of ions, structure disruption, and, eventually, cell death. This cytotoxicity enables EO applications against human pathogens or parasites and for the preservation of vegetal and marine products. Due to their large number of constituents, EOs affect several targets simultaneously, and this may be the reason for the lack of microorganism resistance development or adaptation. Besides cytotoxicity, the antioxidant properties of EOs are generally invoked as an indication of their potential benefits for human health. This is related to the notion that many diseases are due to high oxidative stress generated by diet, environmental contaminants, or work habits. However, the prooxidant properties of some EO components can play a protective role by promoting the removal of damaged cells. The mitochondria produce reactive oxygen species which can oxidize phenolic compounds (EO components) and give rise to reactive phenoxyl radicals which accelerate the general cell damage [72].

Genus *Lippia* (Verbenaceae family) has been the focus of attention of Colombian researchers (**Figure 9**). *L. alba* and *L. origanoides* EOs have appeared as the prominent representatives of this genus, after the evaluation of various *Lippia* genus EOs for antioxidant [33], antiviral [44], antimicrobial [69], antiprotozoal [71], and antigenotoxic [41, 42] activities.

Lippia origanoides Kunth (mountain oregano) is a good example of the aromatic plant biodiversity found in CENIVAM studies. It is an aromatic shrub found in the wild in northern South America and Central America. At least four different chemotypes have been distinguished according to differences in EO composition [34, 73]. Further research showed notorious differences in the compositions of the various chemotype extracts. Several uses of *L. origanoides* infusions in popular medicine have been related to antimicrobial and analgesic activities due to phenylpropanoids and flavonoids found among its secondary metabolites. The detection of thymol and carvacrol as main EO constituents and pinocembrin, naringenin, quercetin, and luteolin in mg/g amounts in extracts of various *L. origanoides* chemotypes supports the recognition of this species as a promising source of bioactive substances. *L. origanoides* is the second most studied species of the *Lippia* genus, preceded by *L. alba*. The useful bioactive properties found for *L. origanoides* oil have aroused interest in commercial applications such as food additive, preservative, or pest control agent, among others. *L. origanoides* oil is an important ingredient of various current chicken food commercial products. The most widely known sources of thymol and carvacrol are thyme (*Thymus vulgaris*) and oregano (*Origanum vulgare*), both of Eurasian origin. Thymol and carvacrol contents in thyme EO are in the range 37–55% and 0.5–5.5%, respectively (ISO 19817:2017). Oregano EO contains around 22% thymol and 18% carvacrol. Thymol and carvacrol are major components in three L. *origanoides* chemotypes. Several projects conducted in CENIVAM have related the variations in EO composition with the steam distillation conditions, with the phenological stage, the agricultural conditions, and the post-harvest treatment.

Figure 9.
Some tropical aromatic plants of Lippia *genus (Verbenaceae family): L. origanoides, L. americana, L. micromera, and L. dulcis.*

More than 30 botanical outings were organized by CENIVAM to various Colombian regions in the Andes, the Eastern Plains, the Caribbean, and the Pacific coasts. Over 1100 accessions of medicinal and aromatic plants were collected and taxonomically identified, under the due permit of access to the genetic resource. Hydro-distillation and steam distillation of these samples produced close to 1000 different EOs that were subjected to chemical characterization using GC techniques provided with FID and MS systems. Tests for antioxidant, antimicrobial, antiviral, antiparasitic, immunomodulatory, photoprotective, and antigenotoxic activities of the EOs revealed that more than 45% of these oils were highly active in one or two assays. The following sections highlight just a few of the interesting findings obtained in the search for natural ingredients with biological properties that may enable the development of products for the pharmaceutical, cosmetics, hygiene, or food industries.

4.1 Cytotoxicity

Following the recognition of cytotoxicity and antioxidant capacity as main determinants of EO potential pharmaceutical applications, Olivero and co-workers [64] used the brine shrimp assay and the measurement of thiobarbituric acid induced in rat liver microsomes by a Fenton reagent to evaluate 13 EOs from Colombian plants for cytotoxicity and antioxidant capacity, respectively. Mean effective concentrations (EC_{50}) below 100 μg/mL were registered for the *Ocotea* sp., *Tagetes lucida*, and *L. alba* (citral chemotype) EOs. Moderate values (130–174 μg/mL) were obtained for the *Elettaria cardamomum* and *L. alba* (carvone chemotype, Tolima) EOs. No antioxidant activity (EC_{50} > 1 mg/mL) was found for the *Minthostachys mollis*, *L. alba* (carvone chemotype, Cundinamarca), and *Piper sancti-*

felisis EOs. LC_{50} cytotoxicity values between 4.36–64.3 and 1.2–20.8 μg/mL, for 24 and 48 h exposure, respectively, were obtained. Most tested EOs can be considered cytotoxic (LC_{50} < 10 μg/mL), but bioactivity was highly dependent on EO chemotype, extraction mode, and plant growing location.

L. alba EOs of different origins in Colombia showed cytotoxicity in the *Artemia franciscana* assay at concentrations in the range from 7 to 21 μg/mL. The differences were attributed to compositional variations caused by the geographical habitat and environmental factors (temperature, light intensity) [74].

A study of cytotoxic activity of EOs from the Verbenaceae and Asteraceae families included 36 species of various origins. These oils were tested on Jurkat, HeLa, HepG2, and Vero cell lines [75]. None of the tested EOs was cytotoxic, except that from *Ambrosia arborescens*, which showed IC_{50} of 16 ± 3.4 μg/mL and was not active against the tested tumor cell lines. All the Verbenaceae family EOs examined produced dose-dependent inhibition on the growth of HeLa cells with determination coefficient R^2 > 0.7. Four *L. alba* citral chemotype EOs and one oil of the *L. alba*, carvone chemotype, were active against HeLa cells. This activity was attributed to the presence of a target on HeLa that is not present on HepG2 and Jurkat cells.

4.2 Antioxidant activity

A study of 12 EOs of 7 *Lippia* species growing in Colombia employed GC-FID and GC-MS methods for their chemical characterization and the ORAC and ABTS assays for their antioxidant activity evaluation [76]. The ORAC and ABTS methods explore radical scavenging mechanisms in which the fundamental step is either the proton transfer (ORAC) or the electron transfer (ABTS) [77]. The EOs with high phenylpropanoid content showed higher antioxidant capacity in both assays. The ORAC antioxidant activity of these oils was five or more times superior to those of butyl hydroxytoluene (BHT) and α-tocopherol, which are antioxidants used commonly in commercial products. This superiority was maintained, although at a smaller proportion, in the ABTS test. The ORAC and ABTS values measured individually for carvacrol and thymol were close to the values obtained for the EOs that contain them as main components. For the remaining oils, not rich in phenylpropanoids, there was no clear relationship between the oil's antioxidant activity and that of its main constituents (**Table 2**). The *L. americana* EO and the phellandrene-rich *L. origanoides* chemotype EO have mono- and sesquiterpene hydrocarbons as main components. They showed poor antioxidant activity under the ABTS assay conditions but did have higher antioxidant activity than BHT in the ORAC assay. This is consistent with the evaluation of individual unsaturated nonaromatic hydrocarbon terpenes (*trans*-ß-caryophyllene, α-phellandrene, γ-terpinene), which were capable of scavenging radicals by proton transfer (ORAC conditions) but were completely inactive under the ABTS assay conditions (electron transfer). All the 12 EOs examined, those of *L. alba* (carvone), *L. alba* (citral), *L. alba* (carvone+citral), *L. alba* (myrcenone), *L. origanoides* (carvacrol), *L. origanoides* (thymol), *L. origanoides* (phellandrene), *L. citriodora*, *L. micromera*, *L. americana*, *L. graveolens*, and *L. dulcis* exhibited higher antioxidant capacity than BHT and α-tocopherol in the ORAC assay, and this makes them very good candidates to become ingredients of final products in substitution of synthetic antioxidants.

4.3 Antiviral activity

EOs from plants of the Labiatae and Verbenaceae families are considered very useful in folk medicine, as antibacterials, antivirals, antifungals, antioxidants, and

Essential oil (chemotype) or standard compound	Essay (± s, n = 3)			
	Main compounds	Yield, % w/w	ORAC (μmol Trolox®/g compound)	ABTS (μmol Trolox®/g compound)
L. graveolens	6-Methyl-5-hepten-2-one (4.9%), α-terpinene (3.0%), thymol (64 6%), carvacrol (12 2%), *trans*-β-caryophyllene (2.9%)	1.2	3990 ± 58	5410 ± 48
L. citriodora	Limonene (10.7%), 1,8-cineole (5.0%), neral (15.6%), geranial (18.9%), spathulenol (4.7%)	0.1	3630 ± 40	41 ± 2
L. origanoides (carvacrol)	*p*-Cymene (12%), carvacrol (46.2%), γ-terpinene (9.6%)	4.4	3400 ± 120	5200 ± 109
L. origanoides (thymol)	γ-Terpinene (5.0%), thymol (54.5%), thymyl acetate (4.8%)	3.1	2840 ± 72	5090 ± 42
L. origanoides (phellandrene)	*p*-Cymene (11.2%), limonene (7.2%), *trans*-β-caryophyllene (11.3%), α-phellandrene (9.9%)	1.5	1820 ± 82	310 ± 2
L. micromera	*p*-Cymene (13.1%), thymol (29.1%), thymyl methyl ether (14.9%)	1.0	2050 ± 78	2750 ± 80
L. alba (citral)	Geranial (27%), neral (21%), geraniol (6.0%)	0.9	2000 ± 77	24.8 ± 0.4
L. alba (carvona)	Limonene (30.2%), carvone (50.3%)	0.8	1340 ± 54	126.4 ± 0.7
L. americana	Sabinene (7.4%), *trans*-β-caryophyllene (12.2%), germacrene D (16.3%)	0.5	1200 ± 27	239 ± 4
Carvacrol			3410 ± 50	4609 ± 6
Thymol			3000 ± 103	5700 ± 125
Trans-β-caryophyllene			2800 ± 109	N.D.
γ-Terpinene			1766 ± 8	N.D.
α-Phellandrene			1040 ± 18	136 ± 3
α-Tocopherol			550 ± 13	2429 ± 7
BHT			457 ± 9	4760 ± 23
1,8-Cineole			299 ± 5	N.D.
p-Cymene			219 ± 2	N.D.

Table 2.
Main constituents and antioxidant capacity of Lippia essential oils.

insecticides. The antiviral activity of 40 EOs of the Labiatae and Verbenaceae families and some monoterpenes were evaluated on human herpes virus types 1 and 2 using the end point titration technique [78]. Samples that showed reduction factor of viral titer in comparison to control without treatment (*Rf*) at least against one viral type, at concentrations lower than or equal to 100 μg/mL, were considered active. *Hyptis mutabilis* oil showed a high activity against both viruses (HHV-1 and HHV-2), with *Rf* values of 10^3 and 10^2, respectively, at a concentration of 50 μg/mL. *Lepechinia vulcanicola* and *Mintostachys mollis* EOs showed the same reduction factor of viral titer against both viral types at a concentration of 100 μg/mL. *Lepechinia salviifolia* was moderately active against both viruses at the same concentration

(100 μg/mL). *Ocimum campechianum* EO showed relevant (1×10^3) and mild (1×10^1) activity against HHV-1 and HHV-2 at concentrations of 100 and 50 μg/mL, respectively. *Lepechinia salviifolia* and *Rosmarinus officinalis* oils showed moderate anti-herpetic activity against HHV-1 and HHV-2, respectively.

The *in vitro* inhibitory effect of *L. alba*, *L. origanoides*, *Origanum vulgare*, and *Artemisia vulgaris* EOs on yellow fever virus was investigated by exposing African green monkey kidney (Vero) cells to EO prior to virus exposure [79]. None of the EOs studied was cytotoxic on these cells. The minimum concentration of the EO that inhibited virus titer by more than 50% (MIC) was determined by virus yield reduction assay. Preincubation of virus with selected EO for 24 h at 4°C before adsorption on Vero cell inhibited the subsequent extracellular virus titer. Vero cells were exposed to EO 24 h at 37 °C before the adsorption of untreated virus. The presence of EO in the culture medium enhanced the antiviral effect: *L. origanoides* oil at 11.1 μg/mL produced a 100% reduction of virus yield, and the same result was observed with *L. alba*, *O. vulgare*, and *A. vulgaris* oils at 100 μg/mL.

4.4 Antigenotoxicity

Since plants are exposed daily to the sun radiation, they have evolved mechanisms that protect them from the effects of overexposure, such as damages to the DNA. When DNA suffers a damage, the cell responds with a set of actions that was discovered in 1975 by Miroslav Radman [80], who assigned the name of SOS response. The Pasteur Institute developed a colorimetric assay to detect carcinogens, based on this response in which the exposition to UV causes the DNA damage whose extent is associated with the intensity of light absorbance by a chromophore [81]. This SOS chromotest is highly sensitive to UV. A modified version has been used by Fuentes and collaborators [82] to identify plants of Colombian flora that may be a source of genoprotective compounds. Their application of the SOS chromotest in a survey of 50 extracts obtained with supercritical CO_2 from aromatic plants grown in Colombia permitted to identify those that significantly reduced UV-induced genotoxicity depending on their concentration, as follows: *Baccharis nitida*, *Solanum crotonifolium*, *Hyptis suaveolens*, *Persea caerulea*, and *L. origanoides*. Volatile secondary metabolites have been the subject of antigenotoxicity tests. *L. alba*, *L. micromera* and *L. origanoides* EOs were found antigenotoxic, and the evaluation of their main constituents showed that carvacrol, thymol, citral, *p*-cymene, and geraniol inhibited the UV-induced genotoxicity in the SOS chromotest [83].

4.5 Repellence

Synthetic insecticides are the most frequent pest control method in crop production and storage. However, their application has negative effects on environmental resources, elimination of beneficial insects, and toxicity for susceptible species and humans, who represent the last link in the food chain. EOs have attracted attention in recent years as potential pest control agents due to their insecticidal, repellent, and/or antifeedant properties. Stored products of insect pest control are important in managing post-harvest grains, food products and processed goods. *Tribolium castaneum* (Herbst) is one of the most common insect pests worldwide of flourmills, grocery shops and warehouses. Jaramillo *et al.* [84] examined the repellent effect of Colombian *Croton malambo* (Karst) EO against *T. castaneum* using the area preference method. A filter paper was divided in halves. On one half, equal volumes of different concentrations of EO dissolved in acetone (0.00002, 0.0002, 0.002, 0.02, and 0.2 $\mu L/cm^2$) and the other with acetone only as control. These halves were joined, and a fixed number of insects were released on the center. Observations on the number of insects present on both the treated and untreated halves were recorded after 2 and after 4 h. The highest repellent activity was observed at an EO concentration of 0.2 μL/cm^2. Repellence values of 86% ± 5 (2 h) and 92% ± 3 (4 h) were observed, which were higher than those obtained for a commercial repellent at the same concentration and exposure times (78% ± 5 and 76% ± 9, respectively).

Weevils that consume flour (*Tribolium castaneum*), peanuts and wheat bread (*Ulomoides dermestoides*) merit attention alongside other insects of major concern in crop production and storage of cereals and other products. Alcala and co-workers [85] used the area preference method to show that EOs of *Elettaria cardamomum*, *Salvia officinalis* and *L. origanoides* (carvacrol chemotype) had repellent action against both pest insects, while the repellency in the controls was null. This repellency increased when the EO concentrations were higher. None of the EOs presented attractant action for either of the exposure times. A 100% repellency was obtained at the highest concentration tested (1.6% v/v), except for *S. officinalis* against *U. dermestoides* at 2 h of treatment that had a 97% of repellency. Mean repellent concentration (RC50) values showed that *E. cardamomum*, *S. officinalis*, and *L. origanoides* had better repellent properties against *U. dermestoides* than a commercial preparation that contained 15% of ethyl butylacetylaminopropionate. The carvacrol-rich chemotype of *L. origanoides* was the most potent, with RC50 values of 0.220 and 0.207% (v/v), for *T. castaneum* and *U. dermestoides*, respectively.

Several tropical diseases for which there is no vaccine yet (yellow fever, Zika fever, chikungunya, dengue) are transmitted by *Aedes aegypti*. The strategies to prevent these illnesses involve the use of insecticide or repellent agents. Since several pesticides have deleterious environmental effects and affect humans, there is a strong interest in finding EOs and plant extracts that can be effective in controlling *Aedes aegypti*. The guideline that good larvicide candidates are substances with LC_{50} < 100 mg/L [86] shows the importance of the finding that the EOs from *L. origanoides* (LC_{50} = 54 mg/L) and *Swinglea glutinosa* (LC_{50} = 66 mg/L) had an improved performance when used as a mixture (LC_{50} = 38 mg/L). Other EO binary mixtures showed similarly interesting activity (*Turnera diffusa* and *S. glutinosa*, LC_{50} = 64 mg/L; *L. alba* and *S. glutinosa*, LC_{50} = 49 mg/L) [87].

5. Conclusions

Colombia's geographic and botanical conditions favor the development of its EOs agroindustry to convert this country into an important provider to the ever-growing EO world market. The initial offer will consist of EOs from aromatic plants of European and Asian origins, which are well-known and commonly traded in the international market. However, the results from the very small survey of Colombia's biodiversity indicate that there are many promising alternatives for future market expansion. The evaluation of biological activities of EOs obtained from plants growing in Colombia points toward many opportunities to develop a wide range of products that employ them as active ingredients.

Acknowledgements

Financial support from *Patrimonio Autónomo*, *Fondo Nacional de Financiamiento para la Ciencia*, *Francisco José de Caldas*, grants RC-432-2004, RC-0572-2012, and

FP44842-212-2018, is gratefully acknowledged. The authors thank *Ministerio de Ambiente y Desarrollo Sostenible*, through its *Dirección de Bosques, Biodiversidad y Servicios Ecosistémicos* for their permission of access to genetic resources, and derived products for the program ran by the Unión Temporal Bio-Red-CO-CENIVAM (Resolution 0812, June 4, 2014).

Conflict of interest

The authors declare that they do not have conflict of interest.

Author details

Elena Stashenko* and Jairo René Martínez
Research Center for Biomolecules, CENIVAM, Universidad Industrial de Santander, Bucaramanga, Colombia

*Address all correspondence to: elena@tucan.uis.edu.co

References

[1] Departamento Administrativo Nacional de Estadística. Censo Nacional Agropecuario 2014. Available from: https://www.dane.gov.co/index.php/estadisticas-por-tema/agropecuario/censo-nacional-agropecuario-2014 [Accessed 01-05-2019]

[2] Trade map, International Trade Centre. Available at: www.trademap.org. [Accessed: 01-05-2019]

[3] Jirovets L, Buchbauer G, editors. Processing, Analysis and Application of Essential Oils. Dehradun, India: Har Krishan Bhalla & Sons; 2005. p. 322

[4] Collazos EF, Stashenko E, Gelvez OA, Martinez JR. Mobile distiller for the extraction of essential oils by the methods of hydrodistillation and water-vapor. CO6290097A1

[5] Adams P. Identification of Essential Oil Components by Gas Chromatography/Mass Spectrometry. 4th ed. Vol. 804. Carol Stream, IL: Allured Publishing; 2007

[6] Babushok VI, Kinstrom PJ, Zenkevich IG. Retention indices for frequently reported compounds of plant essential oils. Journal of Physical and Chemical Reference Data. 2011;**40**: 043101. DOI: 10.1063/1.3653552

[7] Davies NW. Gas chromatographic retention indices of monoterpenes and sesquiterpenes on methyl silicone and Carbowax 20M phases. Journal of Chromatography. 1990;**503**:1-24. DOI: 10.1016/S0021-9673(01)81487-4

[8] Sandra P, Bicchi C, editors. Capillary Gas Chromatography in Essential Oil Analysis. Chromatographic Methods Series. Heidelberg: Dr. Alfred Huethig Verlag; 1987. p. 436

[9] Mondello L. Comprehensive Chromatography in Combination with Mass Spectrometry. 1st ed. Hoboken, NJ: John Wiley and Sons; 2011. DOI: 10.1002/9781118003466

[10] Mondello L, Tranchida P, Dugo P, Dugo G. Comprehensive two-dimensional gas chromatography-mass spectrometry: a review. Mass Spectrometry Reviews. 2008;**27**: 101-124. DOI: 10.1002/mas.20158

[11] Ryan D, Marriott P. Comprehensive two-dimensional gas chromatography. Analytical and Bioanalytical Chemistry. 2003;**376**:295-297. DOI: 10.1007/s00216-003-1934-x

[12] Stashenko E, Martínez JR. Identification of essential oil components. In: Hashemi SMB, Khaneghah AM, Sant'Ana AS, editors. Essential Oils in Food Processing: Che mistry, Safety and Applications. 1st ed. Chichester: Wiley; 2017. pp. 57-117. DOI: 10.1002/9781119149392.ch3

[13] Stashenko E, Martínez JR. In: Salih B, editor. GC-MS Analysis of Volatile Plant Secondary Metabolites, Gas Chromatography in Plant Science, Wine Technology, Toxicology and Some Specific Applications. Rijeka: InTech; 2012. Available from: http://www.intechopen.com/books/gaschromatography-in-plant-science-wine-technology-toxicology-and-some-specific-applications/gc-ms-analysisof-volatile-plant-secondary-metabolites. ISBN: 978-953-51-0127-7

[14] Stashenko E, Martínez CR, Martínez JR, Shibamoto T. Catalytic transformation of anis oil (*Pimpinella anisum* L.) over zeolite Y. Journal of High Resolution Chromatography. 1995; **18**(8):501-503. DOI: 10.1002/jhrc.1240180810

[15] Puertas-Mejía M, Hillebrand S, Stashenko E, Winterhalter P. *In vitro*

radical scavenging activity of essential oils from colombian plants and fractions from oregano *(Origanum vulgare L.)* essential oil. Flavour and Fragrance Journal. 2002;**17**:380-384. DOI: 10.1002/ffj.1110

[16] Stashenko E, Acosta R, Martínez JR. High-resolution gas chromatographic analysis of the secondary metabolites obtained by subcritical fluid extraction from Colombian rue (*Ruta graveolens* L.). Journal of Biochemical and Biophysical Methods. 2000;**43**:379-390. DOI: 10.1016/S0165-022X(00)00079-8

[17] Stashenko E, Villa H, Combariza MY. Study of compositional variation in Colombian rue oil (*Ruta graveolens*) by HRGC using different detection systems. Journal of Microcolumn Separations. 1995;**7**(2):117-122. DOI: 10.1002/mcs.1220070204

[18] Combariza M, Blanco C, Stashenko E, Shibamoto T. Limonene concentration in lemon (*Citrus volkameriana)* peel oil as a function of ripeness. Journal of High Resolution Chromatography. 1994;**17**(9):643-646. DOI: 10.1002/jhrc.1240170905

[19] Stashenko E, Martínez R, Pinzón ME, Ramírez J. Changes in chemical composition of catalytically hydrogenated orange oil (*Citrus sinensis*). Journal of Chromatography A. 1996;**752**:217-222. DOI: 10.1016/S0021-9673(96)00481-5

[20] Blanco-Tirado C, Stashenko E, Combariza M, Martínez JR. Comparative study of colombian citrus oils by high resolution gas chromatography and gas chromatography-mass spectrometry. Journal of Chromatography A. 1995;***697***:501-513. DOI: 10.1016/0021-9673(94)00955-9

[21] Stashenko E, Martínez JR, Macku C, Shibamoto T. HRGC and GC-MS analysis of essential oil from Colombian ylang-ylang (*Cananga odorata* Hook Fil. et Thomson, forma genuina). Journal of High Resolution Chromatography. 1993;**16**(7):441-444. DOI: 10.1002/jhrc.1240160713

[22] Stashenko E, Torres W, Martínez JR. A study of compositional variation in the essential oil of ylang-ylang (*Cananga odorata* Hook. Fil. et Thomson, forma genuina) during flower development. Journal of High Resolution Chromatography. 1995;**18**(2):101-104. DOI: 10.1002/jhrc.1240180206

[23] Stashenko E, Quiroz N, Martínez JR. HRGC/FID/NPD and HRGC/MSD Study of Colombian ylang-ylang (*Cananga odorata*) oils obtained by different extraction techniques. Journal of High Resolution Chromatography. 1996;**19**(6):353-358. DOI: 10.1002/jhrc.1240190609

[24] Stashenko E, Wiame H, Dassy S, Martínez JR, Shibamoto T. Catalytic transformation of copaiba oil (*Copaifera officinalis*) over zeolite ZSM-5. Journal of High Resolution Chromatography. 1995;**18**(1):54-58. DOI: 10.1002/jhrc.1240180112

[25] Stashenko E, Puertas MA, Combariza MY. Volatile secondary metabolites from *Spilanthes americana* obtained by simultaneous steam distillation-solvent extraction and supercritical fluid extraction. Journal of Chromatography A. 1996;**752**:223-232. DOI: 10.1016/S0021-9673(96)00480-3

[26] Stashenko E, Cervantes M, Combariza MY, Martínez JR. HRGC-FID-MSD Analysis of the secondary metabolites obtained by different extraction methods from *Lepechinia schiedeana*, and evaluation of its antioxidant activity *in vitro*. Journal of High Resolution Chromatography. 1999;**22**(6):343-349. DOI: 10.1002/(SICI)1521-4168(19990601)22:6<343::AID-JHRC343>3.0.CO;2-J

[27] Stashenko E, Jaramillo BE, Martínez JR. HRGC/FID/MSD Analysis of volatile secondary metabolites from *Lippia alba* (Mill.) N.E. Brown grown in Colombia and evaluation of their *in vitro* antioxidant activity. Journal of Chromatography A. 2004;**1025**:99-103. DOI: 10.1016/j.chroma.2003.10.058

[28] Stashenko E, Martínez JR, Jaramillo BE. Analysis of volatile secondary metabolites from Colombian *Xylopia aromatica* (Lamarck) by different extraction and headspace methods and gas chromatography. Journal of Chromatography A. 2004;**1025**:105-113. DOI: 10.1016/j.chroma.2003.10.059

[29] Quintero A, González N, Stashenko E. Aceite esencial de las hojas de *Hyptis umbrosa* Salzm extraído por diferentes técnicas. Acta Científica Venezolana. 2004;**55**:181-187

[30] Güette-Fernández J, Olivero-Verbel J, O'Byrne-Hoyos I, Jaramillo B, Stashenko E. Chemical composition and toxicity against *Artemia franciscana* of the essential oil of *Callistemon speciosus (sims)* DC., collected in Bogota (Colombia). Journal of Essential Oil Research. 2008;**20**(3):272-275. DOI: 10.1080/10412905.2008.9700010

[31] Stashenko E, Martínez JR, Medina JD, Durán DC. Analysis of essential oils isolated by steam distillation from *Swinglea glutinosa* fruits and leaves. Journal of Essential Oil Research. 2015; **27**(4):276-282. DOI: 10.1080/10412905.2015.1045087

[32] Stashenko E, Gutiérrez-Avella D, Martínez JR, Manrique-López DL. Análisis por GC/FID y GC/MS de la composición química y estudio de la actividad antioxidante de los metabolitos secundarios volátiles, aislados por diferentes técnicas, de *Satureja viminea* L. cultivada en Colombia. Scientia Chromatographica. 2017;**9**(1):25-39. DOI: :10.4322/sc.2017.003

[33] Stashenko E, Ruiz C, Muñoz A, Castañeda M, Martínez J. Composition and antioxidant activity of essential oils of *Lippia origanoides* H.B.K. grown in Colombia. Natural Product Communications. 2008;**3**(4):563-566. DOI: 10.1177/1934578X0800300417

[34] Stashenko E, Ruíz CA, Arias G, Durán DC, Salgar W, Cala M, *et al.* *Lippia origanoides* chemotype differentiation based on essential oil GC-MS analysis. Journal of Separation Science. 2010;**33**(1):93-103. DOI: 10.1002/jssc.200900452

[35] Pino N, Melendez E, Stashenko E. Eugenol and methyl eugenol chemotype of essential oil of species *Ocimum gratissimum* L. and *Ocimum campechianum* Mill. from the northwest region of Colombia. Journal of Chromatographic Science. 2009;**47**: 800-803. DOI: 10.1093/chromsci/47.9.800

[36] Stashenko E, Martínez JR, Cárdenas-Vargas S, Saavedra-Barrera R, Durán DC. GC-MS study of compounds isolated from *Coffea arabica* flowers by different extraction techniques. Journal of Separation Science. 2013;**36**(17): 2901-2914. DOI: 10.1002/jssc.201300458

[37] Yáñez X, Betancur L, Solbay L, Agudelo M, Zapata B, Correa J, *et al.* Composición química y actividad biológica de aceites esenciales de *Calycolpus moritzianus* recolectado en el Norte de Santander, Colombia. Revista de la Universidad Industrial de Santander. Salud. 2009;**41**:259-267

[38] Muñoz A, Martínez JR, Stashenko E. Cromatografía de gases como herramienta de estudio de la composición química y capacidad antioxidante de especies vegetales ricas en timol y carvacrol, cultivadas en Colombia. Scientia Chromatographica. 2009;**1**(1):67-78

[39] Caroprese J, Parra M, Arrieta D, Stashenko E. Microscopic anatomy and volatile secondary metabolites at three stages of development of the inflorescences of *Lantana camara* (Verbenaceae). Revista de Biología Tropical. 2011;**59**(1):473-486

[40] Rodríguez R, Ruiz C, Arias G, Castro H, Martínez J, Stashenko E. Estudio comparativo de la composición de los aceites esenciales de cuatro especies del género *Cymbopogon* (Poaceae) cultivadas en Colombia. Boletin Latinoamericano y del Caribe de Plantas Medicinales y Aromaticas. 2012; **11**(1):77-85

[41] Vicuña GC, Stashenko E, Fuentes JL. Chemical composition of the *Lippia origanoides* essential oils and their antigenotoxicity against bleomycin-induced DNA damage. Fitoterapia. 2010;**81(**5:343-349. DOI: 10.1016/j.fitote.2009.10.008

[42] López MA, Stashenko E, Fuentes JL. Chemical composition and antigenotoxic properties of *Lippia* alba essential oils. Genetics and Molecular Biology. 2011;**34**:479-488. DOI: 10.1590/S1415-47572011005000030

[43] Quintero N, Stashenko E, Fuentes JL. The influence of organic solvents on estimates of genotoxicity and antigenotoxicity in the SOS chromotest. Genetics and Molecular Biology. 2012; **35**(2):503-514. DOI: 10.1590/S1415-47572012000300018

[44] Meneses R, Torres FA, Stashenko E, Ocazionez RE. Aceites esenciales de plantas colombianas inactivan el virus del dengue y el virus de la fiebre amarilla. Revista de la Universidad Industrial de Santander. Salud. 2009;**41**: 236-243

[45] Mesa A, Montiel J, Zapata B, Durán C, Betancur L, Stashenko E. Citral and carvone chemotypes from de essential oils of Colombian *Lippia alba* (Mill.) N.E. Brown: Composition, cytotoxicity and antifungal activity. Memórias do Instituto Oswaldo Cruz, Rio de Janeiro. 2009;**104**(6):878-884. DOI: 10.1590/S0074-02762009000600010

[46] Bueno JG, Martínez-Morales JR, Stashenko E. Actividad antimicobacteriana de terpenos. Revista de la Universidad Industrial de Santander. Salud. 2009;**41**:231-235

[47] Zapata B, Durán C, Stashenko E, Correa-Royero J, Betancur-Galvis L. Actividad citotóxica de aceites esenciales de *Lippia origanoides* H.B.K. y componentes mayoritarios. Revista de la Universidad Industrial de Santander. Salud. 2009;**41**:215-222

[48] Zapata B, Durán C, Stashenko E, Betancur L, Mesa AC. Actividad antimicótica y citotoxicidad de aceites esenciales de plantas de la familia Asteraceae. Revista Iberoamericana de Micología. 2010;**27**(2):101-103. DOI: 10.1016/j.riam.2010.01.005

[49] Tangarife V, Correa J, Zapata B, Durán C, Stashenko E, Mesa A. Anti-*Candida albicans* effect, cytotoxicity and interaction with antifungal drugs of essential oils and extracts from aromatic and medicinal plants. Infection. 2011; **15**(3):160-167. DOI: 10.1016/S0123-9392(11)70080-7

[50] Correa J, Tangarife V, Durán C, Stashenko E, Mesa A. *In vitro* antifungal activity and cytotoxic effect of essential oils and extracts of medicinal and aromatic plants against *Candida krusei* and *Aspergillus fumigatus*. Brazilian Journal of Pharmacognosy. 2010;**20**(5): 734-741. DOI: 10.1590/S0102-695X2010005000021

[51] Bueno J, Ericsson DC, Stashenko E. Antimycobacterial natural products—An opportunity for the Colombian biodiversity. Revista Española de Quimioterapia. 2011;**24**(4):175-183

[52] Bueno J, Escobar P, Martínez JR, Leal SM, Stashenko EE. Composition of three essential oils, and their mammalian cell toxicity and antimycobacterial activity against drug resistant-tuberculosis and nontuberculous mycobacteria strains. Natural Product Communications. 2011; **6**(11):1567-1798. DOI: 10.1177/1934578X1100601143

[53] Caballero-Gallardo K, Olivero-Verbel J, Stashenko E. Repellent activity of essential oils and some of their individual constituents against *Tribolium castaneum* Herbst. Journal of Agricultural and Food Chemistry. 2011; **59**:1690-1696. DOI: 10.1021/jf103937p

[54] Caballero-Gallardo K, Olivero-Verbel J, Stashenko E. Repellency and toxicity of essential oils from *Cymbopogon martinii, Cymbopogon flexuosus* and *Lippia origanoides* cultivated in Colombia against *Tribolium castaneum*. Journal of Stored Products Research. 2012;**50**:62-65. DOI: 10.1016/j.jspr.2012.05.002

[55] Jaramillo GI, Ramirez G, Logan J, Loza-Reyes E, Stashenko E, Moores G. *Repellents Inhibit P450 Enzymes in Stegomyia (Aedes) aegypti*. PLoS One. 2012;**7**(11):e48698. DOI: 10.1371/journal.pone.0048698

[56] Nerio LS, Olivero-Verbel J, Stashenko E. Repellent activity of essential oils from seven aromatic plants grown in Colombia against *Sitophilus zeamais* Motschulsky (Coleoptera). Journal of Stored Products Research. 2009;**45**:212-214. DOI: 10.1016/j.jspr.2009.01.002

[57] Nerio LS, Olivero-Verbel J, Stashenko E. Repellent activity of essential oils: A review. Bioresource Technology. 2010;**101**:372-378. DOI: 10.1016/j.biortech.2009.07.048

[58] Olivero J, Caballero K, Jaramillo B, Stashenko E. Actividad repelente de los aceites esenciales de *Lippia origanoides, Citrus sinensis* y *Cymbopogon nardus* cultivadas en Colombia frente a *Tribolium castaneum*, Herbst. Revista de la Universidad Industrial de Santander. Salud. 2009;**41**:244-250

[59] Olivero J, Neiro L, Stashenko E. Bioactivity against *Tribolium castaneum* Herbst (Coleoptera: Tenebrionidae) of *Cymbopogon citratus* and *Eucalyptus citriodora* essential oils grown in Colombia. Pest Management Science. 2010;**66**(6):664-668

[60] Jaramillo B, Stashenko E, Martínez JR. Composición química volátil de *Satureja brownei* (Sw.) Briq. Colombiana y determinación de su actividad antioxidante. Revista Cubana de Plantas Medicinales. 2010;**15**:52-63

[61] Jaramillo BE, Duarte E, Muñoz K, Stashenko E. Composición química volátil del aceite esencial de *Croton malambo* H. Karst. colombiano y determinación de su actividad antioxidante. Revista Cubana de Plantas Medicinales. 2010;**15**(3):133-142

[62] Muñoz A, Kouznetsov VV, Stashenko E. Composición y capacidad antioxidante *in-vitro* de aceites esenciales ricos en timol, carvacrol, *trans*-anetol o estragol. Revista de la Universidad Industrial de Santander. Salud. 2009;**41**:287-294

[63] Muñoz-Acevedo A, Vargas LY, Stashenko E, Kouznetsov VV. Improved Trolox® equivalent antioxidant capacity assay for efficient and fast search of new antioxidant agents. Analytical Chemistry Letters. 2011;**1**(1):86. DOI: 10.1080/22297928.2011.10648207

[64] Olivero-Verbel J, González-Cervera T, Güette-Fernandez J, Jaramillo-Colorado B, Stashenko E. Chemical composition and antioxidant activity of essential oils isolated from Colombian plants. Brazilian Journal of

Pharmacognosy. 2010;**20**(4):568-574. DOI: 10.1590/S0102-695X2010000400016

[65] Stashenko E, Martínez JR, Ramírez E, Arias AJ, Vásquez EG. Rendimiento y capacidad antioxidante de extractos de *Rosmarinus officinalis, Salvia officinalis* y *Psidium guajava* obtenidos con CO_2 supercrítico. Revista de la Academia Colombiana de Ciencias Exactas. 2012; **36**(140):305-315

[66] Olivero-Verbel J, Güette-Fernandez J, Stashenko E. Acute toxicity against *Artemia franciscana* of essential oils isolated from plants of the genus *Lippia* and *Piper* collected in Colombia. Boletin Latinoamericano y del Caribe de Plantas Medicinales y Aromaticas. 2009;**8**(5): 419-427

[67] Vera AP, Olivero JT, Jaramillo BE, Stashenko E. Efecto protector del aceite esencial de *Lippia alba* (Mill.) N.E. Brown sobre la toxicidad del mercurocromo en raíces de *Allium cepa* L. Revista Cubana de Plantas Medicinales. 2010;**15**(1):27-37

[68] Zapata B, Durán C, Stashenko E, Betancur L, Mesa AC. Actividad antimicótica, citotoxicidad y composición de aceites esenciales de plantas de la familia Labiatae. Revista de la Universidad Industrial de Santander. Salud. 2009;**41**:223-230

[69] Bueno-Sánchez J, Martínez-Morales JR, Stashenko E, Ribón W. Anti-tubercular activity of eleven aromatic and medicinal plants occurring in Colombia. Biomédica. 2009;**29**:51-60. DOI: 10.7705/biomedica.v29i1.41

[70] Escobar P, Herrera L, Leal S, Durán C, Stashenko E. Composición química y actividad anti-tripanosomal de aceites esenciales obtenidos de *Tagetes* (Fam. Asteraceae), recolectados en Colombia. Revista de la Universidad Industrial de Santander. Salud. 2009;**41**:280-286

[71] Escobar P, Leal SM, Herrera LV, Martinez JR, Stashenko E. Chemical composition and antiprotozoal activities of Colombian *Lippia* spp essential oils and their major components. Memórias do Instituto Oswaldo Cruz. 2010;**105**: 184-190. DOI: 10.1590/S0074-02762010000200013

[72] Bakkali F, Averbeck S, Averbeck D, Idaomar M. Biological effects of essential oils—A review. Food and Chemical Toxicology. 2008;**46**:446-475. DOI: 10.1016/j.fct.2007.09.106

[73] Ribeiro AF, Andrade EH, Salimena G, FR MJG. Circadian and seasonal study of the cinnamate chemotype from *Lippia origanoides* Kunth. Biochemical Systematics and Ecology. 2014;**55**: 249-259. DOI: 10.1016/j.bse.2014.03.014

[74] Stashenko E. Estudio de la composicion y la actividad biologica de aceites esenciales de *Lippia alba*, de diferentes regiones de Colombia. In: Dellacassa E, org. Normalización de productos naturales obtenidos de especies de la flora aromática latinoamericana: proyecto CYTED IV 20. Porto Alegre—Brasil. EDIPUCRS. 2010. Chapter 8. pp. 173-191

[75] Zapata B, Betancur-Galvis L, Duran C, Stashenko E. Cytotoxic activity of Asteraceae and Verbenaceae family essential oils. Journal of Essential Oil Research. 2014;**26**(1):50-57. DOI: 10.1080/10412905.2013.820674

[76] Stashenko E, Martínez JR, Durán DC, Córdoba Y, Caballero D. Estudio comparativo de la composición química y la actividad antioxidante de los aceites esenciales de algunas plantas del género *Lippia* (Verbenaceae) cultivadas en Colombia. Revista de la Academia Colombiana de Ciencias Exactas. 2014; **38**(Suppl):89-105. DOI: 10.18257/raccefyn.156

[77] Huang D, Ou B, Prior R. The chemistry behind antioxidant capacity

assays. Journal of Agricultural and Food Chemistry. 2005;**53**:1841-1856. DOI: 10.1021/jf030723c

[78] Miranda Y, Roa-Linares VC, Betancur-Galvis LA, Durán-García DC, Stashenko E. Antiviral activity of Colombian Labiatae and Verbenaceae family essential oils and monoterpenes on Human Herpes viruses. Journal of Essential Oil Research. 2016;**28**(2): 130-132. DOI: 10.1080/10412905.2015.1093556

[79] Meneses R, Ocazionez RE, Martínez JR, Stashenko E. Inhibitory effect of essential oils obtained from plants grown in Colombia on yellow fever virus replication *in vitro*. Annals of Clinical Microbiology and Antimicrobials. 2009;**8**:8. DOI: 10.1186/1476-0711-8-8

[80] Radman M. Phenomenology of an inducible mutagenic DNA repair pathway in *Escherichia coli*: SOS repair hypothesis. Basic Life Sciences. 1975;**5A**: 355-367. DOI: 10.1007/978-1-4684-2895-7_48

[81] Ohta T, Watanabe M, Tsukamoto R, Shirasu Y, Kada T. Antimutagenic effects of 5-fluorouracil and 5-fluorodeoxyuridine on UV-induced mutagenesis in *Escherichia coli*. Mutation Research. 1986;**173**:19-24. DOI: 10.1016/0165-7992(86)90005-9

[82] Fuentes JL, García-Forero A, Quintero N, Prada CA, Rey-Castellano N, Franco-Niño DA, *et al.* The SOS Chromotest applied for screening plant antigenotoxic agents against ultraviolet radiation. Photochemical & Photobiological Sciences. 2017;**16**: 1424-1434. DOI: 10.1039/C7PP00024C

[83] Quintero N, Cordoba Y, Stashenko E, Fuentes JL. Antigenotoxic effect against ultraviolet radiation-induced DNA damage of the essential oils from *Lippia* species. Photochemistry and Photobiology. 2017;**93**:1063-1072. DOI: 10.1111/php.12735

[84] Jaramillo-Colorado B, Muñoz K, Duarte E, Stashenko E, Olivero J. Volatile secondary metabolites from Colombian *Croton malambo* (Karst) by different extraction methods and repellent activity of its essential oil. Journal of Essential Oil-Bearing Plants. 2014;**17**(5):992-1001. DOI: 10.1080/0972060X.2014.895185

[85] Alcala-Orozco M, Caballero-Gallardo K, Stashenko E, Olivero-Verbel J. Repellent and fumigant actions of the essential oils from *Elettaria cardamomum* (L.) Maton, *Salvia officinalis* (L.) Linnaeus, and *Lippia origanoides* (V.) Kunth against *Tribolium castaneum* and *Ulomoides dermestoides*. Journal of Essential Oil-Bearing Plants. 2019;**22**(1):18-30. DOI: 10.1080/0972060X.2019.1585966

[86] Dias CN, Moraes DFC. Essential oils and their compounds as *Aedes aegypti* L. (Diptera: Culicidae) larvicides: review. Parasitology Research. 2014;**113**: 565-592. DOI: 10.1007/s00436-013-3687-6

[87] Ríos N, Stashenko E, Duque JE. Evaluation of the insecticidal activity of essential oils and their mixtures against *Aedes aegypti* (Diptera: Culicidae). Revista Brasileira de Entomologia. 2017; **61**:307-311. DOI: 10.1016/j.rbe.2017.08.005

Chapter 9

Comparative Analysis of the Chemical Composition, Antimicrobial and Antioxidant Activity of Essential Oils of Spices Used in the Food Industry in Brazil

Amanda Mara Teles, Adenilde Nascimento Mouchreck, Gustavo Oliveira Everton, Ana Lucia Abreu-Silva, Kátia da Silva Calabrese and Fernando Almeida-Souza

Abstract

There are many food-borne pathogens in the wild and they are considered the cause of serious public health problems in both developed and developing countries. The use of natural products, such as antimicrobial compounds, has been increasing, in an attempt to control bacteria present in foods, mainly pathogens resistant to conventional antibiotics. This chapter is intended to provide the antimicrobial and antioxidant activity of essential oils of *Cinnamomum zeylanicum* (cinnamon), *Origanum vulgare* (oregano), *Zingiber officinale* (ginger), *Rosmarinus officinalis* (rosemary), *Citrus latifolia* (tahiti lemon) and *Curcuma longa* (saffron) as well as to determinate its chemical composition. The oils had been extracted by hydrodistillation with a Clevenger type apparatus and the antimicrobial activity was performed against standard strains *Escherichia coli*, *Pseudomonas aeruginosa* and *Staphylococcus aureus*. The antioxidant activity was carried out using the ABTS [2,2-azinobis-(3-ethylbenzothiazoline-6-sulfonic acid)] method. The essential oils presented a mixture of mono- and sesquiterpenes. The best minimum inhibitory concentration was determined to *C. zeylanicum* against *S. aureus*. *O. vulgare* antioxidant activity presented inhibition of 90.74% and EC_{50} of 14 μg mL^{-1}. These results demonstrate that the essential oils analyzed presented efficient antibacterial activity and antioxidant action being able to satisfy the demand of use as control of microorganisms in the food.

Keywords: essential oils, biological properties, antimicrobial activity, antioxidant activity, chemical composition

1. Introduction

Brazil has an extensive diversity of species in its flora, and great tradition in the use of medicinal plants linked to the popular medicine [1]. Medicinal plants are characterized by common sense within communities as an alternative for nutritional and therapeutic purposes in the prevention and cure of diseases since ancient times. Their therapeutic use has aroused scientific interest, awakening new ways to control several diseases [2].

These species are commonly employed in the commercial sector, such as the food industry. Condiments or spices are used in the preparation of food in order to improve sensory characteristics and as a preservative agent due to its antioxidant and antimicrobial attributes [3]. These types of preservatives are more accepted by the population, mainly due to the search of the industries for healthier products [4].

The antimicrobial and antioxidant activities of various spices, such as *Rosmarinus officinalis* (rosemary) [5, 6], *Cinnamomum verum* (cinnamon) [7]*Curcuma longa* (saffron) [8], *Ocimum basilicum* (basil) [9, 10], *Zingiber officinale* (ginger) [11] and *Origanum vulgare* (oregano) [12, 13] that are widely used in the food industry have such proven biological properties.

The chemical constituents responsible for the antibacterial power of these condiments are named as phenolic compounds, such as carvacrol, linalool, thymol, menthol, limonene and eugenol [14], also including terpene derivatives, such as mono- and sesquiterpenes and phenylpropanoids [15].

These spices are mainly used through the essential oils obtained from these plants. Many of these oils are composed of substances such as those mentioned above and these are related to the permeability of the bacterial cell membrane and through this can act in the control of microbiological growth [16].

A very important factor is the yield of these oils, which are based on the method and extraction time [17] and thus add a higher commercial value associated with cost-benefit. Usually, they are synthesized by extraction techniques, such as distillation [18, 19] that separate it from the water by differences in density and polarity.

Essential oils are composed of a complex mixture of volatile chemicals present in various parts of medicinal plants. They provide the essence of the plant, being responsible for the flavor and aroma of spices [20]. They act in protection against pathogens, in the attraction of pollinators and can be found in leaves, flowers, fruits and even in roots of aromatic plants [21]. These compounds have specific odoriferous and lipophilic characteristics [22] and have received much attention in the last decades due to the antimicrobial activity that they present [23].

These natural products have proven antioxidant and antimicrobial potential and several studies describe the application of these products to prolong the shelf life of food products without risks to the consumer or interference in the natural characteristics of the food [19, 24].

The search for decrease in the use of antioxidants and synthetic antimicrobial agents intensifies studies to demonstrate the promising potential of these compounds [19, 25, 26]. These searches are based on the great risk of contamination through food, and the great resistance of bacteria to antibiotics, appearing the interest of adding natural antimicrobial agents in food as a way to mitigate cases of foodborne diseases [27].

Foodborne diseases can be identified when one or more individuals exhibit similar symptoms after ingestion of food contaminated with pathogenic microorganisms, their toxins, toxic chemicals or objects which constitute a common source. In the case of highly virulent pathogens, such as *Clostridium (C.) botulinum* and *Escherichia (E.) coli* O157: H7, only one case can be considered an outbreak [28]. Bacteria like *Staphylococcus aureus*, *Salmonella*, *Campylobacter* and *Escherichia coli* are important food pathogens, and are among the biggest cause of outbreaks in the word [29].

A polymerase chain reaction (PCR) analysis with samples of beef, sheep and processed chicken showed the presence of *Clostridium perfringens*, *Enterococcus faecalis* and *S. aureus* in 79, 86 and 94%, respectively. In meat samples, *E. coli* and enteric *Salmonella* were also found in respective concentrations of 90 and 91% [30].

Food-borne illness is a real problem in the present scenario as the consumerism of packaged food. Pathogens entering the packaged foods may survive longer. Therefore, antimicrobial agents either alone or in combination are added to the food or packaging materials to eliminate these agents [31].

Treatment in these cases leads to the indiscriminate use of antibiotics. These have provided a growing multidrug resistance of microorganisms, generating public health problems due to the residues in foods [2]. The antibiotics act as an important selective pressure for the emergence and persistence of resistant strains [32].

Exploiting the antimicrobial property, essential oils are considered as a "natural" remedy to this problem. Alternatives to the use of synthetic antimicrobial agents have been proposed in recent years, and some approaches include herbal products [28]. This promising determination of the action of these essential oils on microorganisms using Gram-positive and Gram-negative bacteria should be performed due to its low cost of acquisition, use and therapeutic action, such as the viability of medicinal potential and its use in the food industry, cosmetics and pharmaceuticals, whereas bacterial resistance is one of the most significant challenges to human health [33].

Thus, the objective of this chapter was to provide the antimicrobial and antioxidant activity of *Cinnamomum zeylanicum* (cinnamon), *O. vulgare* (oregano), *Z. officinale* (ginger), *R. officinalis* (rosemary), *C. longa L.* (saffron) and *C. latifolia* (tahiti lemon) essential oils as well as to determine its chemical composition.

2. Essential oils chemical profile

C. zeylanicum leaves, *C. latifolia* barks, *O. vulgare* and *R. officinalis* aerial parts and *Z. officinale* and *C. longa L.* rhizome were collected in the city of São Luis Maranhão, Brazil (latitude: −2.53073, longitude: −44.3068 2° 31′ 51″ South, 44° 18′ 24″ West). The taxonomic identification was performed by Ana Zelia Silva in the Seabra Attic Herbarium of the Department of Botany of the Federal University of Maranhão. All five plants were dried for 48 h and sprayed in an electric knife mill at the Food and Water Quality Control Laboratory of the Federal University of Maranhão.

The essential oil was extracted by hydrodistillation using Clevenger system. A quantity of 300 g of dry plant material diluted in water at a ratio of 1:10 was boiled at 100°C for 3 h. The oil was dried with anhydrous sodium sulfate and kept in an amber bottle under refrigeration. For in vitro biological assay, the essential oils and reference drugs were dissolved in dimethylsulfoxide (DMSO) at 100 times the highest concentration of use, and subsequently diluted in an appropriate medium to a final concentration of DMSO less than 1%.

Chemical characterization of the essential oils was performed by gas chromatography coupled to mass spectrometry (GC-MS). The essential oils under study were dissolved in 1 mg/mL ethyl acetate and analyzed on Shimadzu QP 5000 gas chromatograph with ZB-5 ms capillary column (5% phenyl arylene 95% dimethylpolysiloxane) coupled at 70 eV (40–500 Da) electronic impact detector HP 5MS with a transfer temperature of 280°C. The chromatographic conditions were injection of 0.3 μL of ethyl acetate; helium carrier gas (99.99%); injector temperature: 280°C; split mode (1:10); then an initial temperature of 40°C and a final temperature of 300°C; initial time of 5 min and final time of 7.58min. The results obtained for *C. zeylanicum* leaves are shown in **Table 1**.

A total of 15 compounds were identified and their main constituents such as cinnamic aldehyde (46.30%), α-copaene (16.35%) and trans-β-caryophyllene (8.26%) were identified and quantified. Various researchers have identified and quantified different chemical compounds of *C. zeylanicum* essential oil. A study identified nine

Peak	C. zeylanicum E.O.	
	Compounds	**(%)**
1	α-Pinene	1.47
2	Benzaldehyde	4.16
3	3-Phenylpropionaldehyde	2.95
4	Borneol	1.06
5	α-Terpineol	0.87
6	Cinnamic aldehyde	**46.30**
7	3-Phenyl-1-propanol	1.46
8	α-Copaene	16.35
9	trans-β-Caryophyllene	**8.26**
10	(e)-Cinnamyl acetate	7.54
11	α-Humulene	2.16
12	delta-Cadienene	1.42
13	(−)-Spathulenol	2.09
14	Caryophyllene oxide	2.80
15	Benzyl benzoate	1.12

Table 1.
C. zeylanicum (cinnamon) essential oil chemical composition.

compounds [34] with (E)-cinnamaldehyde as its major component [35, 36]. The essential oils of this plant generally have cinnamaldehyde [37], which corroborates the results obtained in this study. We already presented similar results to chemical composition of *C. zeylanicum* essential oil [38].

The chemical profile obtained for the aerial parts such as *R. officinalis* and *O. vulgare* is shown in **Table 2**. Its total composition presented 17 components with the major constituents being camphor (37.00%), 1,8-cineol (11.32%) and α-terpineol (7.12%). In *O. vulgare* essential oil, 20 compounds were found, represented by the main constituents cis-ρ-menth-2-en-1-ol (33.88%), linalyl acetate (13.90%) and p-cymene (8.29%). Probst also identified camphor as the major component of R. officinalis essential oil [39], being possible to observe similarity with the essential oil composition of this study, while other study obtained 1,8-cineol in a higher percentage, but camphor was present in the second place among the major components [40].

With respect to *O. vulgare* essential oil, while there is a description of similar chemical composition to our study [38, 41], another study found three different chemotypes in 25 samples of this essential oil: linalool/linalyl acetate chemotypes with predominant linalyl acetate; the second major chemotypes rich in carvacrol and c-terpinene; and a third rich chemotype in thymol [42].

The chemical composition of the essential oils of *C. longa* (saffron) and *Z. officinale* rhizomes is shown in **Table 3**. For essential oils obtained from *C. longa* rhizomes, 17 compounds were identified and the major chemical composition is represented by turmerone (55.43%), β-turmerone (12.02%) and γ-curcumene (6.96%). Similar results were described to *C. longa* essential oil [38, 43–45], with the main compounds turmerone and β-turmerone presenting the highest percentages.

In the essential oil of *Z. officinale*, 18 components were identified, constituting α-zingiberene (27.14%), geranial (14.06%) and nerolidol (13.51%) in greater percentage. Diemer studing essential oil of Z. officinale quantified the chemical profile in 12 constituents and concluded that α-zingiberene was the predominant

Peak	***R. officinalis*** **(rosemary) E.O.**		***O. vulgare*** **(oregano) E.O.**	
	Compounds	**(%)**	**Compounds**	**(%)**
1	β-Pineno	2.29	α-Pinene	0.80
2	β-Mirceno	4.36	Bicyclo[3.1.0]hexane	1.73
3	ρ-Cimeno	1.41	(+)-4-Carene	3.08
4	Limoneno	2.02	**p-Cymene**	**8.29**
5	**1,8-Cineol**	**11.32**	Cyclohexene	1.23
6	γ-Terpineno	1.61	β-Phellandrene	2.73
7	Linalol	2.99	p-Menth-2-en-1-ol	4.62
8	**Cânfora**	**37.00**	1,4-Cyclohexadiene	1.21
9	Pinocarvona	217	cis-Sabinene hydrate	1.29
10	Borneol	3.24	Terpinolene	3.11
11	Terpinen-4-ol	4.79	1,6-Octadien-3-ol	5.69
12	**α-Terpineol**	**7.12**	trans-Sabinene hydrate	1.59
13	Verbenona	5.85	**cis-p-Menth-2-en-1-ol**	**33.88**
14	Acetato de bornila	4.28	3-Cyclohexen-1-ol	5.26
15	β-Cariofileno	6.43	(+)-α-Terpineol	2.61
16	α-Humuleno	1.47	Carvacrol methyl ether	0.94
17	α-Bisabolol	1.65	**Linalyl acetate**	**13.90**
18	—		Thymol	2.41
19	—		trans-β-Caryophyllene	2.46
20	—		1H-Cycloprop(E)azulen-7-ol	3.16

Table 2.
Chemical composition of the essential oils of Rosmarinus officinalis (rosemary) and Origanum vulgare (oregano).

Peak	***C. longa*** **(saffron) E.O.**		***Z. officinale*** **(ginger) E.O.**	
	Compounds	**(%)**	**Compounds**	**(%)**
1	α-Pinene	1.15	α-Pineno	1.46
2	Myrcene	0.37	Canfeno	5.02
3	Vinyl propionate	0.20	β-Mirceno	1.29
4	ρ-Cymene	1.01	Sabineno	5.23
5	Bisabolone	0.55	1,8-Cineol	4.35
6	**β-Turmerone**	**12.02**	Linalol	0.50
7	1,8-Cineole	1.01	4,4-Dimetil-2-pentinal	0.80
8	Camphor	1.24	terc-Dodeciltiol	0.71
9	α-Terpineol	4.13	Neral	9.64
10	Terpinolene	0.43	Nerol	1.07
11	α-Zingiberene	0.29	**Geranial**	**14.06**
12	β-Sesquiphellandrene	2.67	2-Undecanona	0.63
13	β-Caryophyllene	1.00	Farnesol	1.27
14	**γ-Curcumene**	**6.96**	1,1-Diciclopropiletileno	0.55

Peak	*C. longa* (saffron) E.O.		*Z. officinale* (ginger) E.O.	
	Compounds	(%)	Compounds	(%)
15	ar-Curcumene	1.58	ar-Curcumeno	3.33
16	**Turmerone**	**55.43**	**α-Zingibereno**	**27.14**
17	β-Sesquiphellandrene	1.10	**Nerolidol**	**13.51**
18	—	-	β-Sesquifelandreno	9.45

Table 3.
Chemical composition of Curcuma longa (saffron) and Zingiber officinale (ginger) essential oils.

Peak	*Citrus latifolia* (tahiti lemon) E.O.	
	Compounds	(%)
1	ρ-Cymene	**10.86**
2	D-Limonene	8.85
3	Cyclooctanone	1.54
4	ρ-Mentha-E-2,8(9)-dien-1-ol	2.47
5	trans-Pinocarveol	3.23
6	14.70 Pinocarvone	2.02
7	ρ-Cymen-8-ol	3.02
8	Bicyclo(3.1.1)hept-2-ene-2-carboxaldehyde,6,6-dimethyl-	5.34
9	Myrtenol	6.31
10	trans-Carveol	1.58
11	cis-Carveol	**11.59**
12	Carvone	1.68
13	19.02 Carvone oxide	3.69
14	Limonene dioxide	**25.92**
15	1,2-Cyclohexanediol 1-methyl-4-(1-methyleth)	8.10
16	7-Oxabicyclo[4.1.0]heptane,1-methyl-4-1-(1-methylethyl)	1.24
17	2,7-Octadiene-1,6-diol,2,6-dimethyl	2.56

Table 4.
Chemical composition of lemon tahiti essential oil.

component [46]. Same results were observed by us in the present study. However, there are also descriptions of geranial as its major constituent in this oil [47].

The chemical composition obtained from *C. latifolia* leaves is presented in **Table 4**. The essential oil obtained presented from 17 components with the main constituents being limonene dioxide (25.92%), cis-carveol (11.59%) and ρ-cymene (10.86%). Similarly, it was found in the researches carried out in *C. latifolia* tanaka, identifying 17 compounds and limonene as the major compound with 46.3% [48] and 58.43% [49].

3. Antimicrobial activity

Escherichia coli (ATCC 25922), *Staphylococcus aureus* (ATCC 12600) and *Pseudomonas aeruginosa* (ATCC 27853) strains were cultured in brain heart infusion broth for 24 h at 37°C and then diluted to 10^8 UFC/mL following the MacFarland scale, recommended by the Clinical and Laboratory Standards Institute [50].

The inoculum (100 μL) of each bacterium was seeded in Mueller-Hinton agar, with filter paper impregnated with 50 μL of essential oil placed on the surface. The plates were incubated at 35°C and after 24 h, the inhibition halo was measured with a millimeter ruler [51]. The minimum inhibitory concentration (MIC) was determined using the broth dilution methodology [52] and performed in triplicates with the same bacterium used in solid media diffusion techniques. Initially, an aliquot of the essential oil prepared in DMSO was transferred to a test tube containing BHI broth. Serial dilutions were then performed resulting in concentrations of 5–2000 μg/mL. Microbial suspensions containing 1.5×10^8 CFU/mL of the bacteria were added at each concentration and incubated at 35°C for 24 h. Tubes without bacteria were reserved for control of broth sterility and bacterial growth. After the incubation period, the minimum essential oil inhibitory concentration was defined as the lowest concentration which visibly inhibited bacterial growth observed by the absence of visible turbidity. To confirm growth inhibition, the BHI broth was subjected to the inoculum microbial seeding test on the surface of the plate-count agar.

The disc diffusion method evaluated the antibacterial activity of *C. zeylanicum*, *O. vulgare*, *Z. officinale*, *R. officinalis*, *C. longa* and *C. latifolia* essential oils to form inhibition halos against the growth of Gram-positive (*S. aureus*) and Gram-negative (*E. coli* and *P. aeruginosa*) bacteria strains. The diameters of the inhibition halos developed by the essential oils are shown in **Table 5**. The halos ranged from 7.67 to 15.33 mm. The largest inhibition halo against Gram-negative *E. coli* bacteria was quantified at 21 mm by *C. latifolia* essential oil. The best bactericidal activity for *S. aureus* was quantified at 15.66 mm by *C. longa*, while the Gram-positive *P. aeruginosa* was strongly inhibited by *C. longa* essential oil of, quantifying a halo of 12 mm.

The minimum inhibitory concentration (MIC) in μg mL^{-1}, the lowest visible concentration that prevents visible microbial growth in the culture medium by the action of the natural product, is being reported in **Table 5**.

The bactericidal activity of *C. zeylanicum* essential oil was demonstrated by larger halos against the Gram-positive bacteria. Similar results were found in *C. zeylanicum* essential oil [34, 38, 53] and the authors reported cinnamaldehyde as responsible for the antimicrobial action. However, lower results were also described when evaluating the antimicrobial activity of this oil against *Salmonella typhimurium* and *E. coli*, being its better inhibition against *S. aureus* [54].

The MIC's of *C. zeylanicum* essential oil were quantified using *S. aureus* strain and obtained a similar inhibitory concentration to that in this research, however *E. coli* and *Salmonella typhi* concentrations were far superior to that described in our study [55]. In another research conducted by Trajano et al. [4], concentrations are relatively lower than those observed in this study.

To the antimicrobial potential of *R. officinalis* essential oil, Cordeiro [56] also obtained inhibition halos for *S. aureus*, as well as Ribeiro [57] for *E. coli*, both using this same essential oil as antimicrobial. The antimicrobial activity of this oil for the strains tested in broth dilution showed MIC's similar to the experiment performed by Silva et al. [58]. However, values for such concentrations have also been reported in smaller units by Thanh et al. [59].

On the other hand, *O. vulgare* essential oil has demonstrated efficiencies for both Gram-positive and Gram-negative strains which can be observed by Stefanakis et al. [60] and Sankar et al. [61] who reported antimicrobial activity of this essential oil similar to this research against the same bacteria tested.

Soković et al. [62] obtained MIC's for *O. vulgare* essential oil smaller than those described in the results of this study for *S. aureus*, *E. coli*, *Salmonella enteritidis*

	E. coli (ATCC 25922)		*S. aureus* (ATCC 12600)		*P. aeruginosa* (ATCC 27853)	
	IH (mm)	**MIC (μg mL^{-1})**	**IH (mm)**	**MIC (μg mL^{-1})**	**IH (mm)**	**MIC (μg mL^{-1})**
C zeylanicum	12.67 (±1.00)	216.67 (±14.43)	15.33 (±0.58)	**83.33** (±28.87)	9.33 (±0.58)	383.33 (±0.01)
O. vulgare	15.33 (±0.58)	**133.33** (±28.87)	14.67 (±0.58)	216.67 (±28.87)	10.33 (±0.58)	550.00 (±28.87)
C. longa	14.33 (±0.58)	266.67 (±28.87)	**15.66** (±0.58)	166.67 (±14.43)	12.00 (±0.58)	483.33 (±57.74)
Z. officinale	10.70 (±0.58)	1000.00	9.70 (±0.58)	200.00	8.67 (±0.58)	1500.0
R. officinalis	9.70 (±0.58)	1700.00	10.70 (±0.58)	1500.00	7.67 (±0.58)	1700.0
C. latifolia	21	250	10	500	11	1000

Table 5.
Diameters of inhibition halos (IH) and minimum inhibitory concentrations (MIC) for the essential oils activity against bacterial strains.

and *Salmonella typhimurium*. Sarikurkcu et al. [38, 63], also performed an assay to determine the MIC against *S. aureus* and *E. coli*, obtaining results similar to those observed.

For *C. longa* essential oil, the largest halos were quantified for Gram-positive bacteria, similar to results found by Gupta et al. [64], Teles et al. [38] and Mishra et al. [65] who reported the formation of halos against the same bacteria in this study submitted to antimicrobial activity assays. Singh et al. [66] also observed satisfactory MIC values for the control of the microorganisms tested in this study.

In relation to the bactericidal effect of the *Z. officinale* essential oil, the values obtained in our disc diffusion test are superior to those obtained in the study by Singh et al. [67], where the authors did not obtain inhibition halos for *E. coli* and *S. aureus*, and the same were reported by Grégio et al. [68]. However, MIC values were similar to those found by Sasidharan and Menon [69].

4. Antioxidant activity

The antioxidant activity by the ABTS method [2,2-azinobis-(3-ethylbenzothiazoline-6-sulfonic acid)] was adapted according to the methodology suggested by [70]. The ABTS$^{\cdot}$+ radical was prepared by the reaction of 5.0 mL of a 3840 μg mL^{-1} of ABTS solution with 88 μL of the 37,840 μg mL^{-1} potassium persulfate solution. The mixture was left in the dark for 16 h. After formation of the radical, the mixture was diluted in ethanol (approximately 1:30 v/v) and absorbance was obtained at 734 nm. From the extracts and essential oils concentrations (5–150 μg mL^{-1}), the reaction mixture was prepared with the ABTS radical cation. In a dark environment, a 30 μL aliquot of each extract and essential oil concentration was transferred into 23 test tubes containing 3.0 mL of the ABTS radical cation and homogenized on a tube shaker. After 6 min, absorbance of the reaction mixture was obtained in a spectrophotometer at 734 nm. The analyzes were performed in triplicate and the capture of the free radical was expressed as percent inhibition (% I) of the ABTS radical cation.

The ABTS method allowed the calculation of the 50% effective concentration of the essential oils, which express the minimum concentration required to reduce the initial concentration of ABTS by 50%, and these are expressed in **Table 6**. The lowest concentration and consequently the best antioxidant activity was observed to oregano, with an EC_{50} quantified in 14 μg mL^{-1} and consequently also the highest percentage of ABTS inhibition.

The antioxidant effect of *C. zeylanicum* essential oil differs from that observed by [71] using the ABTS technique, where these authors verified a lower EC_{50}. This difference is explained by the authors due to variation in the chemical composition of the essential oils studied, which depends on factors such as the geographic location and the time of collection of the plant.

The result for *R. officinalis* essential oil exhibited by [57] shows a fairly high effective concentration comparing that obtained in this article. Also [72], when evaluating the antioxidant activity of this essential oil from five different crop fields, found a lower EC_{50} than this study. According to [73], the main responsible for the free radical stabilization capacity of this species is 1,8-cineol. In this way, it is possible to relate the lower antioxidant potential of the rosemary essential oil obtained in this research with the low content of 1,8-cineol.

On the other hand [72], while evaluating the antioxidant capacity of rosemary essential oil using DPPH, it had a relatively higher concentration than this study *Salmonella thyphimurium*. Also highlighting the difference of methods used, since its concentration, it presented higher concentrations at levels of approximately 700 more units.

Regarding the antioxidant potential of oregano essential oil, the author [74] obtained a higher value of efficient concentration than presented in our study, which highlights the data obtained satisfactory in this research. However [75], while still evaluating the antioxidant activity of *O. vulgare* essential oil, using the ABTS radical discoloration technique, a lower EC_{50} is observed than that observed in this study. According to these authors, the antioxidant potential of this oil is related to the presence of phenolic compounds, but it can also be attributed to a possible synergy between the various constituents.

When checking the antioxidant activity of *C. longa* [76], it is found that concentrations are much higher than those quantified for the essential oil using the DPPH assay, whereas, when using the ABTS assay, we exposed satisfactorily lower concentrations.

When studying the anti-inflammatory and anti-inflammatory activity of ginger essential oil [77], a much higher EC_{50} value is obtained which was presented in this study. These results are lower than that obtained in this study, and the authors attribute to this fact that the low concentration of phenolic compounds is mainly responsible for the antioxidant activity.

Essential oil	Effective concentration 50% EC_{50} (μg mL^{-1})	% ABTS inhibition (50 μg mL^{-1})
C. zeylanicum	215.93	11.11
O. vulgare	**14.00**	**90.74**
Z. officinale	308.16	25.9
R. officinalis	153.7	25.7
C. longa	173.43	14.8
C. latifolia	250	24.89

Table 6.
ABTS free radical sequestering activity by essential oils.

A relatively lower EC_{50} value was found for *Z. officinale* essential oil by the β-carotene/linoleic acid system [78]. The author attributes the antioxidant activity of the oil to the geranium and neral aldehydes, which were found in their essential oil at concentrations much higher than in this work.

5. Conclusions

These studies have shown that the essential oils of *C. zeylanicum* (cinnamon), *O. vulgare* (oregano), *Z. officinale* (ginger), *R. officinalis* (rosemary), *C. longa* L. (saffron) and *C. latifolia* (Tahiti lemon), in the chemical composition, presented a mixture of mono- and sesquiterpenes, with the major constituents being cinnamic aldehyde (46.30%), cis-p-menth-2-en-1-ol (33.8%), α-zingiberene (27.14%), camphor (37%), turmerone (55.43%) and limonene dioxide (25.92%), respectively. Results of disc diffusion showed the essential oil of *C. longa* as the oil with the highest bactericidal action independent of the bacteria strain; however, the lowest bactericidal concentration observed was the lowest concentration of *C. zeylanicum* essential oil (83.33 $\mu g\ mL^{-1}$) in *S. aureus* strains. The antioxidant activity of *O. vulgare* presented the percentage of inhibition of the 90.74% radical by 50 $\mu g\ mL^{-1}$ having the EC_{50} of 14 $\mu g\ mL^{-1}$. These results indicate that bioactive molecules present in the essential oils of the species tested presented antimicrobial activity and antioxidant action. These characteristics contribute to the control of microorganisms and help increase the shelf life of foods.

Acknowledgements

The present study was funded by Coordenacão de Aperfeiçoamento de Pessoal de Nível Superior, Brazil (CAPES) [Finance Code 001]; Fundação de Amparo à Pesquisa do Estado do Rio de Janeiro (E-26/111.252/2014), for Kátia da Silva Calabrese; by the Fundação de Amparo à Pesquisa e Desenvolvimento Científico do Maranhão (APP-00844/09 and Pronex-241709/2014); Conselho Nacional de Desenvolvimento Científico e Tecnológico, (407831/2012.6 and 309885/2017-5) for Ana Lucia Abreu-Silva and by CNPq/SECTI/FAPEMA (DCR03438/16) for Fernando Almeida-Souza; as well as the IOC (article processing charges).

Conflict of interest

The authors declare that there are no conflicts of interest regarding the publication of this article.

Author details

Amanda Mara Teles[1], Adenilde Nascimento Mouchreck[1], Gustavo Oliveira Everton[1], Ana Lucia Abreu-Silva[2], Kátia da Silva Calabrese[3] and Fernando Almeida-Souza[3]*

1 Federal University of Maranhão, São Luís, Brazil

2 State University of Maranhão, São Luís, Brazil

3 Oswaldo Cruz Institute, Rio de Janeiro, Brazil

*Address all correspondence to: fernandoalsouza@gmail.com

References

[1] Silva NCC, Fernandes Júnior A. Biological properties of medical plants: A review of their antimicrobial activity. Journal of Venomous Animals and Toxins Including Tropical Diseases. 2010;**16**:402-413

[2] Souza CN, De Almeida AC, Xavier MTR, Costa JPR, Da Silva LMV, Martins ER. Atividade antimicrobiana de plantas medicinais do cerrado mineiro frente a bacterias isoladas de ovinos com mastite. Unimontes Científica. 2017;**19**:51-61

[3] Teixeira-Loyola ABA, Siqueira FC, Paiva LF, Schreiber AZ. Análise microbiológica de especiarias comercializadas em Pouso Alegre, Minas Gerais. Revista Eletrônica Acervo Saúde. 2014;**6**:515-529

[4] Trajano VN, Lima EDO, Souza ELD, Travassos AER. Propriedade antibacteriana de óleos essenciais de especiarias sobre bactérias contaminantes de alimentos. Ciência e Tecnologia de Alimentos. 2009;**29**:542-545

[5] Morais AM, Centurião LN, De Oliveira LG, De Oliveira Aragão DM. Atividade antioxidante de especiarias in natura e desidratadas no consumo alimentar. Revista de APS—Atenção Primária à Saúde. 2018;**21**:156-157

[6] Silva AC, Iacuzio R, Da Silva Cândido TJ, Rodrigues MX, Silva NCC. Resistência antimicrobiana de *Salmonella* spp., *Staphylococcus aureus* e *Escherichia coli* isolados de carcaças de frangos: Resistência a antibióticos e óleos essenciais. Revista Brasileira de Agropecuária Sustentável. 2018;**8**:95-103

[7] Gomes EMC, Firmino AV, Pena RDCM. Efeito inibitório in vitro de extratos de *Cinnamomum zeylanicum* blume no controle de *Cylindrocladium candelabrum*. Ciência Florestal. 2018;**28**:1559-1567

[8] Oliveira TFVD. Características químicas e microbiológicas do açafrão-da-terra (*Curcuma longa*). Trabalho de Conclusão de Curso. Universidade Tecnológica Federal do Paraná; 2017

[9] Koroch AR, Simon JE, Juliani HR. Essential oil composition of purple basils, their reverted green varieties (*Ocimum basilicum*) and their associated biological activity. Industrial Crops and Products. 2017;**107**:526-530

[10] Lourenço M, Bernardi ACA, Lunardi N, Neto RJB, Bernardi PSM, Boeck EM. In vitro evaluation of the antimicrobial activity of basil (*Ocimum baslicum* L.) and coriander (*Coriandrum satirum* L.) oil extracts on *Streptococus* mutans. Journal of Research in Dentistry. 2018;**5**:40-45

[11] Silva FT. Ação do óleo essencial de gengibre (*Zingiber officinale*) encapsulado em fibras ultrafinas de proteína isolada de soja, poli (óxido de etileno) e zeína no controle antimicrobiano in situ [Dissertação de Mestrado]. Universidade Federal de Pelotas; 2018

[12] Porto LL, LRVD R. Avaliação do Potencial Antimicrobiano de Óleos Essenciais de Coentro (*Coriandrum sativum* L.) e Orégano (*Origanum vulgare* L.). Trabalho de Conclusão de Curso. Universidade Tecnológica Federal do Paraná; 2018

[13] Dutra TV, Castro JC, Menezes JL, Ramos TR, Do Prado IN, Júnior MM, et al. Bioatividade do óleo Essencial de Orégano (*Origanum vulgare*) Contra *Alicyclobacillus* spp. Culturas e Produtos Industriais. Vol. 129. 2019. pp. 345-349

[14] Santos G, Oliveira MC, Silva MM, Castro AA. Estudo comparativo

do coentro (*Coriandum sativum* L.) seco obtido em diferentes métodos de secagem. Revista Geintec-Gestão, Inovação e Tecnologias. 2012;**3**:236-244

[15] Calo JR, Crandall PG, O'bryan CA, Ricke SC. Essential oils as antimicrobials in food systems—A review. Food Control. 2015;**54**:111-119

[16] Santurio DF, Da Costa MM, Maboni G, Cavalheiro CP, De Sá MF, Dal Pozzo M, et al. Atividade antimicrobiana de óleos essenciais de condimentos frente a amostras de *Escherichia coli* isoladas de aves e bovinos. Ciência Rural. 2011;**41**:1051-1056

[17] Oliveira AR, Jezler CN, Oliveira RA, Mielke MS, Costa LC. Determinação do tempo de hidrodestilação e do horário de colheita no óleo essencial de menta. Horticultura Brasileira. 2012;**30**:155-159

[18] Milillo SR, Martin E, Muthaiyan A, Ricke SC. Immediate reduction of salmonella enterica serotype typhimurium viability via membrane destabilization following exposure to multiple-hurdle treatments with heated, acidified organic acid salt solutions. Applied and Environmental Microbiology. 2011;**77**:3765-3772

[19] Solórzano-Santos F, Miranda-Novales MG. Essential oils from aromatic herbs as antimicrobial agents. Current Opinion in Biotechnology. 2012;**23**:136-141

[20] Hintz T, Matthews KK, Rong D. The use of plant antimicrobial compounds for food preservation. BioMed Research International. 2015;ID 246264:1-12

[21] Jorge SSA. Plantas Medicinais: Coletânea de Saberes; 2017

[22] Pereira AIS. Atividade Antibacteriana e Caracterização Físico-Química de Óleos Essenciais Extraídos das Plantas Medicinais Comumente Utilizadas pela População de São Luís do Maranhão; 2017

[23] Bassanetti I, Carcelli M, Buschini A, Montalbano S, Leonardi G, Pelagatti P, et al. Investigation of antibacterial activity of new classes of essential oils derivatives. Food Control. 2017;**73**:606-612

[24] Tajkarimi MM, Ibrahim Salam A, Cliver DO. Antimicrobial herb and spice compounds in food. Food Control. 2010;**21**:1199-1218

[25] Andrade MA, Cardoso MDG, Batista LR, Mallet ACT, Machado SMF. Essential oils of *Cinnamomum zeylanicum*, *Cymbopogon nardus* and *Zingiber officinale*: Composition, antioxidant and antibacterial activities [Óleos essenciais de *Cymbopogon nardus*, *Cinnamomum zeylanicum* e *Zingiber officinale*]. Revista Ciência Agronômica. 2012;**43**:399-408

[26] Miranda CASF, Das Graças Cardoso M, Batista LR, Rodrigues LMA, Da Silva Figueiredo AC. Óleos essenciais de folhas de diversas espécies: Propriedades antioxidantes e antibacterianas no crescimento espécies patogênicas. Revista Ciência Agronômica. 2016;**47**:213-220

[27] Atarés L, Chiralt A. Essential oils as additives in biodegradable films and coatings for active food packaging. Trends in Food Science and Technology. 2016;**48**:51-62

[28] Oliveira ABAD, Paula CMDD, Capalonga R, Cardoso MRDI, Tondo EC. Doenças transmitidas por alimentos, principais agentes etiológicos e aspectos gerais: Uma revisão. Revista do Hospital de Clínicas de Porto Alegre. 2010;**30**:279-285

[29] World Health Organization (WHO). WHO Estimates of the Global Burden of Foodborne Diseases: Foodborne

Disease Burden Epidemiology Reference Group 2007-2015. 2015. p. 255. ISBN 978 92 4 156516 5

[30] Akyol I. Development and application of RTi-PCR method for common food pathogen presence and quantity in beef, sheep and chicken meat. Meat Science. 2018;**137**:9-15

[31] Vergis J, Gokulakrishnan P, Agarwal RK, Kumar A. Essential oils as natural food antimicrobial agents: A review. Critical Reviews in Food Science and Nutrition. 2015;**55**(10):1320-1323.

[32] Gebreyes WA, Wittum T, Habing G, et al. Spread of antibiotic resistance in food animal production systems. In: Dodd C, Aldsworth T, Stein RA, et al., editors. Foodborne Diseases. 3rd ed. Cambridge: Academic Press; 2017. pp. 105-130

[33] Yap PSX, Yiap BC, Ping HC. Essential oils, a new horizon in combating bacterial antibiotic resistance. The Open Microbiology Journal. 2004;**8**:6-14

[34] Unlu M, Ergene E, Unlu GV, Zeytinoglu HS, Vural N. Composition, antimicrobial activity and in vitro cytotoxicity of essential oil from *Cinnamomum zeylanicum* Blume (Lauraceae). Food and Chemical Toxicology. 2010;**48**:3274-3280

[35] Procopio FR, Oriani VB, Paulino BN, do Prado-Silva L, Pastore GM, Sant'Ana AS, et al. Solid lipid microparticles loaded with cinnamon oleoresin: Characterization, stability and antimicrobial activity. Food Research International. 2018; **113**:351-361

[36] Kačániová M, Terentjeva M, Vukovic N, Puchalski C, Roychoudhury S, Kunová S, et al. The antioxidant and antimicrobial activity of essential oils against pseudomonas spp. isolated from fish. Saudi Pharmaceutical Journal. 2017;**25**:1108-1116

[37] Echegoyen Y, Nerin C. Performance of an active paper based on cinnamon essential oil in mushrooms quality. Food Chemistry. 2015;**170**:30-36

[38] Teles AM, Rosa TDS, Mouchrek AN, Abreu-Silva AL, Calabrese KS, Almeida-Souza F. *Cinnamomum zeylanicum*, *Origanum vulgare*, and *Curcuma longa* essential oils: Chemical composition, antimicrobial and antileishmanial activity. Evidence-based Complementary and Alternative Medicine. 2019;ID 2421695:1-12

[39] Probst IS. Dissertação de Mestrado. Universidade Estadual Paulista; 2012

[40] Maia AJ, Schwan-Estrada KRF, Faria CMDR, Oliveira JSB, Jardinetti VA, Batista BN. Óleo essencial de alecrim no controle de doenças e na indução e resistência em videira. Pesquisa Agropecuária Brasileira. 2014;**49**:330

[41] Morshedloo MR, Salami SA, Nazeri V, Maggi F, Craker L. Essential oil profile of oregano (*Origanum vulgare* L.) populations grown under similar soil and climate conditions. Industrial Crops and Products. 2018;**119**:183-190

[42] Mastro G, Tarraf W, Verdini L, Brunetti G, Ruta C. Essential oil diversity of *Origanum vulgare* L. populations from southern Italy. Food Chemistry. 2017;**235**:1-6

[43] Martinez-Correa HA, Paula JT, Kayano ACA, Queiroga CL, Magalhães PM, Costa FT, et al. Composition and antimalarial activity of extracts of *Curcuma longa* L. obtained by a combination of extraction processes using supercritical CO_2, ethanol and water as solvents. The Journal of Supercritical Fluids. 2017;**119**:122-129

[44] Gopalan B, Goto M, Kodama A, Hirose T. Supercritical carbon dioxide extraction of turmeric (*Curcuma longa*). Journal of Agricultural and Food Chemistry. 2000;**48**:2189-2192

[45] Angel GR, Menon N, Vimala B, Nambisan B. Essential oil composition of eight starchy curcuma species. Industrial Crops and Products. 2014;**60**:233-238

[46] Diemer AW. Ação antimicrobiana de *Rosmarinus officinalis* e *Zingiber officinale* frente a *Escherichia coli* e *Staphylococcus aureus* em carne mecanicamente separada de frango [Dissertação de Mestrado]; 2016

[47] Andrade MA, Cardoso MG, Batista LR, Mallet ACT, Machado SMF. Óleos essenciais de *Cymbopogon nardus*, *Cinnamomum zeylanicum* e *Zingiber officinale*: Composição, atividades antioxidante e antibacteriana. Revista Ciência Agronômica. 2012;**43**:399-408

[48] Estevam E, Miranda M, Alves J, Egea M, Pereira P, Martins C, et al. Revista Virtual de Química. Composição Química e Atividades Biológicas dos Óleos Essenciais das Folhas Frescas de Citrus limonia Osbeck e Citrus latifolia Tanaka (Rutaceae). Revista Virtual Química. 2016;**8**:1842-1854

[49] Gargano AC. Dissertação de Mestrado. Brasil: Universidade Estadual Paulista; 2007

[50] Clinical and Laboratory Standards Institute. Performance Standards for Antimicrobial Disk Susceptibility Tests. 8a ed. 2003

[51] Bauer AW. Antibiotic susceptibility testing by a standardized single disk method. American Journal of Clinical Microbiology. 1966;**40**:2413-2415

[52] National Committee for Clinical Laboratory Standard. Methods for Dilution Antimicrobial Susceptibility Tests for Bacteria that Grow Aerobically. 6a ed. 2003

[53] Chao SC, Yong DG, Oberg CJ. Secrening for inibitory activity of essential oils on selected bactéria. Fungi and viroses. Journal of Essential Oil Research. 2000;**12**:639-649

[54] Silveira SMD, Cunha Júnior A, Scheuermann GN, Secchi FL, Vieira CRW. Chemical composition and antimicrobial activity of essential oils from selected herbs cultivated in the south of Brazil against food spoilage and foodborne pathogens. Ciência Rural. 2012;**42**:1300-1306

[55] Boniface Y, Philippe S, de Lima HR, Pierre NJ, Alain AG, Fatiou T, et al. Chemical composition and antimicrobial activities of *Cinnamomum zeylanicum* Blume dry leaves essential oil against food-borne pathogens and adulterated microorganisms. Research Journal of Biological Sciences. 2012;**1**:18-25

[56] Cordeiro TS. Monografia. Universidade do Extremo Sul Catarinense; 2012

[57] Ribeiro DS. Dissertação de Mestrado. Universidade Federal da Bahia; 2011

[58] Silva AA, dos Anjos MM, Ruiz SP, Panice LB, Mikcha JMG, Junior MM, et al. Avaliação da atividade óleos essenciais de *Thimus vulgaris* (tomilho), *Syzygium aromaticum* (cravo-da-india) e *Rosmarinus officinalis* (alecrim) e dos conservantes benzoato de sódio e sorbato de potássio em *Escherichia coli* e *Staphylococcus aureus*. Boletim do Centro de Pesquisa de Processamento de Alimentos. 2015. p. 33

[59] Thanh TT, Lan LX, Thu H, Tam NKM. Isolation by different processes and in vitro bioactivities of rosemary (*Rosmarinus officinalis* L.) essential oil. In: AIP Conference Proceedings. AIP Publishing; 2017. p. 020040

[60] Stefanakis MK, Touloupakis E, Anastasopoulos E, Ghanotakis D, Katerinopoulos HE, Makridis P.

Antibacterial activity of essential oils from plants of the genus Origanum. Food Control. 2013;**34**:539-546

[61] Sankar R, Karthik A, Prabu A, Karthik S, Shivashangari KS, Ravikumar V. Origanum vulgare mediated biosynthesis of silver nanoparticles for its antibacterial and anticancer activity. Colloids and Surfaces, B: Biointerfaces. 2013;**108**:80-84

[62] Soković M, Glamočlija J, Marin PD, Brkić D, Van Griensven LJ. Antibacterial effects of the essential oils of commonly consumed medicinal herbs using an in vitro model. Molecules. 2010;**15**:7532-7546

[63] Sarikurkcu C, Zengin G, Oskay M, Uysal S, Ceylan R, Aktumsek A. Composition, antioxidant, antimicrobial and enzyme inhibition activities of two Origanum vulgare subspecies (subsp. vulgare and subsp. hirtum) essential oils. Industrial Crops and Products. 2015;**70**:178-184

[64] Gupta A, Mahajan S, Sharma R. Evaluation of antimicrobial activity of Curcuma longa rhizome extract against Staphylococcus aureus. Biotechnology Reports. 2015;**6**:51-55

[65] Mishra R, Gupta AK, Kumar A, Lal RK, Saikia D, Chanotiya CS. Genetic diversity, essential oil composition, and in vitro antioxidant and antimicrobial activity of *Curcuma longa* L. germplasm collections. Journal of Applied Research on Medicinal and Aromatic Plants. 2018;**10**:75-84

[66] Singh S, Sankar B, Rajesh S, Sahoo K, Subudhi E, Nayak S. Chemical composition of turmeric oil (*Curcuma longa* L. cv. Roma) and its antimicrobial activity against eye infecting pathogens. Journal of Essential Oil Research. 2011;**23**:11-18

[67] Singh G, Kapoor IPS, Singh P, Heluani CS, Lampasona MP, Catalan CAN. Chemistry, antioxidant and antimicrobial investigations on essential oil and oleoresins of *Zingiber officinale*. Food and Chemical Toxicology. 2008;**46**:3295

[68] Grégio AMT, Fortes ESM, Rosa EAR, Simeoni RB, Rosa RT. Ação antimicrobiana do *Zingiber officinale* frente à microbiota bucal. Estudos de Biologia. 2006;**28**:61

[69] Sasidharan I, Menon AN. Comparative chemical composition and antimicrobial activity fresh & dry ginger oils (*Zingiber officinale* roscoe). International Journal of Current Pharmaceutical Research. 2010;**2**:40-43

[70] Re R, Pellegrini N, Proteggente A, Pannala A, Yang M, Rice-Evans C. Antioxidant activity applying an improved ABTS radical cation decolorization assay. Free Radical Biology & Medicine. 1999;**26**:1231-1237

[71] Wang HF. Comparative study of the antioxidant activity of forty-five commonly used essential oils and their potential active components. Journal of Food and Drug Analysis. 2010;**18**:24-33

[72] Proestos C, Lytoudi K, Mavromelanidou OK, Zoumpoulakis P, Sinanoglou VJ. Antioxidant capacity of selected plant extracts and their essential oils. Antioxidants. 2013;2:11

[73] Ramos RS. Dissertação de Mestrado. Universidade Federal do Amapá; 2014

[74] Alarcon MET. Evaluación de la actividad antioxidante del aceite esencial foliar extraido de especies de oregano (*Origanum vulgare*), oregano "borde blanco" (*Origanum vulgare* ssp.) y oreganito (*Lippia alba*) cultivado en la zona norte del departamento de bolívar (colombia). Dissertação (Mestrado). Universidade Nacional da Colômbia; 2014. p. 131

[75] Babili FE, Bouajila J, Souchard JP, Bertrand C, Bellvert F, Fouraste I, et al. Oregano: Chemical analysis and evaluation of its antimalarial, antioxidant, and cytotoxic activities. Journal of Food Science. 2011;**76**:C512-C518

[76] Stanojević JS, Stanojević LP, Cvetković DJ, Danilović BR. Chemical composition, antioxidant and antimicrobial activity of the turmeric essential oil (*Curcuma longa* L.). Advanced Technologies. 2015;**4**:19-25

[77] Jeena K, Liju VB, Kuttan R. Antioxidant, anti-inflammatory and antinociceptive activities of essential oil from ginger. Indian Journal of Physiology and Pharmacology. 2013;**57**:51

[78] Andrade MA. Dissertação de Mestrado. Lavras: Universidade Federal de Lavras; 2010

Chapter 10

Essential Oils and Microbial Communication

Filomena Nazzaro, Florinda Fratianni, Antonio d'Acierno, Raffaele Coppola, Fernando Jesus Ayala-Zavala, Adriano Gomez da Cruz and Vincenzo De Feo

Abstract

The World Health Organization highlighted the increase in the resistance to conventional antibiotics for most pathogens and observed also a decrease of the threshold for all mechanisms of cell-cell microbial communication, leading to the formation of biofilms and to the increase of microbial pathogenicity. Scientific community is therefore oriented to the identification and study of alternative substances to antibiotics. In such context, substances of vegetal source, such as essential oils (EOs), always used in traditional medicine, stimulated—particularly in recent decades—the scientific world to discover and identify substances, intended as a mixture or single components capable to fight pathogenic microorganisms. From this point of view, the study of plants is very interesting and offers many interesting ideas and results. This brief chapter describes the basis of the microbial communication, until the formation of biofilm, and some mechanisms through which essential oils, or some of their main components, may decrease or inactivate the complex mechanisms that lead to pathogenicity, both of prokaryotes and eukaryotes.

Keywords: bacterial resistance, quorum sensing, biofilm, essential oils

1. Bacterial resistance

In recent years, the World Health Organization repeatedly highlighted with alarm the problem of an increase in the resistance of most pathogens to conventional antibiotics. Several causes determined such alarming picture, not least the lack of availability on the market of "new" molecules (given the low economic appealing that such a study raises on the pharmaceutical industries). Wrong behaviors on the part of man are also included among the recurring causes, such as an unjustified abuse of antibiotics [1], as well as possible incorrect medical prescriptions of the drug or the duration of antibiotic treatment. An extensive and indiscriminate use of antibiotics also in agriculture and livestock breeding can cause an indirect contribution to antibiotic resistance also in humans indeed. The often inconsiderate use of broad-spectrum antibiotics could indiscriminately reduce also the number of the so-called "commensal" microorganisms, favoring the onset of diseases more serious than those for which the use of the drug was initially required, varying and consequently altering the relationship between

 IntechOpen

microorganism and host. Microorganisms that generally do not cause diseases in their natural habitats, due to this new environmental situation, can become highly pathogenic. Normal constituents of the intestinal flora, such as *Escherichia coli*, may therefore become harmful in other districts, such as the urinary bladder, spinal cord, lungs, etc. Other microorganisms can become highly pathogenic under certain conditions: for instance, *Streptococcus viridans* physiologically present in the oropharyngeal tract, in some circumstances can invade different organs through the bloodstream, causing serious diseases (e.g., bacterial endocarditis). Today, there is much talk about the so-called "multidrug-resistant" (MDR), "extensively drug-resistant" (XDR), and "pan-drug-resistant" (PDR) strains. Such microorganisms can be figuratively enclosed in a cluster comprising pathogens of infections that are today intractable [2]. Unfortunately, it is also difficult to fight those pathogens belonging to the so-called "ESKAPE" group, an acronym comprising the microbial species *Enterococcus faecium*, *Staphylococcus aureus*, *Klebsiella pneumoniae*, *Acinetobacter baumannii*, *Pseudomonas aeruginosa*, and *Enterobacter* spp. It is widely believed by the scientific community that the study of alternative strategies to the use of conventional antibiotics could represent an important way to be taken into consideration, to combat this dangerous situation. The use of bacteriophages in phage therapies, known for their high specificity, the development of new vaccines against *P. aeruginosa* [3] and *A. baumannii* [4], and the use of strategies to inhibit the bacterial virulence factors can be considered some of the solutions. Recently, research also focused on the exploitation and identification of new microorganisms, isolated, for example, from the ground, enabling to block the microbial growth of one or more species belonging to the ESKAPE group [5].

2. Mechanisms of cell-cell communication

Multicellular organisms are composed by a rigidly regulated society of individual cells, organized into tissues and organs, which all together collaborate for the functioning of the individual and whose final "purpose," from the biological point of view, is to reproduce (or to allow to the reproduction of a similar genome). The coordinated work of the different cell types that leads to the formation of an adult individual, as well as the cell growth, differentiation, and organogenesis giving rise from a single fertilized cell, requires sophisticated signaling mechanisms. Thus, in the course of evolution, molecular messengers were generated, synthesized, and released in some part of the organism and then specifically recognized by the respective receptors expressed in the target cells. Complex molecular machines were simultaneously selected to transduce the activated receptor signal. While generally the term "extracellular messengers" involves those intercellular communication mechanisms taking place within a multicellular organism, it should be emphasized that even unicellular microorganisms are capable of "social" behaviors that require a coordinated response. This sophisticated cell-to-cell process of communication between microorganisms, the so-called quorum sensing (QS), consists in the synthesis by bacteria, both Gram-positive and Gram-negative, of specific molecules, which are called "autoinducers" or "bacterial pheromones." After production of such molecules, bacteria release them into the extracellular medium to be detected by specific receptors/transducers. Quorum sensing is an extremely important communication system for microorganisms. Through this system, bacteria are in fact capable to measure their concentration and to modulate gene expression in response to population density, which lead to the secretion of virulence factors, biofilm formation, competence, and bioluminescence [6, 7]. When bacteria that generate signals are in close proximity to each other, the concentration of their

QS signal amplifies. This event leads to a boost of the binding of the QS signal to specific receptor proteins, to a consequent activation of the specific receptor, and to the enhanced gene transcription with appropriate promoter sequences. QSs give to bacteria a great evolutionary advantage, allowing them to adapt to the change of the environment. Some authors propose these as neo-Darwinian mechanisms of evolution, which had an important function in the arrival of the first multicellular organisms [8]. The result of this "bacterial communication" can be represented by an increase in virulence (e.g., *Staphylococcus aureus*), by the formation of a biofilm (e.g., *Pseudomonas aeruginosa*), by sporulation, etc. To date, more than 100 different autoinducers are known for bacteria, archaea, and fungi.

Bacteria exhibit two main QS mechanisms, based on distinct signaling pathways, which present a certain analogy with the mechanisms found also in multicellular organisms. The first, generally used by Gram-negative bacteria, is based on the synthesis of a family of small molecules, the so-called AHL (acylated homoserine lactones). These molecules have a similar central structure and differ only in the length of a side chain, which specificity is determined by the length of the acyl chain and the substitution (—H,—OH or =O) on carbon. Generally, every type of bacterium can produce at least one AHL type; however, it can happen that bacteria produce more than one of them. Due to their chemical characteristics, AHL are capable to cross the bacterial membrane and spread outside; in addition, from the extracellular medium, they can freely enter within the cell and bind to specific receptors, called LuxR because the QS phenomenon was described for the first time in a microorganism, *Aliivibrio fischeri*, which is able to emit light *in vitro* only when its concentration exceeds a certain threshold [9]. In this microorganism, the AHL autoinducer is at low concentration and does not induce bioluminescence; when the bacterium is in the luminous organ of the giant squid, its cell density is high, so the transcriptional activator reaches the DNA, binds to the recognition sequence (LuxBox), and activates the transcription of genes for the enzyme luciferase, which produces bioluminescence. The advantage of this mechanism is to save energy, to ensure that bacteria become luminescent only when they are present in large numbers, and to prevent them from wasting energy when the population is toosmall to emit a visible signal. The complex AHL-LuxR in turn binds to the bacterial DNA, thus regulating the transcription of specific genes. In a situation where the concentration of bacteria is low, the level of synthesized AHL is below the threshold required for the LuxR bond. However, as the concentration of bacteria increases, the amount of AHL also increases, and the AHL-LuxR complex is formed accordingly. A gene encoding for the enzyme catalysing the AHL synthesis is present within the genes induced by the AHL-LuxR complex. This gives rise to a positive feedback leading to a rapid and synchronous answer from the whole microbial population. Signal transduction through the AHL-LuxR system, based on an extracellular messenger able to cross the membrane and on a receptor that is also a transcription factor, can be thought to obey the same logic of signaling through steroid hormones. On the other hand, Gram-positive bacteria use autoinducer molecules formed by peptides with a variable length ranging between 5 and 17 amino acids. Such molecules are produced by the processing of precursors and are often subjected to posttranslational modifications. These peptides require special transporters to be secreted in the extracellular environment that, in turn, is detected by a sensor histidine kinase [10] and transduces the signal through phosphorylation of intracellular targets. This mechanism of action is therefore similar to that used by growth factors in multicellular organisms. Not always, the nature of quorum sensing molecules (QSMs) is peptidic: for instance, some Gram-positive bacteria, such as *Streptomyces*, produce γ-butyrolactones as QSMs [11]. Finally, different researches report the autoinducer 2 (AI-2), with a rather unusual cyclic boronic ester, as a QS system common to both Gram-positive and Gram-negative bacteria, although its role as a true QSM has been doubted for some microorganisms [12–15]. This system might give rise to a family of molecules that are supposed to operate as a common language for most bacteria.

The production of the AHL involved in the QS mechanism was recently discovered also in several Gram-negative bacteria, such as *Roseobacter* sp. TB60 and *Psychrobacter* sp. TB67 associated with the Antarctic sponge, *Anoxycalyx joubini* [16], indicating this a certain "universality" of the system. Dong and Zhang [17] and McDougald and co-workers [18] demonstrated the existence of other two novel signaling pathways: hydroxyl-palmitic acid methyl ester and methyl-dodecanoic acid.

2.1 QS in eukaryotes

Some eukaryotic microorganisms monitor their population density through QS mechanisms [19, 20] too. This is not so surprising, taking into account that many bacteria and eukaryotic microorganisms inhabit in common ecological niches and often play similar challenges. In fungi, QS mechanisms are in charge to check and regulate processes such as sporulation and production of some molecules, such as secondary metabolites, as well as to those events giving rise to the morphological transition and enzyme secretion by the cells. Considering this and starting from the assumption that even this type of organisms is extremely varied, we can undoubtedly affirm that fungi exhibit different cell-cell communication mechanisms, using a wide variety of signal molecules [19]. Furthermore, fungi can communicate with bacteria and even with their plant or mammalian hosts. In yeasts and dimorphic fungi, aromatic alcohols originating from amino acids mediate the QS type of regulation [21]. Therefore, yeasts, through the production of tryptophol and phenylethyl alcohol, can manage the formation of pseudohyphae and biofilm [22] and probably trigger the virulence process toward some plants such as *Vitis vinifera* [23]. QSMs stimulate the exit from lag phase inducing germ-tube formation and hyphal development [24]. *Candida albicans* remains the most studied species from this point of view. It produces some QSMs, such as tyrosol, described also in other fungal species, such as *Saccharomyces cerevisiae* [25]. QSMs of *C. albicans* influence the formation and structure of biofilms [26, 27] as well as the dispersal of cells from a biofilm; hence, it, as well as other molecules, plays important roles in pathogenesis. E-farnesol, the other most known QSM produced by *C. albicans* [28], is an exogenous molecule that, on the contrary, inhibits biofilm formation when provided early during adherence; furthermore, it acts as an inhibitor of hyphal formation indeed [29]. Therefore, this organism can modulate its morphology (vegetative/hyphal) and, consequently, all events related, including the pathogenicity, through the modulating production mainly of these two QSMs. Dodecanol and γ-butyrolactone are other molecules identified as mediators of QS processes present in other eukaryotic organisms, such as the filamentous fungi, *Aspergillus* and *Penicillium* spp. Some species of *Penicillium*, such as *P. sclerotinum*, produce sclerotiorin, a secondary metabolite with antibiotic properties, and γ-butyrolactone-containing molecules such as multicolic acid, which act as QSMs [30]. Taking into account that Gram-negative bacteria produced lactones (AHLs) as QSMs and that filamentous fungi produce butyrolactone I [31], γ-heptalactone [32], and γ-butyrolactones [33], the discovery that γ-butyrolactones are produced also by the filamentous bacterium *Streptomyces* [11] suggests a convergent evolution or a horizontal gene transfer occurring during the evolution [30]. At the same time, different fungi, such as basidiomycete *Cryptococcus neoformans*, produce as QSMs some peptides, similar to how Gram-positive bacteria do. This means that fungi use a language "analogous" in some way to that exhibited by other phyla [10, 30].

In some species of *Aspergillus*, such as *A. flavus*, oxylipins were identified as QSMs: these molecules modulate both the morphological differentiation and the production of either asexual spores or sclerotia. Furthermore, oxylipins regulate a QS-dependent pathway controlling development and mycotoxin production [34]. Fungi produce also other QSMs: terpenes, such as farnesol, are produced, for instance, by the dimorphic fungi *C. albicans* [28] and *Ophiostoma piceae* [35]; cyclic sesquiterpenes act as QSMs for the dimorphic fungus *Ophiostoma floccosum* [36]; QS alcohols, including tryptophol and phenylethyl alcohol, are produced by *S. cerevisiae* [21]. It is important to underline that the higher organisms evolved mechanisms with which they are capable to interfere with the quorum sensing process of the bacteria. These mechanisms could play an important role both for peaceful cohabitation of human and microorganisms, such as the case of the intestinal flora, and in processes of resistance to pathogens.

2.2 Biofilm

The term "biofilm" is referred to a structure, enough complex, formed by microbial cells, associated with each other that attach to a surface, which are in a certain sense kept isolated from the external environment (although they exert an important influence on this) through the formation of a sort of "dome" of polysaccharide nature [37]. Biofilm has generally a three-dimensional structure: it contains more or less channels and pores, used as a sort of intercellular communication channel and for the maintenance of the entire bacterial community [38, 39]. Biofilm is thus one of the subsequent mechanisms giving rise from the communication among bacteria, which precisely through the formation of biofilm and other microbial behaviors and social exchange (not only bioluminescence, conjugation, and virulence but also motility, sporulation, competence, etc.) form the social microbial system of interaction, the QS (**Figure 1**) [40, 41], that prokaryotes and some eukaryotes developed many millions of years before the actual human social media, which is certainly more organized and complex. The system is so organized and evolved that allows microorganisms to easily adapt to adverse environmental conditions and to use them even to switch to counterattack, with an action of growth, proliferation, and change in their metabolic pathways and morphology. Biofilms allow the survival of bacterial cells in a hostile environment; the extremely

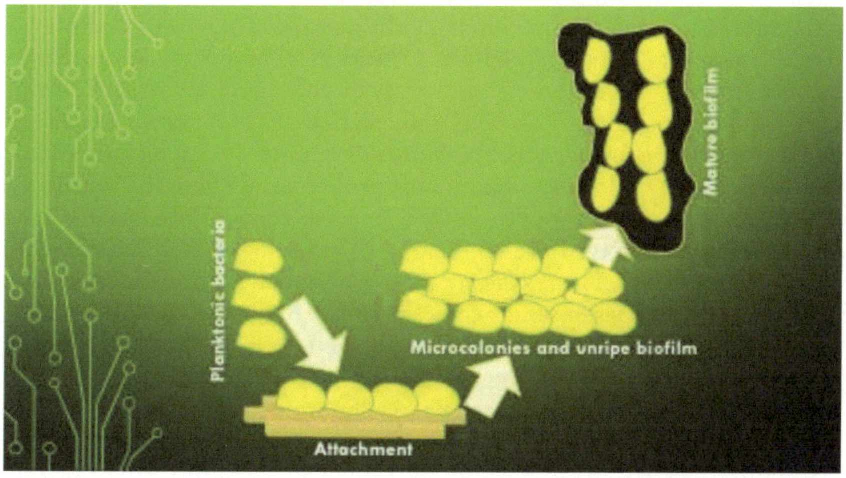

Figure 1.
Mechanism of biofilm formation.

complex structure and the metabolic and physiological heterogeneity that characterize them suggest an analogy between these communities and the tissues of higher organisms. Bacterial biofilms, not easily eradicated with the conventional antibiotic therapies, affect a large number of chronic bacterial infections. Biofilms represent a cohesive matrix of microorganisms and other cellular constituents that can be present in any natural environment; they are also characterized two pints by the ability to adhere to surfaces; by a structural heterogeneity; by a genetic diversity of the components; by complex interactions of communities, even mixed; and by an extracellular matrix of polymeric substances. At the end of the 1990s, it was ascertained that the so called planktonic growthis an artifact and that the type of growth prevalent in natural environments is sessile (fixed to a substrate). When nutrient intake is limited, biofilms tend to adhere to solid supports and remain stable at the solid/liquid interface, where nutrients are concentrated. Once adhered, the biofilms secrete exopolysaccharides that surround them, guaranteeing their cohesion to the support and between them. This creates biofilms, which in most cases are polymicrobial. Bacteria grow slowly inside the biofilm, forming microcolonies. The biofilms are mature when the growth reaches the point where most external cells come off, returning to the planktonic life and then starting the formation of new biofilms. The whole process takes from a few days to a few weeks. In mature biofilms, bacteria are present in different states, depending on the location: the innermost ones are metabolically less active, and the more external have metabolic characteristics similar to those of planktonic growth bacteria. At first, sessile growth attracted attention due to some negative effects of biofilm formation (corrosion of cables and submerged structures but also the dental plaque of many animals) and of the resistance of the bacteria included in the biofilm to the antimicrobials. Studies carried out with the confocal microscope have shown that the biofilm is highly organized within it, with channels through which the surrounding fluid circulates in the matrix carrying the nutrients and removing the toxic products. Maintaining a structure of this type requires complex mechanisms of cell-cell regulation and communication to prevent undifferentiated growth obstructing the canaliculi. The phenomenon of sessile growth has determined, in the last decade, the onset of new pathologies, linked to the colonization of prosthetic implants by bacterial biofilms [42]. Biofilms are present in the most diverse environments, e.g., in thermal springs or on the bottom of lakes and rivers, and can be used not only for the purification of water in an industrial environment but also for the removal of oil or other pollutants from contaminated marine areas. Moreover, it is now established that most bacterial species, when conditions allow it, modify their behavior to find true "microbial cities" in the form of biofilms. These include "fortification walls," consisting of a three-dimensional array of polymeric sugars, and "shipping channels" for the transport of nutrients and catabolites. Two main types of adhesion are involved in the formation of biofilm: *adhesion* of the bacterium to a solid substrate (supra inert) also by attacking host proteins and *intercellular adaption*, which determines the formation of multiple layers of the biofilm. In the biomedical field, biofilms are involved in a wide range of pathologies, involving cochlear, articular, orthopedic, etc. The sessile structures of which biofilms are endowed, the multi-species communities of which they are constituted, and the influence that the dynamics of fluid flows in which they are immersed exert on them are the factors that have contributed to considering biofilms as the core reefs of the microbial world. Obviously, the multicellularity of a biofilm translates into a better defense of microorganisms, contributing substantially to their survival. Nutrient depletion creates some areas of activity alteration; the outer cell layers of the biofilm contribute to the formation of a sort of barrier, capable to absorb external damage. The innermost microorganisms have the task, to some extent, to elaborate a response to intrinsic stress. The biofilm bacteria are 10–10,000 times more resistant to antibiotic treatment than the planktonic phenotype. For this reason, biofilm infections show recurrent symptoms after cycles of antibiotic therapy. This persistence in the heart of the biofilm is linked to the presence of the so-called "sleeper" microorganisms, with low activity and which determine the phenomenon of persistence.

Biofilm adapts to environmental fluctuations, such as temperatures, pH variations, osmolarity, and nutrient availability through multiple gene expression; its resistance is not genotypic. Microbial cells contained within the biofilm are much more difficult to reach; moreover, they have the advantage, compared to the host organism, of being able to communicate outside the biofilm, through the previously indicated system of channels and pores. At the same time, it becomes more difficult for synthetic drugs to "break" the biofilm organization, just for how it is structured and how it is composed. After a certain threshold, bacteria change their life perspective, in the sense that they no longer act as a single cell, but as a component of a microbial team. Such community grows through the recruitment of other cells, which arrive there. In this manner, microbial colony can spread upward of the surface. At the beginning, this gives rise to the formation of small colonies and unripe biofilm. At the end of the process, the biofilm is ripe (**Figure 1**). The production of compounds such as exopolysaccharides determines the embedding of bacteria in a complex matter constituted also by nucleic acids, lipids, and proteins [43]. A so organized matrix supplies bacteria for several advantages: for instance, it can manage the flow of nutrients and protect bacteria against the action of antimicrobial substances and the host immune system, which encounter great difficulty in scratching the structure and organization of the biofilm matrix [44]. So, manipulation and inhibition of the QS system might open new scenarios and improve therapies for chronic bacterial diseases [45, 46], including even cancer [47, 48].

3. Essential oils

Several possible strategies could treat infections associated with biofilms: substances capable of destroying the biofilm matrix (e.g., dispersion B), substances capable of destroying resistant cells, quorum-quenching enzymes that interfere with the quorum sensing phenomenon, substances that cause self-destruction of the biofilm, and then, in particular, strategies to strengthen the action of antimicrobials. The treatment of biofilms with antibiotics often causes only partial killing, allowing the surviving bacteria, present in the depth of the biofilm, to act as a true nucleus of propulsion for the spread of the infection after the interruption of the antibiotic therapy. Antibiotics can be inactivated by the production of specific enzymes within the biofilm. In some extreme cases, even the sessile population must not be surgically removed from the body. Another aspect to take into consideration is the age of biofilm: the younger is the biofilm, the easier is its eradication.

The need to identify substances/active ingredients able to replace synthetic drugs in the fight against pathogens, in particular against those more resistant to conventional treatments, also directed research toward (or better to say, to the rediscovery of) the "natural world," source of bioactive compounds used by traditional medicine since ancient times. Moreover, these substances have always exhibited a great spectrum of action that can be considered of great benefit, also due to the chemical structural differences of the active compounds. In such context, substances of vegetal origin, such as essential oils, have always been successfully used in traditional medicine and stimulated, practically always but particularly in recent decades, the scientific world to discover and identify substances, intended as a mixture or as single components that are able to fight pathogenic microorganisms. From this point of view, the study of plants is very interesting and offers many interesting ideas and results: the same kind of plant can provide a pool of substances with a wide and very diverse spectrum of action [49]. Within the same genus, in fact, there are species with a different chemical composition, which therefore can provide bioactive substances (hydroalcoholic fraction or essential oils) different for the qualitative and quantitative profile. Moreover, the same plant species can diversify and present a different chemical composition depending also on the environmental and climatic conditions in which it grows, the stage of maturity, and method of extraction. Essential oils are substances that appear liquid, aromatic, and limpid and are obtained from different portions of plants through different extraction procedures, such as crushing, distillation, fermentation, the so-called enfleurage, or the use of organic solvents. The International Organization for Standardization (ISO) (ISO/D1S9235.2) defines an essential oil as a product made by distillation with either water or steam or by mechanical processing or by dry distillation of natural materials. About 300 essential oils, within the more than 3000 known types, are available on the market. The antimicrobial and antifungal properties of essential oils have been known since ancient times; however, the first "scientific" demonstrations of this activity date back only to the 1950s, when both Guenther [50] and Boyle [51] described in detail the activity of natural preservatives exhibited by different essential oils derived from plants and spices. The increase in interest from the economic world has meant that, by increasing the research on these substances, other properties were discovered [52], among which, for example, those antivirals [53, 54]. Chemical characterization of essential oils, conducted through chromatographic approaches (GC and GC/MS), has allowed obtaining detailed information on their composition. Essential oils are generally formed by volatile substances, also called volatile organic compounds (VOCs), molecules characterized by a high lipophilicity and a high vapor pressure. Within each essential oil, one can identify one or more quantitatively more abundant molecules and a series (more or less numerous) of other molecules, sometimes present only in traces. Moreover, within the essential oils, other types of molecules can be identified, such as phenolic compounds [55], alkaloids, saponins, and sesquiterpenes, which contribute to the antimicrobial activity of the oil. Further than other properties, EOs protect the plant against some pathogenic microorganisms. Through their smell, they are capable of exercising repulsive action against insect or, concurrently, to attract others to favor the dispersion of pollens and seeds. The same smell can also negatively affect the appetite drive of some herbivores. Some essential oils are reported to be very effective allelopathic agents. Thus, EOs can play a role in mediating the interactions of plants with the environment in a way that, although improperly, we could almost define as similar (however in a certain way opposite) to that exhibited by QSMs that allow these last to communicate in microorganisms with each other and with the environment. Essential oils can be classified according to the chemical constituent contained in greater concentration. Following such criterion, we have, among others:

- EOs with a predominance of mono or sesquiterpene hydrocarbons (e.g., *Citrus*, *Juniperus* L)
- EOs with a prevalence of aldehydes (e.g., cinnamic aldehyde in *Cinnamomum verum*, benzoic aldehyde in *Prunus dulcis* var. *dulcis* and in var. *amara*)
- EOs containing predominantly alcohols (geraniol in *Geranium*, santalol in *Santalum album*, linalool in *Coriandrum sativum*)

- EOs with high content of ketones (carvone in *Carum carvi*, thujone in *Artemisia*, *Thuja*, and in *Salvia officinalis*)
- EOs with a predominant amount of phenols (eugenol in *Dianthus caryophyllus*, carvacrol in *Satureja* and in other Labiatae)
- EOs that have a prevalent content of sulfured compounds (bisulfide, allyl disulfide in *Allium*)
- EOs with a prevalent content in esters and alcohols (linalool and linalyl acetate in *Lavandula angustifolia*)
- EOs having predominantly peroxides (ascaridol in *Chenopodium*)

The composition and the relative differences among the EOs lead to different biological activities that EOs can exhibit [54]. This also means that some species of plants, which exhibit different chemotypes, are characterized by a different composition and rate among the EO components too, to lead a change in EO biological properties. Thus, the final effect of an EO against a specific pathogen can give rise from a synergistic mechanism of its components or from just a unique compound that, although present in less percentage, can enhance the antimicrobial activity of the entire EO. In general, plant EOs and their components have a broad spectrum of inhibitory activities both against Gram-positive and Gram-negative pathogens [56, 57]. Citronellol can exert a broad inhibitory activity against the formation of biofilm. In fact, it acts against the planktonic forms of different Gram-positive (*Listeria monocytogenes*, *Staphylococcus aureus*, *Staphylococcus epidermidis*) and Gram-negative species (*E.coli*, *Pseudomonas aeruginosa*), inhibiting their capability to form biofilm when used at percentage ranging between 3.5% and 7% wt [58, 59]. However, the antibacterial effectiveness can vary depending not only on the EO but also on the bacteria. Therefore, some EOs, such as those of sandalwood (*Santalum album*), manuka (*Leptospermum scoparium*), and vetiver (*Chrysopogon zizanioides*), can act against some Gram-positive bacteria, but result ineffective against Gram-negative [60, 61]. Concurrently, sa microorganism can be more or less to the activity of different EOs. Thus, *Cymbopogon citratus* (lemon grass), *Syzygium aromaticum* (clove), and *Laurus nobilis* (bay laurel) EOs as well as *Thymus vulgaris* (thyme), *Rosmarinus officinalis* (rosemary), and *Mentha piperita* (peppermint) ones are capable to act against *St. aureus* at concentrations of $\leq 0.05\%$. On the other hand, *Ocimum basilicum* (basil) and *Eucalyptus globulus* (eucalyptus) EOs exhibit the same level of activity against this microorganism only if used at 1% concentration [60–62]. Different types of cardamom EOs affect differently the growth, the Qs, and the capability to form biofilm of several bacteria [63]. Thyme, *Origanum vulgare* (oregano), *Melaleuca alternifolia* (tea tree), *Cinnamomum verum* (cinnamon), lemon grass, bay laurel, *Backhousia citriodora* (lemon myrtle), clove, and *Aniba rosaeodora* (rosewood) EOs result the most active antimicrobials, at concentration and MICs also less than 1% [60, 64]. Most of these EOs, in particular bay laurel, clove, lemon grass, oregano, and thyme oils, inhibit growth of *E.coli* at concentrations of 0.02, 0.04, 0.06, 0.05, and 0.05%, respectively. In few cases, a major constituent molecule could exhibit a more effective activity compared to the EO. For example, carvacrol and eugenol, present in clove EO or terpinen-4-ol, which is the main component of tea tree EO, can show greater efficacy than the relative oil. This highlights also that, in the study of a biological activity of an EO, it is important to provide also the chemical composition, so to best argue about [65, 66].

4. Mechanisms of essential oils on microorganisms

As above indicated, the mechanisms which allow EOs to damage bacteria are largely dependent on their composition. Usually, antimicrobial activity can originate from a flow of reactions implicating the total bacterial cell; this is essentially due to the fact that, since the EOs are composed of many groups of chemical compounds, these last act in different ways [30]. Generally, Gram-positive bacteria and Gram-negative bacteria are differently susceptible to the action of EOs, due to the structural differences of their cell wall of these two groups of bacteria. The higher susceptibility of Gram-positive bacteria is caused mainly by the presence of peptidoglycan within their cell wall, which allows more easily the hydrophobic molecules to have access within the cell, acting therein with cytoplasm [30]. The cell wall of Gram-negative bacteria shows an outer membrane, composed of a double layer of phospholipids linked to the peptidoglycan layer by lipopolysaccharides. This allows these bacteria to exhibit greater resistance to the penetration of essential oils and/or their components; in fact, some hydrophobic molecules can be capable to enter into the cell, only through the access given by the porins, proteins that form water-filled channels distributed all over the cell wall. The different compositions of cell wall let it that Gram-negative bacteria are even more resistant to hydrophobic antibiotics [30, 67]. The mechanism through which the EOs or their components act on microbial cell is well known: it includes one or more simultaneous actions, ranging from cell wall degradation, to the damage caused to the cytoplasmic membrane and membrane proteins, as well as to a reduction of the proton-motive force until to damage to the ATP synthesis mechanism. Lipophilic character of EO compounds allows them to penetrate the cell membrane and remain between the phospholipids and/or affect the synthesis of membrane lipids, with a consequent change of membrane structure and with an alteration of its permeability. In addition, EOs can affect directly also the morphology of bacterial cell, altering it even irreversibly, to cause the complete destruction of the entire microbial cell scaffolding [30, 68] (**Figure 2**).

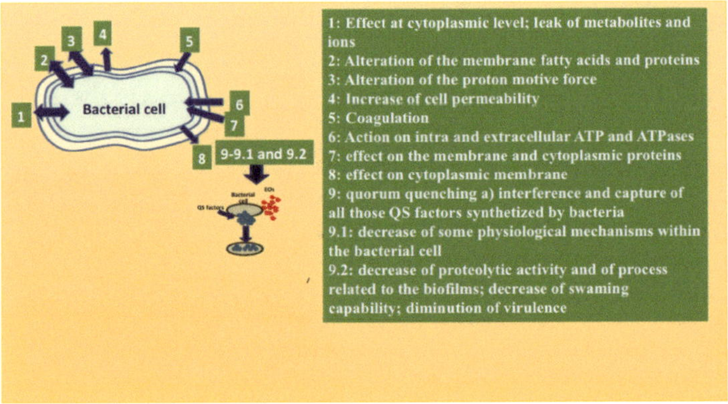

Figure 2.
Effect of essential oils on microbial cells (modified from [65]).

5. The action of the essential oils on quorum sensing system and biofilm formation

EOs can act also on QS systems that coordinate the whole system of pathogenicity of bacteria [30, 69] (**Figure 2**). This property is of noticeable interest, due to the continuous research for new therapeutic and antibacterial agents, which could concurrently act in no toxic manner and without encouraging the development and emergence of resistant bacterial strains [45]. EOs can work on one or more events regulating the entire quorum sensing activity of microorganisms. Summarily, bacterial QS may be inhibited through different mechanisms. Their action against Gram-negative bacteria can be mainly expressed in three basic steps: a first step can block the "upstream" mechanism through the synthesis of AHL; a second mechanism may act further downstream, blocking the AHL transport and/or secretion. If bacteria still manage to produce AHLs and these molecules are still transported and secreted outside, other EO or their components could in any case be able to "capture" these molecules, effectively preventing cell-cell communication between bacteria. Other EOs can therefore act by exhibiting an antagonistic action with respect to AHLs or operating an inhibitory effect downstream of AHL receptor binding [49]. The versatility of action of EOs depends essentially on their chemical composition and the presence of functional groups. EOs containing largely terpenes (*p*-cymene, limonene, terpinene, sabinene, and pinenes) as well as some oxygenated components (for instance, camphor and camphone, borneol and bornyl acetate, 1,8 cineole, α-pinene, and verbenonone) generally do not exhibit a so strong antibacterial activity, which is just more manifest against Gram-positive bacteria. Further than the composition, the antimicrobial activity of EOs is also due to their concentration. In fact, depending on such element, EOs or their components can operate in a different manner on one or more factors that affect the mechanisms of cell-cell communication among bacteria. Thus, some EOs, also at low concentration, are capable to impede the chemical activity of those enzymes involved in the production of energy for the survival and growth of bacteria or, at higher concentration, even to disaggregate and denature microbial proteins [30, 70]. Subinhibitory concentrations of clove EO, tested on *P. aeruginosa* and *Aeromonas hydrophila*, were capable to significantly reduce the las- and rhl-regulated virulence factors such as LasB, total protease, chitinase and pyocyanin production, swimming motility, and exopolysaccharide production. The biofilm-forming capability of these two strains was also reduced in a concentration-dependent manner at all tested sub-MIC values [71]. Peppermint EO at sub-minimum inhibitory concentrations (sub-MICs) strongly can interfere with the production of AHL-regulated virulence factors and biofilm formation in *P. aeruginosa* and *A. hydrophila*. Such effect is mainly due to the presence of menthol, which interferes with QS systems of various Gram-negative pathogens, acting essentially on the las and pqs QS systems [72]. Different bacterial strains are used to test the potential inhibitory effect of essential oils on QS. Apart from the well-known models (*Vibrio harveyi*, *P. aeruginosa*, *S. aureus*, *E. coli*) more recently *Chromobacter violaceum*, in particular the mutant strain CV026, has been also used with this scope. This strain can provide, through the production or not of its purple pigment violacein, directly linked to QS, useful information about the capability of a substance to act or do not act as quorum-quenching agent, respectively. Through the use of such approach, Szabo and co-workers [73] studying several EOs ascertained that, among some EOs, rose, geranium, lavender, and rosemary EOs were the most potent QS inhibitors. Eucalyptus and citrus oils were moderately active, while the chamomile, orange, and juniper EOs which did not show any were ineffective. In several cases, the synergistic effect of more components can enhance the capability of an EO to inhibit the mechanism of communication among bacteria. Khan and co-workers evaluated the capability of different EOs and of their main components to act as quorum-quenching agents. Their study evidenced that clove essential oil showed promising anti-QS activity, followed in activity by cinnamon, lavender, and peppermint oils, and that eugenol, the major constituent of clove oil, could not exhibit anti-QS activity [74]. In other cases, the effectiveness of EOs is related both to their composition and to the bacterium of reference; thus, an EO can act as mixture, better than a singular component on a specific bacterium; therefore, one or more components can act better than parent EOs against another bacterium. The effect of clary sage, juniper, lemon, and marjoram EOs and their major components on the formation of bacterial and yeast biofilms and on the inhibition of AHL-mediated QS, evaluated using *Bacillus cereus*, *Pichia anomala*, *Pseudomonas putida*, and a mix of bacteria containing also *E. coli*, demonstrated that marjoram EO inhibited all these tester strains. However, all components exhibited more strength in limiting the biofilm capacity of *B. cereus* than the parent EOs. Lemon EO was capable to inhibit *E. coli* and mixed-culture biofilms; on the other hand, cinnamon was effective against the mixed forms [75]. Conversely, the entire EO of tangerine (*Citrus reticulata*) is capable to inhibit the *P. aeruginosa* biofilm formation more than its main component limonene, by an inhibition of the QS autoinducer production and elastase activity [76]. This also highlights how, within a same genus, not all the species show the same biological activity. Thus, the EO of *C. reticulata* (tangerine) can be more active in inhibiting the QS system; on the other hand, the EO recovered from orange (*Citrus sinensis*) can be completely ineffective [73]. Some terpenoids, for example, thymol, carvacrol, linalool, menthol, geraniol, linalyl acetate, citronellal, and piperitone, have antibacterial activity mediated by their functional group. Carvacrol is one of the most active components present in different EOs, in particular from Labiatae. Its spectrum of activity is much wide. At sublethal concentrations (<0.5 mM), it is capable to inhibit the formation of biofilms of *C. violaceum*, *Salmonella enterica* subsp. *typhimurium*, and *S. aureus*, while it does not exhibit effects on the formation of *P. aeruginosa* biofilms. In all cases, this concentration seems to not have effects on total bacterial numbers, indicating that carvacrol bactericidal effect could not be also linked to its inhibitory effect on biofilm formation. Sub-MIC concentrations of carvacrol could reduce the expression of cviI (a gene coding for the N-acyl-L-homoserine lactone synthase) and decrease the production of violacein and the activity of chitinase (both regulated by quorum sensing) at concentrations coinciding with carvacrol's inhibiting effect on biofilm formation. These results indicate that carvacrol activity in inhibition of biofilm formation might be also related to the disruption of quorum sensing [77]. Thymol, one of the main constituents of *Thymus vulgaris* EO, can affect (at the same manner of the parent EO) not only the AHL production (acting thus in the blockage of the communication system among bacteria), but it also can suppress flagella gene transcription (reducing the mRNA level of flagella gene), the bacterial motility, and finally the formation of biofilm [78]. Cinnamaldehyde, another widely diffused component, present, for example, in cinnamon EO, can show different mechanisms of action. The use of 60 μM cinnamaldehyde can decrease down to 55% the bioluminescence of *V. harveyi BB886*, which is induced by 3-hydroxy-C4-HSL, and from 60 to 100% that of *V. harvevi* BB170 (mediated by AI-2). This indicates, once again, that the activity of EOs, like all other phytochemicals, can be dependent even on the strain used within the same species [30, 65, 79, 80], further than on the QS molecule involved. Another study showed that cinnamaldehyde particularly directs its action toward the short-chain AHL synthase (RhlI) and inhibits AHL production by RhlI [81]. Also cinnamaldehyde analogs and derivatives are capable to inhibit AI-2-based QS system of *V. harveyi* in a dose-dependent manner [82] and are effective against AI-2-regulated QS of *Vibrio* spp. too [83]. Three other cinnamaldehyde analogs, *trans*-2-nonenal, *trans*-3-decen-2-one, and *trans*-3-nonen-2-one, can interfere with AI-2 QS in different manner. In *Vibrio* spp., *trans*-2-nonenal and *trans*-3-decen-2-one inhibit the AI-2-based QS system by reducing the DNA-binding ability of LuxR, causing a decrease in the production of QS-regulated virulence functions such as biofilm formation, matrix production, and protease production [83]. Therefore, some compounds, such as *p*-anisaldehyde can act as AHL mimics, inhibiting the production of violacein by *C. violaceum* [84]. Eugenol inhibits QS in pathogenic bacteria; this was shown, for example, by Zhou and co-workers [85], evaluating the reduction of violacein production in *C. violaceum* after contact with eugenol. This molecule is also capable to affect lasB and pqsA in *E. coli*. This suggests an inhibitory action of eugenol on Las and pseudomonas quinolone signal (PQS)-controlled transcription. The action of eugenol on pathogenic bacteria at subinhibitory concentrations also considerably translates into a reduction in the QS-regulated production of some molecules/enzymes (elastase, protease, chitinase, pyocyanin, and exopolysaccharides) with a concurrently decreased formation of biofilm EPS in *P. aeruginosa* PAO1 [86]. In the Gram-positive pathogen, *S. aureus*, eugenol exhibited also antivirulence property acting on bacterial capability to produce exotoxin, through the repression of the agrA transcription [86]. Some EOs can effectively act both in preventing the biofilm formation and in disrupting the preformed biofilm. The EOs obtained from *Pogostemon heyneanus* and *Cinnamomum tamala* are capable to reduce the extracellular polymeric substance (EPS) and the synthesis of the two factors of the biofilm assemblage built by methicillin-resistant *S. aureus* (MRSA) strains. These EOs are also effective in reducing some virulence factors, such as staphyloxanthin and hemolysin. In silico docking studies demonstrated that (E)-nerolidol showed better binding affinity toward the enzyme dehydroxysqualene synthase of MRSA which is responsible for the synthesis of staphyloxanthin [87]. Different ratios between two components present in an EO can provide a different effectiveness of the EO as a QS inhibitor. Two among five EOs of *Lippia alba*, in particular one containing a greater prevalence of geranial/neral (the two isomers of the octa-2,6-dienal citral) and the other with an higher limonene/carvone content, were the most effective QS inhibitors and also had small effects on cell growth [88]. The activity of EOs on the cell-cell mechanism of communication could depend also on the chemical organization of one or some of their main components. The (+)-enantiomers of carvone, limonene, and borneol are potentially capable to increase the production of violacein and pyocyanin in *C. violaceum* and *P. aeruginosa*, respectively, while their levorotary analogs inhibit such production [84]. Among phenols present in the EOs, eugenol at subinhibitory concentrations is capable of inhibiting the production of virulence factors, involving production of violacein and pyocyanin, synthesis and expression of elastase, and finally the organization of the biofilm. In fact, using two *E. coli* biosensors, MG4/pKDT17 and pEAL08-2, Zhou and co-workers demonstrated that also this compound could act of one or more QS systems, in particular on las and pqs QS systems [85]. The process applied for the extraction of the EO may affect its biological activity indeed. The inhibitory effect of *Citrus medica* L. var. *sarcodactylis* EO obtained by hydrodistillation extraction (HE), microwave-assisted hydrodistillation extraction (MHE), and ultra-microwave-assisted hydro distillation extraction (UMHE) on biofilm formation by *S. aureus* and *S. typhimurium* was significantly higher than that of the essential oil obtained by standard extraction. This could also be related to the different chemical compositions of the different EOs, which elements can differ in terms of quality and rate [89]. Therefore, the diurnal variation can affect the chemistry of the essential oils, affecting their biological properties, including the capability to inhibit the biofilm [90]. New EOs are exhibiting interesting action against the formation of biofilm by microorganisms. *Cannabis sativa* EO is receiving particular attention because, further than other well-known properties, it showed a certain capability to attenuate the virulence of *Listeria monocytogenes* [91], with downregulation of flagella motility genes and of the regulatory gene prfA and a decreasing ability to form biofilm and to invade Caco-2 cells.

6. Action of EOs on biofilm of eukaryotes

More recently, the role of EOs and their components was studied for their potential capability to block the formation of biofilms in eukaryotes [64]. Terpenes are capable to inhibit the formation of biofilm through different mechanisms of action. Thymol, for instance, can affect the envelope of the planktonic form of *C. albicans*: it reaches to alter its membrane permeability [92] by infiltrating between the fatty acyl chains of the membrane lipid bilayers, with subsequent disruption of the lipid organization and damage to membrane fluidity. These events led to important alterations of yeast cell and can also reduce its adherence capability, which represents a major step in biofilm formation. Eugenol acts as a potent agent in blocking the biofilms associated with polystyrene too. Also in the case of *C. albicans*, the action of terpenes on the biofilm formation depends on the concentration of the compound used. Thus, a decrease of approximatively 50% of the metabolic pathway linked to biofilm is observed with 0.016% of carvacrol, geraniol, or thymol; however, a higher concentration is requested when we want to use citral, and even a percentage > or equal to 0.25% is necessary to decrease at 50% the biofilm of *C. albicans* if we want use 1,8-cineole, eugenol, farnesol, linalool, menthol, and α-terpinene [93]. Also using the same terpenes at the same concentration, the capability to inhibit the biofilm formation is related to the strain within the same species of the yeast [94], to the different species belonging to the same genus [95] or even to the genera considered [96]. It should not be overlooked the fact that much often, due to the different mechanisms of communication between bacteria, between fungi and among different bacteria and fungi, the own nature led to the formation of complex biofilm containing mixed population. Even, we can find such very frequent situation. In this case, it could be also more difficult to eradicate a biofilm, being the strength of more than a unique type of microorganisms against whom to combat. For instance, *C. albicans* may be associated with mixed infections of *Streptococcus mutans* to form plaque biofilms [97]. The chemical interaction between these two pathogens results in mixed biofilm development, mainly at oral level; therefore, there are no effective treatments in preventing or inhibiting the formation of mixed biofilms or in preventing inter-microbial communication. Eugenol, at sub-MICs, inhibits single and mixed biofilms of *C. albicans* and *S. mutans*. Also in this case, such capability cannot be expanded to every strain of *C. albicans* neither to all strains of *S. mutans*. In fact, Jafri and co-workers [98] ascertained that eugenol is effective against the biofilm formed by two of more than ten strains of *C. albicans* (in particular, the strain CAJ-01) and *S. mutans* (strain MTCC497), studied singularly on in mixed form, with a concurrent reduction of cell viability and a disruption of cell membrane.

7. New opportunities from the use of essential oils

In view of the importance of quorum sensing on the physiology regulation of microorganisms, research is also addressing toward the identification of new prospects for the use of essential oils in the blocking of cellular communication mechanisms. In recent years, the opportunity to associate essential oils or their main components with synthetic substances (drugs, enzymes, etc.) has been evaluated, to identify the associations that improve their performance in this sector. In particular, emerging resistance to last-resort antibiotics, such as carbapenems and polymixin B, led to theory of the so-called post-antibiotic era [99–101], and now research is moving to identify new compounds with stronger activity also against the most resistant pathogens. Different research groups are thus carrying a screening of "no-drug compounds," including the EOs, in association with nonsteroidal anti-inflammatory drugs (the so-called NSAIDs) for inhibition of quorum sensing and biofilm formation in pathogens. Some NSAIDs, such as Z-phytol and lonazolac, can block the QS system of *Salmonella enteritidis*, acting as antagonists with respect to the AHLs [102]. The association of EOs with azolic drugs gives interesting challenges in the treatment of fungal infections determined by fungi particularly difficult to be eradicated [103]. Thymol in combination with the azolic drug, fluconazole, is capable to act against several species of *Candida* isolated from clinical specimens [104]. Chitosan nanoparticles containing miconazole and farnesol can inhibit *Candida* proliferation. The presence of farnesol is also capable of decreasing the pathogenicity of infection, demonstrated through the absence of inflammation [105]. The combination between the EO of *Cinnamomum tamala*, or its main component cinnamaldehyde, and linalool, with a commercial DNase I and marine bacterial DNase, might offer an alternative strategy to fight the biofilm formation and quorum sensing-mediated virulence factors in the aquaculture pathogen *Vibrio parahaemolyticus*, acting as effective food-preserving agent too [106]. The synergistic association between some EOs and DNase can decrease the biofilm-forming ability of *S. aureus* [87]. The association between EOs and/or their main components with conventional antibiotics can allow to eradicate also some particularly resistant biofilm, decreasing concurrently their resistance threshold. This allows also to minimize the antibiotic concentration, decreasing also its potential accompanying toxic side effects [107]. Synergistic interactions between EOs and their components with antibiotics are recognized, including several instances of antibiotic re-sensitization in resistant isolates, in support of this strategy to control antibiotic resistance, although synergistic effects are not well explored outside a preliminary identification of antibacterial interactions and mechanism of action is seldom defined, despite many hypotheses and recommendations for future studies [108]. The possibility of using essential oils has also been evaluated in the prevention of biofilm development by microorganisms (*Klebsiella*, *S. aureus*, *Staphylococcus haemolyticus*, *E. coli*, *Enterococcus cloacae*, *Enterococcus faecalis*), and the EO of *Matricaria chamomilla* (Asteraceae) could be considered as a good candidate, encouraging research related to the application of essential oils to fight diabetic complications [109]. The nanomaterials including EOs, at subinhibitory concentration, can inhibit QS and prevent biofilm formation and virulence development in pathogens [110]. The capability of blocking the quorum sensing mechanisms and biofilm formation by EOs can be combined with conventional drugs for a better treatment efficacy, as well as to design new more effective drugs, capable of acting also against those particularly resistant bacteria and fungi. The possibility to use nanotechnology, through the production of nano-vesicles containing EO, can result useful in a variety of applications including medical and pharmaceutical recipients and in home products for treating or preventing microbial colonization, as well as avoiding biofilm development [111, 112], also in food technology [113, 114]. Nano-coating with different inorganic and organic materials also supported by EOs proves to be particularly useful in the treatment of chronic wounds, such as venous or arterial ulcers, diabetic foot ulcers, pressure sores, and non-healing surgical wounds, all scenarios associated with chronic mono- or polymicrobial biofilm infections, formed by different bacteria, such as *S. aureus* and *P. aeruginosa*, followed by various species of *Enterobacteriaceae* such as *E. coli*, *Klebsiella* spp., *Proteus* spp., *Enterobacter* spp., *Morganella morganii*, *Citrobacter freundii*, *Serratia marcescens*, *Providencia* spp., *Enterococcus* spp., *Streptococcus* spp., and rarely *Corynebacterium* spp. or *Acinetobacter baumannii* [115]. In agro-food industry, the increasing demand for eco-friendly material stimulated the study of the elaboration of complex bio-nanocomposite films containing also EOs, such as that of rosemary, which can act as effective eco-friendly nanocomposite films in packaging industries [116]. Algae are beginning the new frontier for the supplement of new bioactive compounds with antimicrobial activity and represent a unique opportunity for the science. Some investigations are exploring the therapeutic potential of algal extracts and their chemical groups, including terpenes, capable to have antimicrobial activity and to block the mechanism of communication between microorganisms. The alga *Pithophora oedogonium* targets *Staphylococcus* and *Salmonella*. The algae *Rivularia bullata*, *Nostoc spongiaeforme*, *Codium fragile*, *Colpomenia peregrina*, *Cystoseira barbata*, and *Zanardinia typus* have already demonstrated activity against many Gram-negative and Gram-positive bacteria [117, 118]. Finally, innovative techniques, such as the optical technique of bio-speckle [119] and the biofilm electrostatic test (BET) [120], will support the research in the near future to have a very fast scenario of EO biological activity. Speckle decorrelation can lead us to visualize the effect of essential oil on the decrease of the usual self-propelling movement of microorganisms taking place when they interact with coherent light. BET is as a simple, rapid, and highly reproducible tool for evaluating *in vitro* the ability of bacteria to form biofilms through electrostatic interaction with a pyro-electrified carrier and for ascertaining the impeding effect of an EO on the microbial capability to form biofilm in just 3 h.

8. Conclusion

Essential oils can represent a precious mine to fight pathogenic microorganisms, in particular to counteract the communication mechanisms that allow them to trigger those processes leading to their greater virulence and danger, in the hospital, environmental, and food sectors. However, *in vivo* studies remain crucial to evaluating the potential of this strategy as a mean to treat antibiotic-resistant infections, and profounder understanding of the mechanism of action is required. Finally, in the case of the association EOs-antibiotics, the preliminary evaluation of the *in vitro* toxicity of EO-antibiotic combinations is needed prior to in vivo studies indeed.

Conflict of interest

The authors declare no conflict of interest.

Author details

Filomena Nazzaro[1]*, Florinda Fratianni[1], Antonio d'Acierno[1], Raffaele Coppola[2], Fernando Jesus Ayala-Zavala[3], Adriano Gomez da Cruz[4] and Vincenzo De Feo[5]

1 Institute of Food Science, CNR-ISA, Avellino, Italy

2 DiAAA, Department of Agricultural, Environmental and Food Sciences, University of Molise, Campobasso, Italy

3 Center for Research in Nutrition and Development (CIAD AC), Hermosillo, Mexico

4 Department of Food Engineering, State University of Ponta Grossa (UEPG), Ponta Grossa, Brazil

5 Department of Pharmacy, University of Salerno, Fisciano (Salerno), Italy

*Address all correspondence to: filomena.nazzaro@cnr.it

References

[1] Lee VC. The antibiotic resistance crisis. Part 1: Causes and threats. Pharmacy and Therapeutics. 2015;**40**:277-283. PMID: 25859123

[2] Ruiz J, Castro I, Calabuig E, Salavert M. Non-antibiotic treatment for infectious diseases. Revista Espanola De Quimioterapia. 2017;**30**:66-71. PMID: 28882020

[3] Westritschnig K, Hochreiter R, Wallner G, Firbas C, Schwameis M, Jilma B. A randomized, placebo-controlled phase I study assessing the safety and immunogenicity of a *Pseudomonas aeruginosa* hybrid outer membrane protein OprF/I vaccine (IC43) in healthy volunteers. Human Vaccines & Immunotherapeutics. 2014;**10**:170-183. DOI: 10.4161/hv.26565

[4] Garcia-Quintanilla M, Pulido MR, McConnell MJ. First steps towards a vaccine against *Acinetobacter baumannii*. Current Pharmaceutical Biotechnology. 2013;**14**:897-902. DOI: 10.2174/1389201014666131226123511

[5] Terra L, Dyson PJ, Hitchings MD, Thomas L, Abdelhameed A, Banat IM, et al. A novel alkaliphilic *Streptomyces* inhibits ESKAPE pathogens. Frontiers in Microbiology. 2018;**9**:2458. DOI: 10.3389/fmicb.2018.02458

[6] Camilli A, Bassler BL. 2006 bacterial small-molecule Signaling pathways. Science. 2006;**311**:1113-1116. DOI: 10.1126/science.1121357

[7] Albuquerque P, Casadevall A. Quorum sensing in fungi—A review. Medical Mycology. 2012;**50**:337-345. DOI: 10.3109/13693786.2011.652201

[8] Miller MB, Bassler BL. Quorum sensing in bacteria. Annual Review in Microbiology. 2001;**55**:165-199. DOI: 10.1146/annurev.micro.55.1.165

[9] Nealson KH, Platt T, Hastings JW. Cellular control of the synthesis and activity of the bacterial luminescent system. Journal of Bacteriology. 1970;**104**:313-322

[10] Hawver LA, Jung SA, Ng WL. Specificity and complexity in bacterial quorum-sensing systems. FEMS Microbiology Reviews. 2016;**40**:738-752. DOI: 10.1093/femsre/fuw014

[11] Biarnes-Carrera M, Breitling R, Takano E. Butyrolactone signalling circuits for synthetic biology. Current Opinion in Chemical Biology. 2015;**28**:91-98. DOI: 10.1016/j.cbpa.2015.06.024

[12] Chen X, Schauder S, Potier N, Van Dorsselaer A, Pelczer I, Bassler BL, et al. Structural identification of a bacterial quorum-sensing signal containing boron. Nature. 2002;**415**:545-549. DOI: 10.1038/415545a

[13] Winzer K, Hardie KR, Williams P. Bacterial cell-to-cell communication: Sorry, can't talk now-gone to lunch! Current Opinion in Microbiology. 2002;**5**:216-222. DOI: 10.1016/S1369-5274(02)00304-1

[14] Bandara HMHN, Lam OLT, Jin LJ, Samaranayake L. Microbial chemical signaling: A current perspective. Critical Review in Microbiology. 2012;**38**: 217-249. DOI: https://doi.org/10.3109/1040841X.2011.652065

[15] Papenfort K, Bassler BL. Quorum sensing signal-response systems in gram-negative bacteria. Nature Reviews Microbiology. 2016;**14**(9):576-588. DOI: 10.1038/nrmicro.2016.89

[16] Mangano S, Caruso C, Michaud L, Lo Giudice A. First evidence of quorum sensing activity in bacteria associated with Antarctic sponges. Polar Biology.

2018;**41**:1435-1445. DOI: 10.1007/s00300-018-2296-3

[17] Dong HY, Zhang HL. Quorum sensing and quorum-quenching enzymes. Journal of Microbiology. 2005;**43**:101-109. PMID: 15765063

[18] McDougald D, Rice AS, Kjelleberg S. Bacterial quorum sensing and interference by naturally occurring biomimics. Analytical and Bioanalytical Chemistry. 2007;**387**:445-453. DOI: 10.1007/s00216-006-0761-2

[19] Barriuso J, Hogan DA, Keshavarz T, Martínez MJ. Role of quorum sensing and chemical communication in fungal biotechnology and pathogenesis. FEMS Microbiology Reviews. 2018;**42**:627-638. DOI: 10.1093/femsre/fuy022

[20] Hogan DA. Talking to themselves: Autoregulation and quorum sensing in fungi. Eukaryotic Cell. 2006;**5**:613-619. DOI: 10.1128/EC.5.4.613-619.2006

[21] Hogan DA. Quorum sensing: Alcohols in a social situation. Current Biology. 2006;**16**:R457-R458. DOI: 10.1016/j.cub.2006.05.035

[22] Chen H, Fink GR. Feedback control of morphogenesis in fungi by aromatic alcohols. Genes Development. 2006;**20**:1150-1161. DOI: 10.1101/gad.1411806

[23] Gognies S, Barka EA, Gainvors-Claisse A, Belarbi A. Interactions between yeasts and grapevines: Filamentous growth, endopolygalacturonase and phytopathogenicity of colonizing yeasts. Microbial Ecology. 2006;**5**:09-116. DOI: 10.1007/s00248-005-0098-y

[24] Chen H, Fujita M, Feng Q, Clardy J, Fink FR. Tyrosol is a quorum sensing molecule in *Candida albicans*. Proceeding of the National Academy of Sciences of United States of America. 2004;**101**:5048-5052. DOI: 10.1073/pnas.0401416101

[25] Avbelj M, Zupan J, Kranjc L, Raspor P. Quorum-sensing kinetics in *Saccharomyces cerevisiae*: A symphony of ARO genes and aromatic alcohols. Journal of Agricultural and Food Chemistry. 2015;**63**:8544-8550. DOI: 10.1021/acs.jafc.5b03400

[26] Ramage G, Saville SP, Wickes BL, López-Ribot JL. Inhibition of *Candida albicans* biofilm formation by farnesol, a quorum-sensing molecule. Applied and Environmental Microbiology. 2002;**68**:5459-5463. DOI: 10.1128/AEM.68.11.5459-5463.2002

[27] de A, Cordeiro R, Teixeira CEC, Brilhante RSN, Castelo-Branco DSCM, Alencar LP, et al. Exogenous tyrosol inhibits planktonic cells and biofilms of *Candida* species and enhances their susceptibility to antifungals. FEMS Yeast Research. 2015;**15**:12. DOI: 1093/femsyr/fov012

[28] Hornby JM, Jensen EC, Lisec AD, Tasto JJ, Jahnke B, Shoemaker R, et al. Quorum sensing in the dimorphic fungus *Candida albicans* is mediated by farnesol. Applied and Environmental Microbiology. 2001;**67**:2982-2992. DOI: 10.1128/AEM.67.7.2982-2992.2001

[29] Peleg AY, Hogan DA, Mylonakis E. Medically important bacterial-fungal interactions. Nature Review Microbiology. 2010;**8**:340-349. DOI: 10.1038/nrmicro2313

[30] Nazzaro F, Fratianni F, De Martino L, Coppola R, De Feo V. Effect of the essential oils on pathogenic bacteria. Pharmaceuticals. 2013;**6**:1451-1474. DOI: 10.3390/ph6121451

[31] Raina S, De Vizio D, Palonen EK, Odell M, Brandt AM, Soini JT, et al. Is quorum sensing involved in lovastatin production in the filamentous fungus *Aspergillus terreus*? Process Biochemistry. 2012;**47**:843-852. DOI: 10.1016/j.procbio.2012.02.021

[32] Williams HE, Steele JCP, Clements MO, Keshavarz T. ɣ-Heptalactone is an endogenously produced quorum-sensing molecule regulating growth and secondary metabolite production by *Aspergillus nidulans*. Applied Microbiology and Biotechnology. 2012;**96**:773-781. DOI: 10.1007/s00253-012-4065-5

[33] Raina S, Odell M, Keshavarz T. Quorum sensing as a method for improving sclerotiorin production in *Penicillium sclerotiorum*. Journal of Biotechnology. 2010;**148**:91-98. DOI: 10.1016/j.jbiotec.2010.04.009

[34] Affeldt KJ, Brodhagen M, Keller NP. *Aspergillus* oxylipin signaling and quorum sensing pathways depend on G protein-coupled receptors. Toxins. 2012;**4**:695-717. https://doi.org/10.3390/toxins4090695

[35] De Salas F, Martínez MJ, Barriuso J. Quorum-sensing mechanisms mediated by farnesol in *Ophiostoma piceae*: Effect on secretion of sterol esterase. Applied and Environmental Microbiology. 2015;**81**:4351-4357. DOI: 10.1128/AEM.00079-15

[36] Berrocal A, Oviedo C, Nickerson KW, Navarrete J. Quorum sensing activity and control of yeast mycelium dimorphism in *Ophiostoma floccosum*. Biotechnology Letters. 2014;**36**:1503-1513. DOI: 10.1007/s10529-014-1514-5

[37] Costerton WJ, Cheng KJ, Geesey GG, Ladd TI, Nickel JC, Dasgupta M, et al. Bacterial biofilms in nature and disease. Annual Review of Microbiology. 1987;**41**:435-464. DOI: 10.1146/annurev.micro.41.1.435

[38] O' Toole G, Kaplan BH, Kolter R. Biofilm formation as microbial development. Annual Review in Microbiology. 2000;**54**:49-79. DOI: 10.1146/annurev.micro.54.1.49

[39] de Kievit RT. Quorum sensing in *Pseudomonas aeruginosa* biofilms. Environmental Microbiology. 2009;**11**:279-288. DOI: 10.1111/j.1462-2920.2008.01792.x

[40] Goldstone RJ, Popel R, Fletcher MP, Cruisz SA, Diggie SP. Quorum sensing and social interaction in microbial biofilm. In: Lear G, Lewis GD, editors. Microbial Biofilms: Current Research and Application. Poole: Caister Academic Press; 2012. pp. 1-24. ISBN: 978-1-904455-96-7S

[41] Scutera S, Zucca M, Savoia D. Novel approaches for the design and discovery of quorum-sensing inhibitors. Expert Opinion on Drug Discovery. 2014;**9**:353-366. DOI: 10.1517/17460441.2014.894974

[42] Li J, Helmerhorst EJ, Leone CW, Troxler RF, Yaskell T, Haffajee AD, et al. Identification of early microbial colonizers in human dental biofilm. Journal of Applied Microbiology. 2004;**97**:1311-1318. DOI: 0.1111/j.1365-2672.2004.02420.x

[43] Czaczyk K, Myszka K. Biosynthesis of extracellular polymeric substances [EPS] and its role in microbial biofilm formation. Polish Journal of Environmental Studies. 2007;**16**:799-806

[44] Donlan RM, Costerton JW. Biofilms: Survival mechanisms of clinically relevant microorganisms. Clinical Microbiology Review. 2002;**15**:167-193. DOI: 10.1128/CMR.15.2.167-193.2002

[45] Hentzer M, Givskov M. Pharmacological inhibition of quorum sensing for the treatment of chronic bacterial infections. Journal of Clinical Investigation. 2003;**112**: 1300-1307. DOI: 10.1172/JCI20074

[46] Ishida T, Ikeda T, Takiguchi N, Kuroda A, Ohtake H, Kato J. Inhibition of quorum sensing in *Pseudomonas aeruginosa* by N-acyl cyclopentylamides. Applied and Environmental

Microbiology. 2007;**73**:3183-3188. DOI: 10.1128/AEM.02233-06

[47] Shiner EK, Terentyev D, Bryan A, Sennoune S, Martinez-Zaguilan R, Li G, et al. *Pseudomonas aeruginosa* autoinducer modulates host cell responses through calcium signalling. Cellular Microbiology. 2006;**8**:1601-1610. DOI: 10.1111/j.1462-5822.2006.00734.x

[48] Hickson J, Yamada DS, Berger J, Alverdy J, O'Keefe J, Bassler B, et al. Societal interactions in ovarian cancer metastasis: A quorum-sensing hypothesis. Clinical and Experimental Metastasis. 2009;**26**:67. DOI: 10.1007/s10585-008-9177-z

[49] Nazzaro F, Fratianni F, Coppola R. Quorum sensing and phytochemicals. International Journal of Molecular Sciences. 2013;**14**:12607-12619. DOI: 10.3390/ijms140612607

[50] Guenther E. The Essential Oils: History—Origin in Plants. Production—analysis. New York: D. Van Nostrand company Inc.; 1948. pp. 1-427. DOI: 10.1002/jps.3030370518

[51] Boyle W. Spices and essential oil as preservatives. The American Perfumer and Essential Oil Review. 1955;**66**:25-28

[52] Raut JS, Karuppayil SM. A status review on the medicinal properties of essential oils. Industrial Crops and Products. 2014;**62**:250-264. DOI: 10.1016/j.indcrop.2014.05.055

[53] Flechas MC, Ocazionez RE, Stashenko EE. Evaluation of *in vitro* antiviral activity of essential oil compounds against dengue virus. Pharmacognosy Journal. 2018;**10**:55-59. DOI: 10.5530/pj.2018.1.11

[54] Aziz Zarith AA, Akil A, Setapar SHM, Karakucuk A, Azim MM, Lokhat D, et al. Essential oils: Extraction techniques, pharmaceutical and therapeutic potential—A review. Current Drug Metabolism. 2018;**19**:1100-1110. DOI: 10.2174/1389200219666180723144850

[55] Cosentino S, Tuberoso CIG, Pisano B, Satta M, Mascia V, Arzedi E, et al. *In-vitro* antimicrobial activity and chemical composition of Sardinian *Thymus* essential oils. Letters in Applied Microbiology. 1999;**29**:130-135. DOI: 10.1046/j.1472-765X.1999.00605.x

[56] Edris AE. Pharmaceutical and therapeutic potentials of essential oils and their individual volatile constituents: A review. Phytotherapy Research. 2007;**21**:308-323. DOI: /10.1002/ptr.2072

[57] Lang G, Buchbauer G. A review on recent research results (2008-2010) on essential oils as antimicrobials and antifungals: A review. Flavour and Fragrance Journal. 2012;**27**:13-39. DOI: 10.1002/ffj.2082

[58] dos Santos JFS, Rocha JER, Fonseca Bezerra C, do Nascimento Silva MK, Soares YML, Sampaio de Freitas MT, et al. Chemical composition, antifungal activity and potential anti-virulence evaluation of the *Eugenia uniflora* essential oil against *Candida* spp. Food Chemistry. 2018;**261**:233-239. DOI: 10.1016/j.foodchem.2018.04.01

[59] Nostro A, Scaffaro R, D'Arrigo M, Botta L, Filocamo A, Marino A, et al. Development and characterization of essential oil component-based polymer films: A potential approach to reduce bacterial biofilm. Applied Microbiology and Biotechnology. 2013;**97**:9515-9523. DOI: 10.1007/s00253-013-5196-z

[60] Hammer KA, Carson CF, Riley TV. Antimicrobial activity of essential oils and other plant extracts. Journal of Applied Microbiology. 1999;**86**:985-990. DOI: 10.1046/j.1365-2672.1999.00780.x

[61] Hammer KA, Carson CF. In: Thormar H, editor. Lipids and

Essential Oils as Antimicrobial AgentsAntibacterial and Antifungal Activities of Essential Oil. Chichester: Wiley; 2010. pp. 1-336

[62] Smith-Palmer A, Stewart J, Fyfe L. Antimicrobial properties of plant essential oils and essences against five important food-borne pathogens. Letters in Applied Microbiology. 1998;**26**:118-122. DOI: 10.1046/j.1472-765X.1998.00303.x

[63] Noumi E, Snoussi M, Alreshidi MM, Rekha PD, Saptami K, Caputo L, et al. Chemical and biological evaluation of essential oils from cardamom species. Molecules. 2018;**3**:1-18. DOI: 10.3390/molecules23112818

[64] Oussalah M, Caillet S, Lacroix M. Mechanism of action of Spanish oregano, Chinese cinnamon, and savory essential oils against cell membranes and walls of *Escherichia coli* O157:H7 and *Listeria monocytogenes*. Journal of Food Protection. 2006;**69**:1046-1055. DOI: 10.4315/0362-028X-69.5.1046

[65] Nazzaro F, Fratianni F, Coppola R, De Feo V. Essential oils and antifungal activity. Pharmaceuticals. 2017;**10**:1086-1105. DOI: 10.3390/ph10040086

[66] da Cunha JA, Heinzmann BM, Baldisserotto B. The effects of essential oils and their major compounds on fish bacterial pathogens–a review. Journal of Applied Microbiology. 2018;**125**: 328-344. DOI: https://doi.org/10.1111/jam.13911

[67] Nikaido H. Prevention of drug access to bacterial targets: Permeability barriers and active efflux. Science. 1994;**64**:382-388. DOI: 10.1126/science.8153625

[68] Di Pasqua R, Betts G, Hoskins N, Edwards M, Ercolini D, Mauriello G. Membrane toxicity of antimicrobial compounds from essential oils. Journal of Agricultural and Food Chemistry. 2007;**55**:4863-4870. DOI: 10.1021/jf0636465

[69] Bouyahya A, Dakka N, El-Boury H, Bakri Y. Correlation between phenological changes, chemical composition and biological activities of the essential oil from Moroccan endemic oregano (*Origanum compactum* Benth). Industrial Crops and Products. 2017;**108**:29-737. DOI: 10.1016/j.indcrop.2017.07.033

[70] Tiwari R, Karthik K, Rana R, Singh Malik Y, Dhama K, Joshi SK. Quorum sensing inhibitors/antagonists countering food spoilage bacteria-need molecular and pharmaceutical intervention for protecting. Current issues of food safety. International Journal of Pharmacology. 2016;**12**: 262-271. DOI: 10.3923/ijp.2016.262.271

[71] Husain FM, Ahmad I, Asif M, Tahseen Q. Influence of clove oil on certain quorum-sensing-regulated functions and biofilm of *Pseudomonas aeruginosa* and *Aeromonas hydrophila*. Journal of Biosciences. 2013;**38**:835-844. DOI: 10.1007/s12038-013-9385-9

[72] Husain FM, Ahmad I, Khan MS, Ahmad E, Tahseen Q, Khan MS, et al. Sub-MICs of *Mentha piperita* essential oil and menthol inhibits AHL mediated quorum sensing and biofilm of gram-negative bacteria. Frontiers in Microbiology. 2015;**6**:1-12. article 420. DOI: 10.3389/fmicb.2015.00420

[73] Szabó MA, Varga GZ, Hohmann J, Schelz Z, Szegedi E, Amaral L, et al. Inhibition of quorum-sensing signals by essential oils. Phytotherapy Research. 2010;**24**:782-786. DOI: 10.1002/ptr.3010

[74] Khan MSA, Zahin M, Hasan S, Husain FM, Ahmad I. Inhibition of quorum sensing regulated bacterial functions by plant essential oils with special reference to clove oil. Letters in Applied Microbiology. 2009;**49**:354-360. DOI: 10.1111/j.1472-765X.2009.02666.x

[75] Kerekes EB, Deák E, Takó M, Tserennadmid R, Petkovits T, Vágvölgyi C, et al. Anti-biofilm forming and anti-quorum sensing activity of selected essential oils and their main components on food-related microorganisms. Journal of Applied Microbiology. 2013;**115**: 933-942. DOI: 10.1111/jam.12289

[76] Luciardic MC, Blázquez MA, Cartagena E, Bardóna A, Arena ME. Mandarin essential oils inhibit quorum sensing and virulence factors of *Pseudomonas aeruginosa*. LWT—Food Science and Technology. 2016;**68**: 373-380. DOI: 10.1016/j.lwt.2015.12.056

[77] Burt SA, Ojo-Fakunle VTA, Woertman J, Veldhuize EJA. The natural antimicrobial carvacrol inhibits quorum sensing in *Chromobacterium violaceum* and reduces bacterial biofilm formation at sub-lethal concentrations. PLoS One. 2014;**9**:e93414. DOI: 10.1371/journal.pone.0093414

[78] Myszka K, Schmidt MT, Majcher M, Juzwa W, Olkowicza M, Czaczyka K. Inhibition of quorum sensing-related biofilm of *Pseudomonas fluorescens* KM121 by *Thymus vulgare* essential oil and its major bioactive compounds. International Biodeterioration & Biodegradation. 2016;**114**:252-259. DOI: 10.1016/j.ibiod.2016.07.006

[79] Niu S, Afre S, Gilbert ES. Sub-inhibitory concentrations of cinnamaldehyde interfere with quorum sensing. Letters in Applied Microbiology. 2006;**43**:489-494. DOI: 10.1111/j.1472-765X.2006.02001.x

[80] Noumi E, Merghni A, Alreshidi MM, Haddad O, Akmadar G, De Martino L, et al. Chromobacterium violaceum and Pseudomonas aeruginosa PAO1: Models for evaluating anti-quorum sensing activity of *Melaleuca alternifolia* essential oil and its main component terpinen-4-ol. Molecules. 2018;**23**:1-16. DOI: 10.3390/molecules23102672

[81] Chang CY, Krishnan T, Wang H, Chen Y, Yin WF, Chong YM, et al. Non-antibiotic quorum sensing inhibitors acting against N-acyl homoserine lactone synthase as druggable target. Scientific Reports. 2014;**4**:7245. DOI: 10.1038/srep07245

[82] Brackman G, Defoirdt T, Miyamoto C, Bossier P, Van Calenbergh S, Nelis H, et al. Cinnamaldehyde and cinnamaldehyde derivatives reduce virulence in *Vibrio* spp. by decreasing the DNA-binding activity of the quorum sensing response regulator LuxR. BMC Microbiology. 2008;**8**:149. DOI: 10.1186/1471-2180-8-149

[83] Brackman G, Celen S, Hillaert U, Van Calenbergh S, Cos P, Maes L, et al. Structure-activity relationship of cinnamaldehyde analogs as inhibitors of AI-2 based quorum sensing and their effect on virulence of *Vibrio* spp. PLoS One. 2011;**6**:e16084. DOI: 10.1371/journal.pone.0016084

[84] Ahmad A, Viljoen AM, Chenia HY. The impact of plant volatiles on bacterial quorum sensing. Letters in Applied Microbiology. 2015;**60**:8-19. DOI: 10.1111/lam.12343

[85] Zhou L, Zheng H, Tang Y, Yu W, Gong Q. Eugenol inhibits quorum sensing at sub-inhibitory concentrations. Biotechnology Letters. 2013;**35**:631-637. DOI: 10.1007/s10529-012-1126-x

[86] Al-Shabib NA, Husain FM, Ahmad I, Baig MH. Eugenol inhibits quorum sensing and biofilm of toxigenic MRSA strains isolated from food handlers employed in Saudi Arabia. Biotechnology and Biotechnological Equipment. 2017;**31**:387-396. DOI: 10.1080/13102818.2017.1281761

[87] Rubini D, Banu SF, Prakash N, Murugan NR, Thamotharan S, Percino MJ, et al. Essential oils from unexplored aromatic plants quench biofilm

formation and virulence of methicillin resistant *Staphylococcus aureus*. Microbial Pathogenesis. 2018;**122**:162-173. DOI: 0.1016/j.micpath.2018.06.028

[88] Olivero-Verbel J, Barreto-May A, Bertel-Sevilla A, Stashenko EE. Composition, anti-quorum sensing and antimicrobial activity of essential oils from *Lippia alba*. Brazilian Journal of Microbiology. 2014;**45**:759-767. DOI: 10.1590/S1517-83822014000300001

[89] Zhang H, Lou Z, Chen X, Cui Y, Wang H, Kou X, et al. Effect of simultaneous ultrasonic and microwave assisted hydrodistillation on the yield, composition, antibacterial and antibiofilm activity of essential oils from *Citrus medica* L. var. *sarcodactylis*. Journal of Food Engineering. 2019;**244**:126-135. DOI: 0.1016/j.jfoodeng.2018.09.014

[90] Dos Santos Icon RC, De Melo Filho AA, Chagas EA, Montero Fernández I, Aparecida J, Icon T, et al. Influence of diurnal variation in the chemical composition and bioactivities of the essential oil from fresh and dried leaves of *Lantana camara*. Journal of Essential oil Research. 2019;**31**:228-234. DOI: 10.1080/10412905.2018.1555102

[91] Marini E, Magi G, Ferretti G, Bacchetti T, Giuliani A, Pugnaloni A, et al. Attenuation of *Listeria monocytogenes* virulence by *Cannabis sativa* L. essential oil. Frontiers in Cellular and Infection Microbiology. 2018;**8**:293. DOI: 10.3389/fcimb.2018.00293

[92] Braga PC, Ricci D. Thymol-induced alterations in *Candida albicans* imaged by atomic force microscopy. In: Braga P, Ricci D, editors. Atomic Force Microscopy in Biomedical Research. Methods in Molecular Biology (Methods and Protocols). New York: Humana Press; 2011. p. 736. DOI: 10.1007/978-1-61779-105-5_24

[93] He M, Du M, Fan M, Bian Z. *In vitro* activity of eugenol against *Candida albicans* biofilms. Mycopathologia. 2007;**163**:137-143. DOI: 10.1007/s11046-007-0097-2

[94] Sajjad M, Khan A, Ahmad I. Biofilm inhibition by *Cymbopogon citratus* and *Syzygium aromaticum* essential oils in the strains of *Candida albicans*. Journal of Ethnopharmacology. 2012;**140**: 416-423. DOI: 10.1016/j.jep.2012.01.045

[95] Dalleau S, Cateau E, Bergès T, Berjeaud JM, Imbert C. *In vitro* activity of terpenes against *Candida* biofilms. International Journal of Antimicrobial Agents. 2008;**31**:572-576. DOI: 10.1016/j.ijantimicag.2008.01.028

[96] Bennis S, Chami F, Chami N, Bouchikhi T, Remmal A. Surface alteration of *Saccharomyces cerevisiae* induced by thymol and eugenol. Letters in Applied Microbiology. 2004;**38**:454-458. DOI: 10.1111/j.1472-765X.2004.01511.x

[97] Sampaio AA, Souza SE, Ricomini-Filho AP, Del Bel Cury AA, Cavalcanti YW, Cury JA. *Candida albicans* increases dentine demineralization provoked by *Streptococcus mutans* biofilm. Caries Research. 2019;**53**:322-331. DOI: 10.1159/000494033

[98] Jafri H, Sajjad M, Khan A, Ahmad I. *In vitro* efficacy of eugenol in inhibiting single and mixed-biofilms of drug-resistant strains of *Candida albicans* and *Streptococcus mutans*. Phytomedicine. 2019;**54**:206-213. DOI: 10.1016/j.phymed.2018.10.005

[99] Laxminarayan R, Duse A, Wattal C, Zaidi AKM, Wertheim HFL, Sumpradit N, et al. Antibiotic resistance—The need for global solutions. Lancet Infectious Diseases. 2013;**13**:1057-1098. DOI: 10.1016/S1473-3099(13)70318-9

[100] Olaitan AO, Morand S, Rolain JM. Mechanisms of polymyxin resistance: Acquired and intrinsic resistance in bacteria. Frontiers in

Microbiology. 2017;**23**:1-18. DOI: 10.339/fmicb.2014.00643

[101] Langeveld WT, Veldhuizen EJA, Burt SA. Synergy between essential oil components and antibiotics: A review. Critical Reviews in Microbiology. 2014;**40**:1-19. DOI: 10.3109/1040841X.2013.763219

[102] de Almeida FA, Giraldo Vargas EL, Guimarães Carneiro D, Pinto UM, Dantas Vanetti MC. Virtual screening of plant compounds and nonsteroidal anti-inflammatory drugs for inhibition of quorum sensing and biofilm formation in *Salmonella*. Microbial Pathogenesis. 2018;**121**:369-388. DOI: 10.1016/j.micpath.2018.05.014

[103] de Oliveira Santos GC, Vasconcelos CC, Lopes AJO, de Sousa Cartágenes M, Filho AKDB, do Nascimento FRF, et al. *Candida* infections and therapeutic strategies: Mechanisms of action for traditional and alternative agents. Frontiers in Microbiology. 2018;**9**:1-23. DOI: 10.3389/fmicb.2018.01351

[104] Sharifzadeh A, Khosravi AR, Shokri H, Shirzadia H. Potential effect of 2-isopropyl-5-methylphenol (thymol) alone and in combination with fluconazole against clinical isolates of *Candida albicans*, *C. glabrata* and *C. krusei*. Journal de Mycologie Médicale. 2018;**28**:294-299. DOI: 10.1016/j.mycmed.2018.04.002

[105] Fernandes Costa A, Araujo DE, Cabral MS, Teles Brito I, de Menezes Leite LB, Pereira M, et al. Development, characterization, and *in vitro–in vivo* evaluation of polymeric nanoparticles containing miconazole and farnesol for treatment of vulvovaginal candidiasis. Medical Mycology. 2019;**57**:52-62. DOI: 10.1093/mmy/myx155

[106] Farisa S, Ramar BDR, Vellingiri M, Shanmugaraj V, Shunmugiah G, Pandiand K, et al. Exploring the antivirulent and sea food preservation efficacy of essential oil combined with DNase on *Vibrio parahaemolyticus*. LWT. 2018;**95**:107-115. DOI: 10.1016/j.lwt.2018.04.070

[107] Mohamed SH, Mohamed MSM, Khalil MS, Azmy M, Mabrouk MI. Combination of essential oil and ciprofloxacin to inhibit/eradicate biofilms in multidrug-resistant *Klebsiella pneumoniae*. Journal of Applied Microbiology. 2018;**125**:84-95. DOI: 10.1111/jam.13755

[108] Owen L, Laird K. Synchronous application of antibiotics and essential oils: Dual mechanisms of action as a potential solution to antibiotic resistance. Critical Reviews in Microbiology. 2018;**44**:414-435. DOI: 10.1080/1040841X.2018.1423616

[109] Nikolayeva AB, Akhmetova SB, Gazaliyeva MA, Sirota VB, Zhumaliyeva VA, Zhakenova SR. Effects of essential oils on biofilm-forming microorganisms from patients with diabetic foot. Diabetes. 2018;**67**. DOI: 10.2337/db18-2223-PUB

[110] Bai AJ, Rai R. Nanotechnological approaches in quorum sensing inhibition. In: Kaliapp VC, editor. Biotechnological Applications of Quorum Sensing Inhibitors. Berlin: Springer; 2018. pp. 245-261. DOI: 10.1007/978-981-10-9026-4_12

[111] Perez AP, Perez N, Suligoy Lozano CM, Altube MJ, Marcelo Alexandre de Farias MA, Villares Portugal R, et al. The anti MRSA biofilm activity of *Thymus vulgaris* essential oil in nanovesicles. Phytomedicine. 2019;**57**:339-351. DOI: 10.1016/j.phymed.2018.12.025

[112] Liakos IL, Iordache F, Carzino R, Scarpellini A, Oneto M, Bianchini P, et al. Cellulose acetate—Essential oil nanocapsules with antimicrobial activity for biomedical applications. Colloids and Surfaces B: Biointerfaces.

2018;**172**:471-479. DOI: 10.1016/j.colsurfb.2018.08.069

[113] Prakash B, Kujur A, Yadav A, Kumar A, Singh PP, Dubey NK. Nanoencapsulation: An efficient technology to boost the antimicrobial potential of plant essential oils in food system. Food Control. 2018;**89**:1-11. DOI: 10.1016/j.foodcont.2018.01.018

[114] Omonijo FA, Nia L, Gong J, Wang Q, Lahaye L, Yang C. Essential oils as alternatives to antibiotics in swine production. Animal Nutrition. 2018;**4**:126-136. DOI: 10.1016/j.aninu.2017.09.001

[115] Mihai MM, Preda M, Lungu I, Cartelle Gestal M, Popa MI, Holban AM. Nanocoatings for chronic wound repair—Modulation of microbial colonization and biofilm formation. International Journal of Molecular Sciences. 2018;**19**:1179. DOI: 10.3390/ijms19041179

[116] Alizadeh-Sani M, Arezou Khezerlou A, Ehsani A. Fabrication and characterization of the bionanocomposite film based on whey protein biopolymer loaded with TiO_2 nanoparticles, cellulose nanofibers and rosemary essential oil. Industrial Crops and Products. 2018;**124**:300-315. DOI: 10.1016/j.indcrop.2018.08.001

[117] Barbosa JP, Pereira RC, Abrantes JL Cirne dos Santos CC, Rebello MA, de Palmer Paix Frugulhetti IC, et al. *In vitro* antiviral diterpenes from the Brazilian brown alga *Dictyota pfaffii*. Planta Medica. 2004;**70**:856-860. DOI: 10.1055/s-2004-827235

[118] Jena J, Subudhi E. Microalgae: An untapped resource for natural antimicrobials. In: Sukla L, Subudhi E, Pradhan D, editors. The Role of Microalgae in Wastewater Treatment. Singapore: Springer; 2019. pp. 99-114. DOI: 10.1007/978-981-13-1586-2_8

[119] Bianco V, Mandracchia B, Nazzaro F, Marchesano V, Gennari O, Paturzo M, et al. Food quality inspection by speckle decorrelation properties of bacteria colonies. Optical Methods for Inspection, Characterization, and Imaging of Biomaterials III. 2017:103331N. DOI: 10.1117/12.2272945

[120] Gennari O, Marchesano V, Rega R, Mecozzi L, Nazzaro F, Fratianni F, et al. Pyroelectric effect enables simple and rapid evaluation of biofilm formation. ACS Applied Material Interfaces. 2018;**10**:15467-15476. DOI: 10.1021/acsami.8b02815

Chapter 11

Application of Essential Oils for Shelf-Life Extension of Seafood Products

Marzieh Moosavi-Nasab, Armin Mirzapour-Kouhdasht and Najme Oliyaei

Abstract

This chapter will discuss the antimicrobial and antioxidant activities of various essential oils on possible shelf-life extension of different seafood products. Furthermore, the effect of antimicrobial coatings incorporated with various essential oils on the shelf-life of seafood products will be investigated. Microbiological and physico-chemical properties such as total count, psychrophilic and lactic acid bacterial count, peroxide test, thiobarbituric acid (TBA) test, total volatile basic nitrogen (TVB-N) test, and pH, also sensory evaluations of seafood products will be included. During this chapter the effect of chemical composition of some essential oils on the antimicrobial and antioxidant activities will be discussed briefly.

Keywords: essential oils, shelf-life, algae, seafood, antimicrobial, antioxidant

1. Introduction

The safety and quality of food is one of the most important factors which concerns the related industries as consumers prefer fresh and minimally/not processed products. Using various technical preservation methods have been reported in order to improve the shelf-life extension of seafood. Generally, these techniques are including simple methods like salting and freezing as well as more complicated methods such as chemical preservation and modified atmosphere packaging. Application of chemical and synthetic preservatives in seafood is globally common and convenient. During the last decades, antimicrobial and antioxidant additives, principally synthetic origin, are added to refrigerated seafood products for shelf-life extension. Nonetheless, consumers are interested in the use of natural origin material as alternative preservatives in food, since the safety risks of synthetic preservatives, excess antioxidants added to food might produce toxicities or mutagenicities, has been proved [1, 2].

Essential oils (EOs) are aromatic oily liquids including terpenoids, sesquiterpenes and possibly diterpenes with different groups of aliphatic hydrocarbons, acids, alcohols, aldehydes, acyclic esters or lactones which obtain from plants [3, 4] and algae extract [5, 6]. EOs are also known for their antioxidant, antimicrobial and pharmaceutical properties, thus, they can use as natural additives or preservatives in foods [7–9]. Moreover, EOs extracted from various plants have shown to possess several biological activities and potential health benefits including antidiabetic, anti-inflammatory, anti-viral activities and antiprotozoal agent [10]. Among various techniques for extending the shelf-life of refrigerated seafood products, the application of biopolymer-based edible coatings and films are regularly the method of choice. Edible coatings from polysaccharides, proteins, and lipids can extend the shelf life of foods by functioning as a solute, gas, and vapor barriers [11]. Thus, essential oil incorporation into edible coatings or packaging can prevent the food spoilage and extend the food shelf life in particular fish products [12].

Therefore, great attention has been arisen to identified and used EOs in the food industry. This chapter provides an overview of antioxidant and antimicrobial activities of EOs derived from different sources and their potential organoleptic beneficial and applications in shelf life extension of raw fishes.

2. Chemical composition of EOs

2.1 EOs from algae

Algae extracts are proven to be rich sources of metabolites with a wide range of biological activities such as anti-microbial, anti-oxidant, and pharmaceutical activities [13], thus, several extraction methods have been performed to preparation of algal extract [14] and evaluated their nutritional and pharmacological applications, however, a few number of studies focused on the characterization and composition of EOs from algal extracts. Hence, some scientific efforts have been dedicated to study essential oil composition of algae extracts. The GC-MS analysis of chemical composition shows the presence of different groups of essential oil in micro and macroalgae.

Asparagopsis taxiformis is species of red algae (Rhodophyta) which its EOs consist of bromine and iodine-containing haloforms with the smaller amount of other halogenated methanes and several halogenated ethanes, ethanols, formaldehydes, acetaldehydes, acetones, 2-propanols, 2-acetoxypropanes, propenes, epoxypropanes, acroleins, butenones, halogenated acetic and acrylic acids [15]. Two other red seaweeds (Rhodophyta) *Laurencia obtusa* and *Laurencia obtusa* var. pyramidata are also rich in EOs and 28 components in the oil of *L. obtusa* and 27 components in the oil of *L. obtusa* var. *pyramidata* were identified and 2,6-dimethyl-4-oxa-endo-tricyclodecane was the highest account in both red algae [16].

In addition, the brown macroalgae (Phaeophyta) such as *Colpomenia sinuosa*, *Dictyota dichotoma*, *Dictyota dichotoma* var. *implexa*, *Petalonia fascia* and *Scytosiphon lomentaria* are rich in the EOs. The GC/MS analysis discovered the components including hydrocarbons, terpenes, acids, phenols, sulfur-containing compound, aldehydes, naphthalene skeleton and alcohols in *C. sinuosa*, *D. dichotoma*, *D. dichotoma* var. *implexa*, *P. fascia* and *S. lomentaria*. Among these brown seaweeds, *S. lomentaria* is rich in crown ether (18-crown-6-ether). Moreover, the presence of dihexylsulfide in essential oil profile of *C. sinuosa* revealed the potential of *C. sinuosa* for supplying the rare sulfur-containing compound in seaweeds [17]. Ref. [17] discovered the eight (58.41%) for *D. dichotoma* var. *implexa*, 12 (83.53%) for *D. dichotoma*, 4 (91.71%) for *P. fascia*, 6 (87.89%) for *S. lomentaria* and 14 (74.17%) compounds for *C. sinuosa* in total composition of their essential oil.

Recently, there is interest in the microalgae as well as macroalgae for development of EOs. For this respect, the 50 total compositions of the EOs from *Dunaliella salina* extract were identified and octadecanoic acid, methyl ester (27.43%), hexadecanoic acid, methyl ester (Cas) methyl palmitate (24.82%), 9,12,15-octadecatrienoic acid, ethyl ester, (Z,Z,Z) (7.39%), octadecanoic acid (5.03%), pentadecanoic acid (3.60%) were detected as major compounds [18].

Furthermore, the other various microalgae such as *Stichococcus bacillaris*, *Phaeodactylum tricornutum*, *Microcystis aeruginosa* and *Nannochloropsis oculata* extracts exhibited the antileukemic effects which was related to their EOs. According to [6] findings, the essential oil profile of *S. bacillaris* consist of 5,6-dihydroergosterol, ergost-7-en-3-ol, (3beta)-(CAS)5,6,22,23-tetrahydroergosterol, N-methoxy-N-methylacetamide, 9-octadecenamide, (Z)- and pentan-1,3-dioldiisobutyrate, 2,2,4-trimethyl-compounds. While, *P. tricornutum* essential oil include pentan-1,3-dioldiisobutyrate, 2,2,4-trimethyl-compound. Tricosane (CAS) n-Tricosane was detected only in *P. tricornutum* extract. Further, cyclopropanecaronic acid,-2-phenyl, ethyl ester (E-), molybdenum, bis[(1,2,3,4,5-eta)-1,3-bis(1,1-dimethylethyl)-2,4-cyclopentadien-1-yl]di-mu-carbonyldicarbonyldi-, (mo-mo), 9-octadecenoic acid (Z)-, methyl ester and 9-octadecenoic acid, methyl ester (CAS) methyl octadec-9-enoate were detected only in *M. aeruginosa* extract. Acetic acid 3-isopropyl-8,10,14-trimethyl-16-phenyl-1,2,3,5,6,7,8,9,10,11,12,14- and 2,6-dihydroxybenzoic acid 3TMS were detected only in *N. oculata* extract.

2.2 EOs from other plants

So many researches inquired into the chemical composition of the EOs obtained from various sources including *Thymus ulgaris, Nigella sativa, Achillea millefolium, Curcuma zedoaria*, *Rosmarinus officinalis* etc. A summary of these investigations is reported in **Table 1**. In an outstanding study, the essential oil composition of thyme (*Thymus ulgaris* L.) was investigated by capillary GC/MS evaluation method. The effect of vegetative cycle on the variation of EOs chemical composition was looked over, as well. Generally, the oil was had high amounts of monoterpene

EOs sources	Major components	References
A. taxiformis	Bromine and iodine-containing haloforms	[15]
L. obtusa and *L. obtusa* var. pyramidata	2,6-Dimethyl-4-oxa-endo-tricyclo decane	[16]
S. lomentaria	Crown ether (18-crown-6-ether)	[17]
C. sinuosa	Dihexylsulfide	[17]
D. salina	Octadecanoic acid, methyl ester, hexadecanoic acid, methyl ester (Cas) methyl palmitate, 9,12,15-octadecatrienoic acid, ethyl ester	[18]
N. oculata	Acetic acid 3-isopropyl-8,10,14-trimethyl-16-phenyl-1,2,3,5,6,7,8,9,10,11,12,14- and 2,6-dihydroxybenzoic acid 3TMS	[6]
Thyme (*Thymus ulgaris* L.)	Carvacrol, thymol, *p*-cymene, and γ-terpinene	[19]
Flowering Thyme (*Thymis vulgaris* L.)	Camphor, camphene, α-pinene, 1, 8-cineole, borneol, and β-pinene	[20]
Nigella sativa	Thymoquinone, *p*-cymene, carvacrol, 4-terpineol, t-anethole, and sesquiterpene longifolene	[21]
Turmeric (*Curcuma longa* L.)	*ar*-turmerone, turmerone, and curlone	[22]
Rosemary (*Rosmarinus officinalis* L.)	1,8-cineole, α-pinene, camphor, and camphene	[24]

Table 1.
Some investigations performed to investigate the chemical composition of EOs.

phenols (carvacrol and thymol) and their related monoterpene hydrocarbon precursors (*p*-cymene and γ-terpinene), that demonstrated integrated effects of the different collection periods and seasons on the chemical composition of EOs. The EOs obtained from old plant included much lower quantities of monoterpene hydrocarbones (mostly γ-terpinene) and the highest quantities of the oxygenated monoterpenes (linalool and borneol), monoterpene phenols (mostly thymol) and their derivatives (mostly carvacrol methyl ether), sesquiterpenes (mostly β-caryophyllene) and their oxygenated derivatives (e.g., caryophyllene oxide). A characteristic presence of camphor and thymodihydroquinone was also discovered in the old plant EOs [19].

The EOs obtained by hydrodistillation from flowering Thyme (*Thymis vulgaris* L.) was investigated by GC/FID and GC/MS. The yield of extraction in this study was reported as 1%, in which 43 chemical compounds (97.85% of total constituents) were identified. The EOs extracted from flowering Thyme were mainly consisted of camphor (38.54%), camphene (17.19%), α-pinene (9.35%), 1, 8-cineole (5.44%), borneol (4.91%) and β-pinene (3.90%) [20].

In an another research, seven EOs of *N. sativa*, which were all extracted by soxhlet extraction and steam distillation, were analyzed by GC/MS. A total of 32 compounds were identified. The major fraction of every EOs was a mixture of monoterpenes. The major components were thymoquinone (30–48%), *p*-cymene (7–15%), carvacrol (6–12%), 4-terpineol (2–7%), t-anethole (1–4%) and the sesquiterpene longifolene (1–8%). Very small quantities of the esters of special un/saturated fatty acids were also detected [21].

Curcumin, the yellowish pigment of turmeric, is generated from turmeric oleoresin. In a study performed in order to investigate the antibacterial activity of turmeric oil extracted by hexane and fractionated by silica gel column chromatography, GC/MS analysis identified 13 major components in turmeric oil, fraction I, and fraction II. *ar*-turmerone (62.0%), *trans-â*-farnesene (6.6%), turmerone (5.1%), and curlone (3.9%) were found to be the major compounds in turmeric oil whereas fraction II contained *ar*-turmerone (77.9%), curlone (5.3%), and turmerone (5.2%) [22].

Rosemary (*Rosmarinus officinalis* L.), a member of mint family, is an ordinary aromatic shrub grown in various places around the world [23]. Some researchers has assessed the chemical composition of rosemary EOs to understand the reason of biological activities such as antimicrobial activity. In an experimentation 22 components were identified from this plant by GC/MS. The major constituents were 1,8-cineole (26.54%), α-pinene (20.14%), Camphor (12.88%), and camphene (11.38%) [24].

3. Antioxidant activity of EOs

There are many EOs which have antioxidant activity, and their application as natural antioxidants has been increasingly interested due to harmful effects to human health that some synthetic antioxidants (e.g., BHA and BHT) are faced. The antioxidant activity of EOs is due to their potential ability to cease or suspend the oxidation reaction of organic materials in the presence of oxygen which is a result of some special components including phenols. There are EOs which lack of phenolic compounds also show antioxidant activity. Some constituents including terpenoids and other volatile constituents (such as sulfur-containing components) have special radical chemistry which capable them to express antioxidant activity [25, 26].

As it was discussed earlier (Section 2.2), the major constituents of many EOs can be categorized in two specific structural families: terpenoids (monoterpene, sesquiterpene, and diterpene) and phenylpropanoids, which both comprise phenolic compounds. Some phenolic compounds are demonstrated in **Figure 1**.

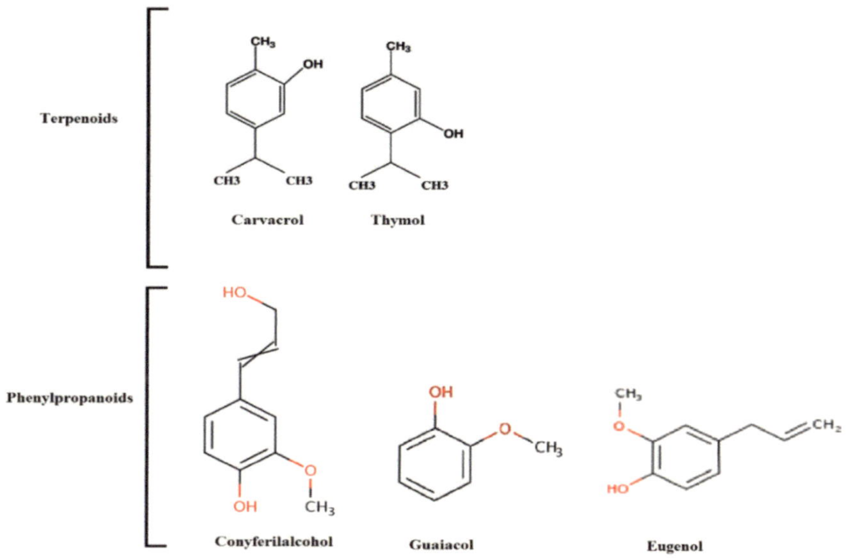

Figure 1.
Some phenolic compounds present in EOs.

$$PhOH + ROO^{\cdot} \longrightarrow PhO^{\cdot} + ROOH \qquad \textbf{a}$$

$$PhO^{\cdot} + ROO^{\cdot} \xrightarrow{\text{Quick}} \text{Non-radical products} \qquad \textbf{b}$$

Figure 2.
Mechanism of antioxidant activity of phenolic compounds. The reaction between phenolic compounds and peroxyl radicals (a), quenching the second peroxyl group by phenoxyl radical (b).

Generally, phenolic compounds can potentially react with peroxyl radicals and transfer the H atom (**Figure 2a**). Due to the stability of phenoxyl radical, it will not continue the radical chain reactions. Instead it will quench the second peroxyl radical quickly (**Figure 2b**).

In contrast with phenolic compounds present in EOs, unsaturated non-phenolic terpenoids such as α-pinene (**Figure 3**) can autoxidize similarly to unsaturated lipids [27].

Many researchers have investigated the antioxidant activity of EOs. A potential antioxidant essential oil was extracted from *Achillea millefolium subsp. millefolium* Afan, which significantly reduced DPPH radical (IC50 = 1.56 μg/ml) and showed lipid peroxidation (IC50 = 13.5 g/ml). The authors demonstrated that the polar phase of the extract exhibited antioxidant activity [28]. The essential oil extracted from dried rhizome *Curcuma zedoaria* (Berg.) Rosc. (Zingiberaceae) showed a moderate to good antioxidant activity at 20 mg/ml. This activity was measured by three different methods, reducing power (good activity), DPPH radical scavenging (excellent activity), and ferrous ion chelating (low activity) [29]. In another research, the antioxidant activity of *Rosmarinus officinalis* essential oil were determined against gastric injury caused by ethanol. Results showed that the *R. officinalis* essential oil (50 mg/kg) ingestion can make gastro-protective influences by reducing the ethanol induced ulcers. The resultant data suggested possible antioxidant mechanism induced by *R. officinalis* essential oil. The major components of chemical composition of this essential oil were cineole (28.5%), camphor (27.7%), and α-pinene (21.3%).

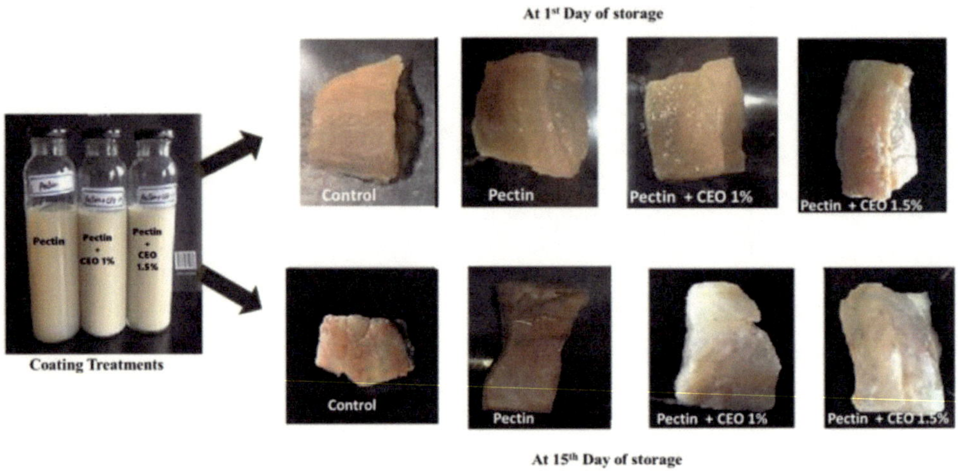

Figure 3.
The effect of pectin-CEO coating on the bream fillets at day 1 and day 15 of the storage (adopted from Nisar et al. [42]).

The lipid oxidation is one of the most important limiting factors for the shelf-life seafood products. For this purpose the antioxidant activity of the EOs of five Mediterranean spices (*Origanum vulgare*, *Thymus vulgaris*, *Rosmarinus officinalis*, *Salvia officinalis*, and *Syzygium aromaticum*) was analyzed. The *S. aromaticum* essential oil, which comprised the highest level of total phenols (898.89 mg/l GAE), demonstrated the highest antioxidant activity (98.74% for DPPH radical inhibition and 1.47 TEAC for FRAP value). The EOs extracted from *T. vulgaris* and *R. officinalis* showed the highest TBARS inhibition (89.84%) and iron (II) chelating (76.06%) activities, respectively [30].

4. Antimicrobial activity of EOs

The antimicrobial effect of essential oils is attributed to actions including alteration of the permeability, and disruption of lipophilic cell membrane. The antimicrobial potential of essential oils can be completely associated with their constituents. Phenolic compounds with their hydrophobicity inherent, breakdown the lipid of cell membrane and mitochondria and enhance the permeability [31, 32]. The inherent of cellular energy generation system (ATP) and damage of proton motive force is a result of changing cell and cytoplasmic membrane permeability [33]. In addition, leakage of internal contents of the cell during the disruption of the membrane is another mechanism which causes cell death [32]. It is generally believed that Gram-negative bacteria are more resistance to essential oils because of their outer hydrophilic cell wall which exhibits inhibitory activity against the penetration of phenolic components [34].

Algae extracts and their components have displayed antimicrobial activity against organisms found in foods. Algae extracts and their components have displayed antimicrobial activity against organisms found in foods. The antibacterial activities of essential oils from two red seaweed *Laurencia obtuse* and *L. obtusa* var. *Pyramidata* showed that the essential oils from Laurencia obtuse exhibited the strong antimicrobial effect against Gram-positive (*Bacillus subtilis*, *Staphylococcus aureus*, methicillin-oxacillin resistant *Staphylococcus aureus*, *Enterococcus faecalis* and *Staphylococcus epidermidis*), Gram-negative bacteria (*Enterobacter cloaceae*) and yeast (*Candida albicans*) [16].

Terpenoids, thymol, carvacrol, β-cubebene, β-eudesmol, β-ionone, dactylol and pachydictol A are the usual volatile compounds in seaweeds and it is known that there is correlation between β-ionone and antibacterial and antifungal activity of seaweeds [17]. However, two known sesquiterpenes (1R*,2S*,3R*,5S*,8S*,9R*)-2,3,5,9-tetramethyltricyclo(6.3.0.0^{1,5})undecan-2-ol and (1S*,2S*,3S*,5S*,8S*,9S*)-2,3,5,9-tetramethyltricyclo-(6.3.0.0^{1,5})undecan-2-ol were isolated from the red macroalgae *Laurencia dendroidea* had no antibacterial effect on eight bacteria strains (*Staphylococcus aureus*, *Bacillus subtilis*, *Enterococcus faecalis*, *Streptococcus pneumonia*, *Klebsiella pneumoniae*, *Salmonella typhi*, *Escherichia coli* and *Pseudomonas aeruginosa*) and the yeast *Candida albicans* [5].

5. Shelf-life extension of seafood products

The short shelf-life of fresh seafood is a practical issue in the industries and distribution chain systems. Short shelf-life caused by chemical and microbial spoilage reactions can be stopped by traditional preservation methods but there is increasing interest in natural preservation methods. EOs are natural antioxidants and antimicrobials by which the shelf-life of seafood can be extended alone or in combination with other techniques. However, the reduction of antimicrobial effect of EOs in a food system due to some components of food and also the reverse action of EOs as antioxidant agents in some cases, has slowed down the use of them in practical systems.

Combination of EOs exhibit the synergistic antimicrobial activity. Thus, using of EOs into packaging can be the safe approach for food preservation technology [35]. The antimicrobial activity of gelatin-chitosan films incorporated with organo essential oil exhibited the great inhibitory effect through reducing the *E. coli*, *S. aureus*, *B. subtilis* and *B. entritidis* growth. Its inhibition zone was larger for Gram-positive bacteria compared with Gram-negative bacteria. Furthermore, the lower total aerobic plate count and total volatile basic nitrogen values that can extend the shelf life of grass carp muscle was recorded in fish muscle packed with film containing the 4% organo essential oil. It seems the high percentage of carvacrol, eugenol, and thymol as phenolic components are responsible for this antimicrobial activity by damaging the cell membrane or interfere the enzyme functionality located on the cell wall. Moreover, the TVB-N value of sample packaged with gelatin-chitosan-EOs film was lower compared with control and the shelf life of grass carp muscle packaged with the film containing EOs was extended to 12 days [36]. The same observation was gained from the gelatin-chitosan film incorporated with other EOs including clove, fennel, cypress, lavender, thyme, herb-of-cross, pine and rosemary for cod fillet preservation. Among all EOs, the high antimicrobial effect was obtained from clove against a wide range of food pathogen and spoilage bacteria such as *Salmonella*, *Lactobacillus*, *Listeria*, *Citrobacter*, *Escherichia*, *Yersinia*, *Brochothrix*, *Staphylococcus*, *Bacillus*, *Listeria*, *Clostridium*, *Aeromonas*, *Shewanella*, *Vibrio* and *Photobacterium*. In addition, the film containing the clove essential oil used for preservation of cod fillets, lowered the microorganisms in particular, *Entrobacteria*. Further, by delaying the formation of TVB-N, can extend the shelf life of chilled stored fish [37].

Immersion of salmon in marinade solution containing 1 w/w% essential oil from organo, cinnamon and thyme revealed that the antimicrobial effect, however, organo and cinnamon essential oil caused to enhance the shelf life of salmon and scampi. In addition, reduction of yeast and mold was observed by cinnamon (1%) addition in marinade for 6 days. Moreover, salmon treated with marinade containing 1% essential oil, shoed appropriate sensorial properties and high hedonic score rather than 5% essential oil [38]. Combination of EOs with different types of packaging is another approach for enhancement of shelf life. For instance, the combination of cinnamon essential oil (1 w/w%) and MAP/vacuum packaging extend the shelf life of salmon. However, the MAP+ cinnamon had a better effect on salmon shelf life and the microbial shelf life reach nine or more days. While it was 6 days for vacuum packaged salmon treated with cinnamon. Moreover, cinnamon had no additional antimicrobial effect on LAB, when salmon stored vacuum or MAP [12].

Furthermore, the vacuum packaged common carp (*Cyprinus carpio*) stored 4°C had high quality with the combination of cinnamon essential oil. In addition, cinnamon essential oil inhibited the *Aeromonas* and *Lactococcus* on day 10. *Pseudomonas* and H2S-producing bacteria count was lower in treated fillets and did not exceed the microbial level of 7 log CFU/g at the end of the fillet's shelf life. Moreover, was effective in inhibition the increase of TVB-N and the accumulation of biogenic amines. TVB-N value fluctuated in 6.21 and 9.90 mg/100 g before 12 days treated sample contained and the highest value (15.15 mg/100 g) occurred at day 14. Moreover, crap fillets treated with cinnamon essential oil exhibited the acceptance longer sensorial shelf life (14 days) [39]. The Flounder fillet covered with clove essential oil agar films (0.5 g clove essential oil/g agar) had high low microbial count because of the great antimicrobial activity of clove essential oil against pathogens such as *Staphylococcus aureus*, *Yersinia enterocolitica*, *Aeromonas hydrophila*, *Debaryomyces hansenii* and *Listeria innocua*. The chemical indicators such as TVB-N was 25.83 mg TVB-N/100 g after 15 days of storage. The low total volatile bases and pH values and inhibitory effect on H_2S producing microorganisms suggested clove essential oil could be suitable biopreservative for the flour fillet shelf life extension [40].

In another study, the active films accommodated by poly lactic acid enriched with ZnO nanoparticles (1.5%w/w) and *Zataria multiflora* Boiss (0.5, 1, 1.5%w/w) and the effect of this film on shelf-life extension of refrigerated *Otolithes ruber* fillet during 16 days was investigated. One aspect of shelf-life extension effect is the antibacterial activity of the films which in this case was conducted against *Escherichia coli*, *Salmonella enterica*, *Pseudomonas aeruginosa*, *Bacillus cereus* and *Staphylococcus aureus* by disc diffusion procedure. PLA/ZnO/ZEO and PLA/ZnO/MEO [*Zataria multiflora* Boiss. essential oil (ZEO) and *Menta piperita* L. essential oil] films demonstrated magnified antibacterial (691 and 513.33 mm^2, respectively, against *S. aureus*). The authors expressed that according to the microbial count, the active film remarkably enhanced the shelf-life extension from 8 to 16 days. Chemical factors such as TBARS and TVB-N were also determined. The fillet wrapped with PLA/ZnO containing 1.5% ZEO, showed the lowest TBARS (0.8 mg MA/kg muscle) and TVB-N (21.23 mg/100 g muscle). GC/MS analysis of EOs showed that the carvacrol and menthone were the major components of ZEO and MEO, respectively [41].

A new edible coating of pectin containing clove essential oil (CEO), was assessed to extension of bream (*Megalobrama ambycephala*) fillets shelf-life during 15 days. Physicochemical (pH, PV, TBA and TVB-N), microbiological (Total viable count, Psychrophilic bacteria, Lactic acid bacteria, *Enterobacteriaceae*, *Pseudomonas* spp., H_2S producing bacteria) and organoleptic characteristics were analyzed to determine the influences of the pectin-CEO coating. Physicochemical analysis revealed that lipid oxidation decreased. Some other factors such as weight loss, water holding capacity, color, and texture of the fillets were improved as a result of coating with pectin-CEO (**Figure 3**). During 15 days, lactic acid bacteria were not affected by coating. However, the effects of coating on bacterial growth, especially on Gram-negative bacteria, was observed [42].

The effect of chitosan, thyme essential oil and their combination, on the shelf-life of vacuum packaged smoked eel fillets at 4°C, was investigated and according to sensory odor analysis the shelf-life of chitosan/thyme and chitosan-thyme combination treated samples extended 1 and >2 weeks, respectively, compared with than control sample (35 days for control). The control sample showed a significantly higher thiobarbituric acid value compared chitosan-thyme combination treated sample. Control, thyme, chitosan, and chitosan-thyme combination treated samples showed TVB-N values below the maximum permissible level (35 mg N/100 g) in fish and fishery products which was 31.5, 18.1, 14.9, and 13.1 mg N/100 g in 35 and 42, 49 days of storage, respectively [43]. The maximum permissible level of TVB-N in fish and fishery products is 35 mg N/100 g [44].

6. Conclusions

During the past decades, EOs have achieved great attention due to their food preservation effects, particularly for the antimicrobial and antioxidant effects.

The EOs of different sources from land and the seas, have variety of phenolic and non-phenolic components which the most actives are low molecular weight terpenoids, terpenes, and aliphatic chemicals (obtained data from analysis by GC-MS and GC/FID in literature). These EOs have shown significant antioxidant, antimicrobial activities which can extent the shelf-life of seafood products. However, it is still mandatory to inquire into cytotoxicity and toxicity of these EOs.

Conflict of interest

The authors declare no conflict of interest.

Author details

Marzieh Moosavi-Nasab[1,2]*, Armin Mirzapour-Kouhdasht[1,2] and Najme Oliyaei[1,2]

1 Department of Food Science and Technology, School of Agriculture, Shiraz University, Shiraz, Iran

2 Seafood Processing Research Group, School of Agriculture, Shiraz University, Shiraz, Iran

*Address all correspondence to: marzieh.moosavi-nasab@mail.mcgill.ca

References

[1] Williams GM. Inhibition of chemical-induced experimental cancer by synthetic phenolic antioxidants. Toxicology and Industrial Health. 1993;**9**(1-2):303-308

[2] Williams G. Interventive prophylaxis of liver cancer. European journal of cancer prevention: The official journal of the European Cancer Prevention Organisation (ECP). 1994;**3**(2):89-99

[3] Tavakolpour Y, Moosavi-Nasab M, Niakousari M, Haghighi-Manesh S, Hashemi SMB, Mousavi Khaneghah A. Comparison of four extraction methods for essential oil from thymus daenensis subsp. Lancifolius and chemical analysis of extracted essential oil. Journal of Food Processing and Preservation. 2017;**41**(4):e13046.a

[4] Tajkarimi M, Ibrahim SA, Cliver D. Antimicrobial herb and spice compounds in food. Food Control. 2010;**21**(9):1199-1218

[5] Gressler V et al. Sesquiterpenes from the essential oil of *Laurencia dendroidea* (*Ceramiales*, *Rhodophyta*): Isolation, biological activities and distribution among seaweeds. Revista Brasileira de Farmacognosia. 2011;**21**(2):248-254

[6] Atasever-Arslan B et al. Screening of new antileukemic agents from essential oils of algae extracts and computational modeling of their interactions with intracellular signaling nodes. European Journal of Pharmaceutical Sciences. 2016;**83**:120-131

[7] Tavakolpour Y, Moosavi-Nasab M, Niakousari M, Haghighi-Manesh S. Influence of extraction methods on antioxidant and antimicrobial properties of essential oil from thymua danesis subsp. Lancifolius. Food Science & Nutrition. 2016;**4**(2):156-162

[8] Moosavi-Nasab M, Jamalian J, Heshmati H, Haghighi-Manesh S. The inhibitory potential of zataria multiflora and syzygium aromaticum essential oil on growth and aflatoxin production by aspergillus flavus in culture media and Iranian white cheese. Food Science & Nutrition. 2018;**6**(2):318-324

[9] Popović-Djordjević J et al. *Calamintha incana*: Essential oil composition and biological activity. Industrial Crops and Products. 2019;**128**:162-166

[10] Raut JS, Karuppayil SM. A status review on the medicinal properties of essential oils. Industrial Crops and Products. 2014;**62**:250-264

[11] Volpe M et al. Active edible coating effectiveness in shelf-life enhancement of trout (*Oncorhynchus mykiss*) fillets. LWT-Food Science and Technology. 2015;**60**(1):615-622

[12] Van Haute S et al. Combined use of cinnamon essential oil and MAP/vacuum packaging to increase the microbial and sensorial shelf life of lean pork and salmon. Food Packaging and Shelf Life. 2017;**12**:51-58

[13] Kim S-K, Chojnacka K. Marine Algae Extracts: Processes, Products, and Applications. Germany: John Wiley & Sons; 2015

[14] Michalak I, Chojnacka K. Production of seaweed extracts by biological and chemical methods. In: Marine Algae Extracts: Processes, Products, and Applications. Germany: John Wiley & Sons; 2015. pp. 121-144

[15] Kladi M, Vagias C, Roussis V. Volatile halogenated metabolites from marine red algae. Phytochemistry Reviews. 2004;**3**(3):337-366

[16] Demirel Z et al. Antimicrobial and antioxidant activities of solvent extracts and the essential oil composition of

Laurencia obtusa and *Laurencia obtusa var. pyramidata*. Romanian Biotechnological Letters. 2011;**16**(1):5927-5936

[17] Demirel Z et al. Antimicrobial and antioxidant activity of brown algae from the Aegean Sea. Journal of the Serbian Chemical Society. 2009;**74**(6):619-628

[18] Atasever-Arslan B et al. Cytotoxic effect of extract from *Dunaliella salina* against SH-SY5Y neuroblastoma cells. General Physiology and Biophysics. 2015;**34**:201-207

[19] Hudaib M et al. GC/MS evaluation of thyme (*Thymus vulgaris* L.) oil composition and variations during the vegetative cycle. Journal of Pharmaceutical and Biomedical Analysis. 2002;**29**(4):691-700

[20] Imelouane B et al. Chemical composition and antimicrobial activity of essential oil of thyme (*Thymus vulgaris*) from eastern Morocco. International Journal of Agriculture and Biology. 2009;**11**(2):205-208

[21] Burits M, Bucar F. Antioxidant activity of *Nigella sativa* essential oil. Phytotherapy Research. 2000;**14**(5):323-328

[22] Negi P et al. Antibacterial activity of turmeric oil: A byproduct from curcumin manufacture. Journal of Agricultural and Food Chemistry. 1999;**47**(10):4297-4300

[23] Moss M et al. Aromas of rosemary and lavender essential oils differentially affect cognition and mood in healthy adults. International Journal of Neuroscience. 2003;**113**(1):15-38

[24] Jiang Y et al. Chemical composition and antimicrobial activity of the essential oil of rosemary. Environmental Toxicology and Pharmacology. 2011;**32**(1):63-68

[25] Golmakani MT, Moosavi-Nasab M, Keramat M, Mohammadi MA. Arthrospira platensis extract as a natural antioxidant for improving oxidative stability of common kilka (Clupeonella cultriventris caspia) oil. Turkish Journal of Fisheries and Aquatic Sciences. 2018;**18**(11):1315-1323

[26] Valgimigli L et al. Essential oils: An overview on origins, chemistry, properties and uses. Essential Oils as Natural Food Additives. New York: Nova Science Publishers; 2012. pp. 1-24

[27] Neuenschwander U, Guignard F, Hermans I. Mechanism of the aerobic oxidation of α-Pinene. ChemSusChem: Chemistry & Sustainability Energy and Materials. 2010;**3**(1):75-84

[28] Candan F et al. Antioxidant and antimicrobial activity of the essential oil and methanol extracts of *Achillea millefolium* subsp. *millefolium* Afan. (Asteraceae). Journal of Ethnopharmacology. 2003;**87**(2-3):215-220

[29] Mau J-L et al. Composition and antioxidant activity of the essential oil from *Curcuma zedoaria*. Food Chemistry. 2003;**82**(4):583-591

[30] Viuda-Martos M et al. Antioxidant activity of essential oils of five spice plants widely used in a Mediterranean diet. Flavour and Fragrance Journal. 2010;**25**(1):13-19

[31] Moosavi-Nasab M, Jamal Saharkhiz M, Ziaee E, Moayedi F, Koshani R, Azizi R. Chemical compositions and antibacterial activities of five selected aromatic plants essential oils against food-borne pathogens and spoilage bacteria. Journal of Essential Oil Research. 2016;**28**(3):241-251

[32] Bajpai VK, Baek K-H, Kang SC. Control of salmonella in foods by using essential oils: A review. Food Research International. 2012;**45**(2):722-734

[33] Li M et al. Use of natural antimicrobials from a food safety

perspective for control of *Staphylococcus aureus*. Current Pharmaceutical Biotechnology. 2011;**12**(8):1240-1254

[34] Soković M et al. Antibacterial effects of the essential oils of commonly consumed medicinal herbs using an in vitro model. Molecules. 2010;**15**(11):7532-7546

[35] Moosavi-Nasab M, Shad E, Ziaee E, Yousefabad SHA, Golmakani MT, Azizinia M. Biodegradable chitosan coating incorporated with black pepper essential oil for shelf life extension of common carp (Cyprinus carpio) during refrigerated storage. Journal of Food Protection. 2016;**79**(6):986-993

[36] Wu J et al. Properties and antimicrobial activity of silver carp (*Hypophthalmichthys molitrix*) skin gelatin-chitosan films incorporated with oregano essential oil for fish preservation. Food Packaging and Shelf Life. 2014;**2**(1):7-16

[37] Gómez-Estaca J et al. Biodegradable gelatin-chitosan films incorporated with essential oils as antimicrobial agents for fish preservation. Food Microbiology. 2010;**27**(7):889-896

[38] Van Haute S et al. The effect of cinnamon, oregano and thyme essential oils in marinade on the microbial shelf life of fish and meat products. Food Control. 2016;**68**:30-39

[39] Zhang Y et al. Effect of cinnamon essential oil on bacterial diversity and shelf-life in vacuum-packaged common carp (*Cyprinus carpio*) during refrigerated storage. International Journal of Food Microbiology. 2017;**249**:1-8

[40] da Rocha M et al. Effects of agar films incorporated with fish protein hydrolysate or clove essential oil on flounder (*Paralichthys orbignyanus*) fillets shelf-life. Food Hydrocolloids. 2018;**81**:351-363

[41] Heydari-Majd M et al. A new active nanocomposite film based on PLA/ZnO nanoparticle/essential oils for the preservation of refrigerated *Otolithes ruber* fillets. Food Packaging and Shelf Life. 2019;**19**:94-103

[42] Nisar T et al. Physicochemical responses and microbiological changes of bream (*Megalobrama ambycephala*) to pectin based coatings enriched with clove essential oil during refrigeration. International Journal of Biological Macromolecules. 2019;**124**:1156-1166

[43] El-Obeid T et al. Shelf-life of smoked eel fillets treated with chitosan or thyme oil. International Journal of Biological Macromolecules. 2018;**114**:578-583

[44] EEC D. 95/149/EC, Total volatile basic nitrogen TVBN limit values for certain categories of fishery products and specifying the analysis methods to be used. Official Journal L. 1995;**97**:84-87

Chapter 12

Seed Propagation and Constituents of the Essential Oil of *Stevia serrata* Cav. from Guatemala

Juan Francisco Pérez-Sabino, Max Samuel Mérida-Reyes, José Vicente Martínez-Arévalo, Manuel Alejandro Muñoz-Wug, Bessie Evelyn Oliva-Hernández, Isabel Cristina Gaitán-Fernández, Daniel Luiz Reis Simas and Antonio Jorge Ribeiro da Silva

Abstract

Stevia serrata Cav. (Eupatorieae, Asteraceae) grows in Central America and Mexico usually over 1500 m. In this study, essential oils of aerial parts from three populations of western Guatemala were obtained yielding 0.17–0.27% of oil by hydrodistillation. Chamazulene (42–62%) was the most abundant compound in the oil analyzed GC/MS, also presenting germacrene D (4.4–15.3%), caryophyllene oxide (3.2–11.8%), (*E*)-nerolidol (3.9–7.1%), spathulenol (2.3–7.9%), and (*E*)-caryophyllene (2.5–6.6%). Besides, a propagation trial was carried out on seeds of plants collected in Santa Lucía Utatlán, as the first step for the domestication of the plant, obtaining approximately 75% survival in the transplanting of the germinated seedlings. After the flowering of the individuals, a greenish essential oil was obtained from the roots yielding 0.2% of oil. This oil did not present chamazulene, but α-longipinene (23.5%), germacrene D (22.2%), santolina triene (12.6%), and (*E*)-caryophyllene (8.1%) as major components. As conclusion, it was confirmed that the aerial parts of the essential oil of *S. serrata* from western Guatemala presents a high content of chamazulene and that there is feasibility for the domestication of the plant through the germination of seeds.

Keywords: α-longipinene, chamazulene, Guatemala, sesquiterpenes, *Stevia serrata*

1. Introduction

The high biodiversity of Guatemala, caused by the great variety of microclimates and the convergence of the flora of North and South America, presents plants that have developed a large number of secondary metabolites to fulfill functions of defense and interaction with the environment. Many of these metabolites have biological and pharmacological activities that are used by communities, through the use of plants for the treatment of different diseases [1]. In this way, many investigations have been carried out aimed at determining the composition and biological activity of the metabolites of different medicinal plants used in Guatemala [2–5].

One of the biodiverse plants of Guatemala, which also grows in neighboring countries and for which no medicinal uses have been reported in Guatemala, is *Stevia serrata* Cav. [6] whose essential oil presents chamazulene in high proportions. Chamazulene is a substance of intense blue color of high economic value, which has been shown to have a high anti-inflammatory activity [7].

The genus *Stevia* belongs to the Asteraceae family within the Eupatorieae tribe [8]. It is a New World genus distributed from the south of the United States to Argentina and the highlands of Brazil, passing through Mexico, the Central American countries, and the South American Andes [9, 10]. The records indicate that the genus is not represented in the Antilles or the Amazon. The members of the *Stevia* genus are found mainly at altitudes between 500 and 3500 m. Although they usually grow in semidry mountainous terrain, their habitats range from meadows, leafy forests, forested mountain slopes, coniferous forests, to subalpine vegetation [8].

The genus *Stevia* consists of between 220 and 230 accepted species. Of these, only about 34 (15%) have some type of ethnobotanical record that relate uses with common names of the species. Of these 34 species, only the South American species *Stevia rebaudiana* (Bertoni) Bertoni presents records of outstanding use because its sweet leaves are used for imparting sweetness to beverages and foods [8, 12]. Due to this, *S. rebaudiana* is of great economic importance internationally, given its intensive commercialization due to its use as a natural low calorie sweetener [8].

The sesquiterpenoids are by far the majority and characteristic constituents of the aerial parts and roots of the *Stevia* genus. The overwhelming majority of these compounds belong to the guaiane, longipinane, and germacrene groups [8]. Derivatives of longipinene have been isolated and elucidated mainly in roots of *S. eupatoria*, *S. porphyria*, and *S. pilosa* in Mexico, in *S. triflora* from Venezuela, and in *S. lucida* of Colombia [13–18]. Diterpene glycosides have been isolated from commercial extracts of *S. rebaudiana* leaves in Malaysia [19, 20]. The composition of the essential oil of plants of the genus *Stevia* has been determined in leaves of *S. urticifolia* in Brazil being the main components found the oxygenated sesquiterpene α-cadinol (8.6%) and the sesquiterpene hydrocarbon germacrene D (10.4%) [21].

On the other hand, the composition of the essential oil of *S. rebaudiana* leaves analyzed in Nigeria showed carvacrol (67.89%), caryophyllene oxide (23.50%), spathulenol (15.41%), cardinol (5.59%), α-pinene (3.75%), ibuprofen (1.79%), isopinocarveol (1.26%), and α-caryophyllene (1.15%) as the main components found [22].

Other types of compounds isolated in plants of this genus include four flavonoids isolated from the aerial parts of *S. urticifolia* in Brazil [23], two triterpenes isolated from the roots of *S. viscida* and *S. eupatoria* from Mexico [24], the breviarolide and guaianolide isolated from the aerial parts of *S. breviaristata* from Argentina [25], and the stephalic acid isolated from the whole plant of *S. polycephala* from Mexico [26]. Nineteen hydroxycinnamic acid derivatives were successfully characterized in *S. rebaudiana* leaves: three monocaffeoylquinic acids (Mr354), seven dicaffeoylquinic acids (Mr516), one *p*-coumaroylquinic acid (Mr338), one feruloylquinic acid (Mr368), two caffeoyl-feruloylquinic acids (Mr530), three caffeoylshikimic acids (Mr336), and two tricaffeoylquinic acids (Mr678) [12].

Likewise, two new *stevia* amino acid sweeteners have been synthesized from natural stevioside: *stevia* glycine ethyl ester and *stevia* L-alanine methyl ester. The sweetness intensity rate of the new sweeteners was higher than sucrose, and they also had a clean sweetness without the unpleasant bitter aftertaste of stevioside [27]. *Stevia* products have been elaborated as an infusion with suitable organoleptic characteristics using a formulation of 80–85% of leaves + dried flowers of anise (*Tagetes filifolia* Lag.) and 15–20% of dried *stevia* leaves (*S. rebaudiana*) [28].

As for the *S. serrata* plant, it is distributed from southern Arizona, New Mexico and Texas to northern Oaxaca, and from Chiapas to Honduras, Colombia, Venezuela, and Ecuador. In Guatemala it is found in the departments of Chimaltenango, Guatemala, Huehuetenango, Quetzaltenango, El Quiché, Sacatepéquez, and Sololá [6, 11].

The species grows along pastures and roadsides in various habitats from *Yucca-Opuntia* scrub, sand pine woods, steep rock outcrops in *Quercus-Acacia* grasslands, and pastured slopes, usually between 900 and 2800 m. The plants prefer sunny, stony, well-drained places but also grow in moist pastures and other flat areas [6, 11]. They grow as erect perennial herbs to 0.6–1 m, the stems single to many, puberulent to densely pilose. Leaves alternate, scattered and often crowded, sessile to subsessile, serrate toward the apex, 2.5–6.5 cm long, 0.2–1.5 cm wide, apex rounded to acute. Capitula 5–9 mm, phyllaries 3.5–6 mm long, 0.7–1 mm wide, puberulent with numerous glandular dots. Corollas white, 3–5 mm long, often gland-dotted, lobes 1–1.5 mm long, puberulent. Achenes are usually heteromorphic, 2.2–4.2 mm long, hispid. Pappus of the four adelphocarps with 3–5 awns equaling the corolla and alternating with 3–5 scales, 0.2–0.7 mm long [6, 11].

As for the chemistry of *S. serrata*, five new derivatives of longipinene have been isolated and elucidated from the roots of the plant in Mexico, these being 7β,9α-diangeloyloxy-8α-hydroxylongipinan-1-one; 8β,9α-diangeloyloxy-9α-hydroxylongipinan-1-one; 7β,9α-diangeloyloxy-8α-acetyloxylongipinan-1-one; 7β,9α-diangeloyloxy-8α-acetyloxylongipin-2-en-1-one; and 7β-angeloyloxy-8α-isobutyryloxylongipin-2-en-1-one [29]. Likewise, in Mexico, two new prochamazulene sesquiterpene lactones from the dried leaves of *S. serrata* from Mexico were isolated and identified: steviserrolide A and steviserrolide B [30]. The presence of the R enantiomer of chamazulene carboxylic acid (**Figure 1**) of *S. serrata* from Central America was determined [31].

Regarding studies of the essential oil of the plant, the distillation of 178 g of flowers of *S. serrata* from Mexico provided 700 mg of the blue essential oil, which yielded 320 mg of chamazulene [32]. The compounds found in highest concentration in the essential oil of *S. serrata* from Guatemala were the sesquiterpenes chamazulene (60.1%), (*E*)-nerolidol (7.3%), caryophyllene oxide (6.3%), and germacrene D (5.4%) [33], which are shown in **Figure 2**. Chamazulene is produced from prochamazulenic sesquiterpenlactones. Among these precursors, matricine (**Figure 2**) and the carboxylic acid of chamazulene, among others, have been identified, which are present in the plant and are transformed into chamazulene by the action of the temperature during the steam extraction process [31]. Other compounds isolated from the plant include the methyl-ripariochromene A from the dried leaves of *S. serrata* cultivated in Japan [34].

The plant, known in Mexico as "tlachichinole," was used in decoction of the aerial parts for the washing of infected pimples [8], while the "donkey chili" or "sheep tail" is used as medicine to treat intestinal discomforts in Honduras [35]; the

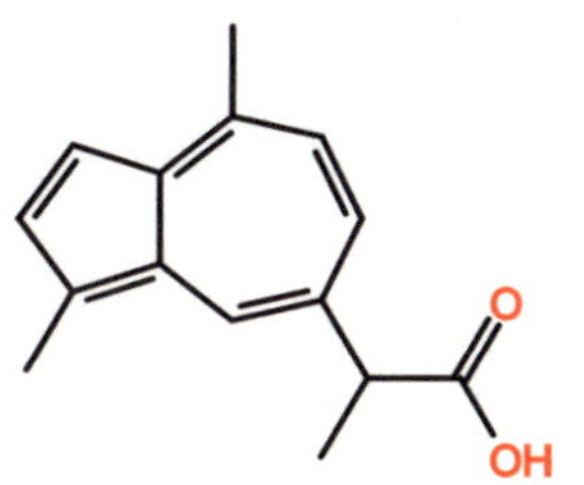

Figure 1.
Chamazulene carboxylic acid.

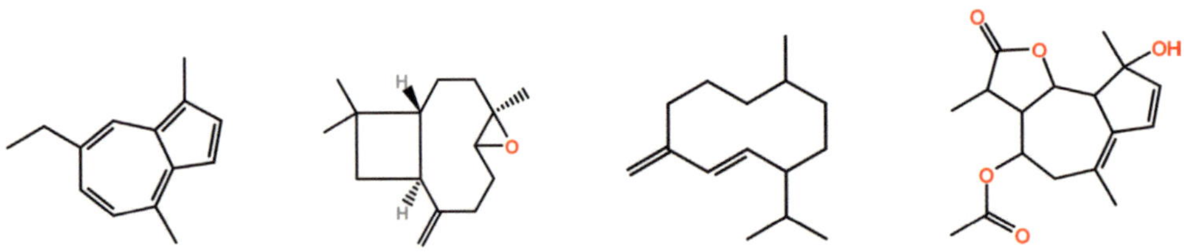

Figure 2.
From left to right, structures of chamazulene, caryophyllene oxide, germacrene D, and matricine.

decoction of the "October flower" is used by the midwives to accelerate the contractions of the parturients during childbirth [36]. Oral administration of *S. serrata* essential oil from Guatemala produced a marked antinociceptive activity in mice in the formalin test [33].

The purpose of the study was to determine the composition of the essential oil of aerial parts of *S. serrata* from different localities of the Guatemalan highlands, to evaluate the variability of the content of chamazulene. The capability of propagation of plants of *S. serrata* was also determined by a seed propagation trial. Finally, the composition of the essential oil of the roots of the propagated plants was determined to compare it with the composition of the oil extracted from aerial parts of the plant.

2. Methodology

2.1 Collection and preparation of plant material

Aerial parts of *S. serrata* were collected from populations in different localities (**Table 1**) during 2018. The plant material was dried in a solar dryer at a temperature between 30 and 35°C and immediately extracted. **Figure 3** shows pictures of the population in Santa Cruz del Quiché, Quiché, and details of floral button of the plant.

2.2 Seed germination

Seeds of *S. serrata* were collected in the surroundings of Santa Lucia Utatlán, Sololá (N 14° 46 40.4″ W 091° 14 41.5″/2430 m), in December 2015. Seeds were stored in trays inside a solar dryer at a temperature between 30 and 35°C for 2 months.

After drying, seeds were manually removed from the flower receptacles and subsequently placed for germination in peat moss previously moistened into plastic strainers (**Figure 4**).

2.3 Transplantation of seedlings and root obtention

The seedlings obtained were transplanted to 4-gallon flowerpots containing potting soil. The plants were placed in direct sunlight and watered daily. After the seed production by the individuals grown in pots, their roots were removed, washed, and dried in a solar dryer. Then, the roots were pulverized in a forage mill for the extraction of the essential oil.

2.4 Extraction of essential oil

The oil from 50.0 g of aerial parts of *S. serrata* was extracted by hydrodistillation using a Clevenger-type apparatus for 2 h. It was then weighed with an analytical scale. The extraction of the essential oil of 100 g of powdered roots was carried out

Localities and dates of collection of individuals of *S. serrata*

Locality	Sample code	Organ	Geographic position	Altitude (m)	Collection date	Phenological stage
San Miguel Ixtahuacán, San Marcos.	SS3	aerial parts	N 15° 14'21.7" W 091° 41'31.4"	2093	21/08/2018	Flowering
Santa Cruz del Quiché, Quiché	SS4	aerial parts	N 14° 59'03.7" W 091° 07' 15.0"	2013	10/07/2018	Floral button
Santa María Chiquimula. Totonicapán	SS5	aerial parts	N 14° 58'30.4" W 091° 25'59.2"	2830	13/06/2018	Vegetative
Santa Lucía Utatlán, Sololá*	S-SLU	roots	N 14° 46'40.4" W 091° 14'41.5"	2430	/05/2017	Vegetative

*The roots were obtained from the first generation of plants cultivated in Guatemala city using seeds from this locality.

Table 1.
Localities and dates of collection of individuals of S. serrata.

Figure 3.
Population of S. serrata in Santa Cruz del Quiché, Quiché, on the left and details of S. serrata in floral button stage on the right.

Figure 4.
Germinated seeds of S. serrata on the left, seedlings in peat moss in the middle, and transplanted plants on the right.

in the same Clevenger-type apparatus for 2 h. The essential oils of the aerial parts and of the roots were collected in pentane which was later removed in a rotatory evaporator at 40°C. All the extractions were made in triplicate, and the reported yield corresponds to the average of the three extractions.

2.5 Gas chromatography coupled to mass spectrometry analyses (GC/MS)

GC/MS analyses were performed using a chromatograph Shimadzu 2010 Plus system coupled with a Shimadzu QP-2010 Plus selective detector (MSD) and equipped with a DB5-MS capillary fused silica column (60 m, 0.25 mm I.D., 0.25 μm film thickness). The oven temperature program initiated at 60°C, then was raised by 3°C/min to 246°C, and then was held for 20 min. Other operating conditions were as follows: carrier gas, He (99.999%), with a flow rate of 1.03 mL/min; injector temperature, 220°C; split ratio of 1:50; and injection volume of 1 μL. Mass spectra were taken at 70 eV. The m/z values were recorded in the range of m/z 40–700 Da.

3. Results

Tables 2 and **3** present the results of yields and chemical composition of the essential oils of the three sampled populations of *S. serrata* and roots of plants obtained by seed propagation, respectively. Chamazulene was the major component of the essential oils of the aerial parts meanwhile α-longipinene was the compound found in major proportion in the essential oil of the roots.

4. Discussion

4.1 Essential oil of aerial parts of *S. serrata*

Table 2 shows the yield and composition results of the intense blue essential oil obtained from the aerial parts of individuals of *S. serrata* collected in three different populations distinct of the population sampled in a previous study of the chemical composition of oil of *S. serrata* from Guatemala [33]. The three populations are located in the highlands of western Guatemala. Extraction yields were between 0.2 and 0.3% (w/w) (**Table 3**), corresponding the highest yield to the SS4 oil from Santa Cruz del Quiché. A probable explanation for the difference in yields among the sampled populations is that the production of essential oil depends on the phenological stage, so that there is a greater production of oil in the flowering stage and lower production in the fruiting stage.

Another probable explanation could be edaphic factors affecting the production of secondary metabolites in general, but only after new investigations could the relationship between these factors and the production of essential oil and other metabolites be determined.

Regarding the chemical composition analyzed by GC/MS, 22 compounds were identified in the SS3 (94.7% of the total area) and SS4 (97.6% of the total area) oils and 18 compounds in the SS5 oil (98.4% of the total area). A chromatogram of the essential oil of SS4 is shown in **Figure 5**. The most abundant compound was the chamazulene in area percentages between 42 and 62%, with the highest percentage corresponding to the SS5 essential oil. The mass spectrum of chamazulene from the essential oil of sample SS4 is shown in **Figure 6**. The other compounds found in high percentage in the oil were germacrene D (4.4–15.3%), caryophyllene oxide (3.2–11.8%), (*E*)-nerolidol (3.9–7.1%), spathulenol (2.3–7.9%) and (*E*)-caryophyllene (2.5–6.6%). The α-longipinene, frequently found in *Stevia* genus plants [8] that had not been reported in the essential oil of *S. serrata*, was found in the SS4 oil in 0.4%.

RI	**Compound**	**SS3**	**SS4**	**SS5**
	yield	**0.2**	**0.29**	**0.17**
939	α-pinene	0.6	--	--
1353	α -longipinene	--	0.4	--
1388	β-bourbonene	0.4	--	--
1419	*(E)*-caryophyllene	6.6	2.5	4.0
1441	aromadendrene	--	0.2	--
1455	α -humulene	1.2	0.6	0.6
1480	γ-muurolene	0.7	1.1	0.4
1485	germacrene D	4.4	8.7	15.3
1493	bicyclogermacrene	1.7	0.6	0.2
1500	α -muurolene	1.6	0.8	2.3
1502	epizonarene	0.5	--	--
1512	NI	0.3	0.3	--
1514	γ-cadinene	0.6	0.7	0.3
1523	δ-cadinene	1.8	2.4	1.0
1539	α -cadinene	--	0.2	--
1555	NI	0.5	0.2	--
1563	*(E)*-nerolidol	7.1	4.1	3.9
1572	aromadendrene oxide	0.4	0.8	0.3
1575	NI	0.4	--	--
1578	spathulenol	7.9	6.0	2.3
1583	caryophyllene oxide	11.8	9.0	3.2
1587	isoaromadendrene epoxide	--	0.3	--
1601	guaiol	0.6	--	--
1608	humulene epoxide II	0.5	0.4	0.2
1616	NI	--	0.3	--
1624	10-epi-γ-eudesmol	1.0	--	--
1634	NI	0.3	--	--
1640	epi- α -cadinol	0.6	0.5	0.2
1646	NI	--	0.3	0.3
1648	NI	--	0.2	--
1651	NI	0.4	--	--
1654	α -cadinol	1.1	1.3	1.0
1660	caryophyllene<14-hydroxy-9-epi-*(E)*->	--	0.2	0.4
1672	NI	0.9	0.6	--
1685	NI	--	0.3	0.2
1693	NI	--	0.4	--
1698	Eudesm-7(11)-en-4-ol	0.7	0.7	0.3
1732	chamazulene	42.9	56.1	62.5
1780	NI	--	0.6	0.6
		94.7	97.6	98.4

NI: Not identified

Table 2.
Composition of the essential oil of the aerial parts of S. serrata from three localities.

The results confirm that essential oil of *S. serrata* with high content of chamazulene can be obtained from the different populations of the Guatemalan highlands. The authors consider that although the extraction yield in all the samples has been lower than 0.3%, the plant presents economic potential for its domestication for oil production in view of its high content of chamazulene and the presence in it of other components for which pharmacological activity has been reported.

RI	Compound	Area %
	yield	0.2
909	santolina triene	12.6
939	α-pinene	1.7
979	β-pinene	0.2
1003	α-phellandrene	0.6
1030	β-phellandrene	0.7
1037	NI	0.2
1238	trans-chrysanthenyl acetate	0.3
1261	NI	0.2
1267	NI	0.3
1290	lavandulyl acetate	4.9
1337	NI	1.1
1353	α-longipinene	23.5
1362	neryl acetate	1.9
1374	longicyclene	1.4
1375	α-ylangene	4.8
1380	NI	0.2
1401	β-longipinene	1.4
1408	longifolene	0.5
1419	*(E)*-caryophyllene	8.1
1451	α-himachalene	0.6
1457	*(E)*-*b*-farnesene	3.5
1483	γ-himachalene	1.3
1485	germacreno D	22.2
1500	bicyclogermacrene	1.0
1505	β-himachalene	1.7
1506	β-bisabolene	0.3
1583	caryophyllene oxide	1.2
1609	NI	1.1
1623	NI	0.5
1637	NI	0.3
1651	vulgarone B	0.5
1654	himachalol	0.3
1730	NI	0.4
1743	Cedr-8(15)-en-9-alpha-ol acetate	0.6
		95.8

NI: Not identified

Table 3.
Composition of the essential oil of roots of propagated S. serrata.

When comparing this source of essential oil with chamazulene content in the oil of *Matricaria recutita* L. (Asteraceae), which is obtained only from the flowers of this species [31], *S. serrata* is shown as a promising species because all aerial parts (leaves, stems, and flowers) produce essential oil with high chamazulene content.

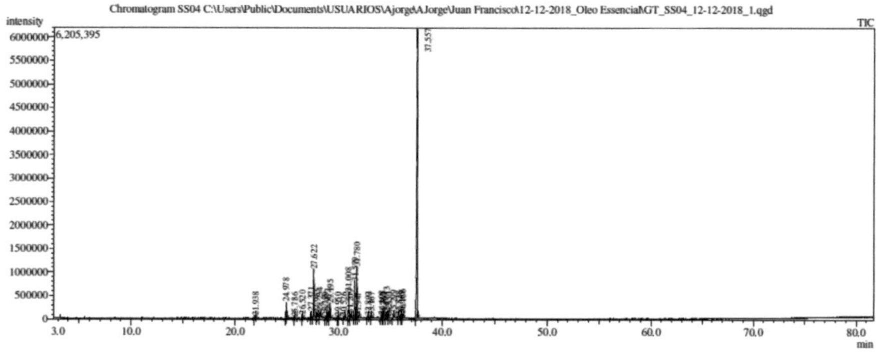

Figure 5.
Chromatogram of the essential oil of S. serrata from SS4 sample obtained by GC/MS.

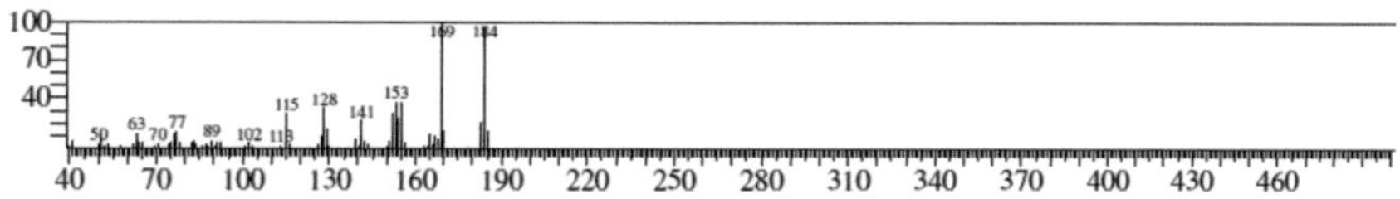

Figure 6.
Spectrum of chamazulene corresponding to the essential oil of sample SS4.

It is worth noting that the composition of the three oils is in congruence with the composition obtained by Simas et al. [33] of *S. serrata* from a population in the department of Sololá, presenting the same major compounds with some percentage variations and the majority of compounds such as sesquiterpenoids.

4.2 Essential oil of roots of propagated plants of *S. serrata*

A seed propagation trial was carried out with seeds of plants of *S. serrata* collected from a population of Santa Lucía Utatlán, Sololá, from where the composition of essential oil with a high content of chamazulene had been previously reported [33]. The purpose of the trial was to evaluate the capability of propagation of the plants, generate new seeds, and extract and analyze the essential oil from the root. The interest in analyzing the root oil was due to the fact that in interviews with residents of the region, the authors had received information that previously the root of the plant had been used in traditional medicine for the treatment of stomach pain [33]. The seeds were germinated in peat moss, and then seedlings were transplanted to pots where they developed well with approximately 75% survival reaching 1 m height after 6 months. It is important to note that the cultivation experiment was carried out in Guatemala City, at an altitude of 1495 m, this being a lower altitude than in the region where the plant grows naturally.

After obtaining the seeds during a plant vegetative stage, the roots were collected from which an essential oil with a light green color was obtained with a yield of 0.2% (w/w), and 25 compounds representing 95.8% of the total chromatographic area were identified (**Table 3**). The chromatogram of the essential oil of the roots is shown in **Figure 7**. Due to the green coloration of the oil, it was supposed that the chamazulene was absent in the oil, which was confirmed after the analysis by GC/MS. The major components of the root oil corresponded to α-longipinene (23.5%), germacrene D (22.2%), santolina triene (12.6%), and (*E*)-caryophyllene (8.1%). The mass spectrum of α-longipinene is shown in **Figure 8**.

The common components between the root and the aerial parts oils were germacrene D and (*E*)-caryophyllene. The α-longipinene (**Figure 9**) was only found in one of the oils of the aerial parts in low percentage (0.4%), while the santolina triene (**Figure 9**) was not found in any of the oils of the aerial parts. As in the oil of aerial parts, sesquiterpenoids predominated in the root oil. Since the plant has been used in the past for the treatment of stomach pain, the authors consider it of value to carry out pharmacological activity tests with this oil in the near future.

5. Conclusions

It was found in this study that the essential oil of aerial parts of wild *S. serrata* from different populations of the highlands of Guatemala showed high concentrations of chamazulene. In addition, the essential oil of roots of the plant was analyzed for the first time, which presented a composition very different from that of the aerial parts, as it did not present chamazulene and presented α-longipinene as the major component. It was also verified that the seeds of *S. serrata* present a high viability and that the seedlings obtained from seeds also have a high percentage of survival. Therefore, *S. serrata* can be considered as a plant with high potential for domestication and cultivation for the production of essential oil with high content of chamazulene.

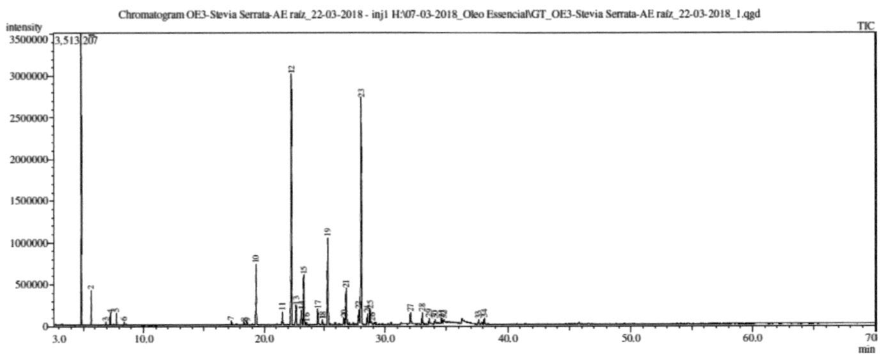

Figure 7.
Chromatogram of the essential oil of roots of S. serrata.

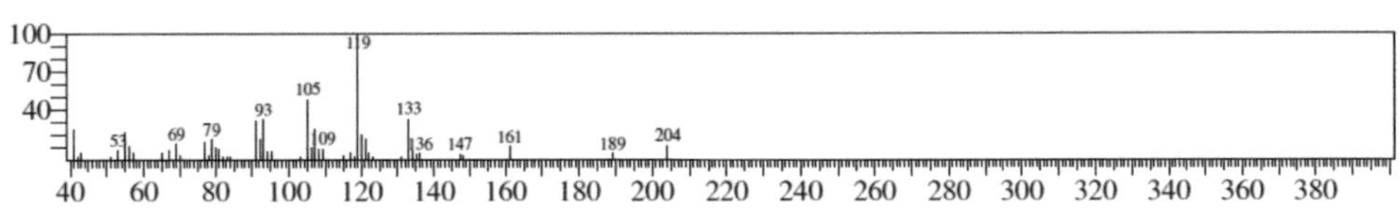

Figure 8.
Mass spectrum of α-longipinene corresponding to the essential oil of roots of S. serrata.

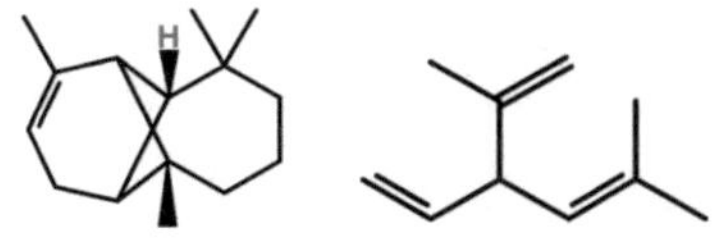

Figure 9.
Structures of α-longipinene and santolina triene, major components of the essential oil of roots of S. serrata.

Acknowledgements

The present research was partially funded by the General Directorate of Research of the University of San Carlos of Guatemala, project 4.8.63.1.06, within the framework of the University Program of Interdisciplinary Research in Health. The authors would like to agree to CAPES, CNPq, and FAPERJ from Brazil.

Conflict of interest

The authors declare that they have no conflict of interest with respect to this publication.

Author details

Juan Francisco Pérez-Sabino[1*], Max Samuel Mérida-Reyes[1], José Vicente Martínez-Arévalo[2], Manuel Alejandro Muñoz-Wug[1], Bessie Evelyn Oliva-Hernández[1], Isabel Cristina Gaitán-Fernández[3], Daniel Luiz Reis Simas[4] and Antonio Jorge Ribeiro da Silva[5]

1 School of Chemistry, University of San Carlos of Guatemala, Guatemala City, Guatemala

2 Faculty of Agronomy, University of San Carlos of Guatemala, Guatemala City, Guatemala

3 School of Biological Chemistry, University of San Carlos of Guatemala, Guatemala City, Guatemala

4 Institute of Biomedical Sciences, Federal University of Rio de Janeiro, Brazil

5 Research Institute of Natural Products, Federal University of Rio de Janeiro, Brazil

*Address all correspondence to: fpsabino@usac.edu.gt

References

[1] MSPAS-USAC. In: Ministerio de Salud Pública y Asistencia Social, editor. Vademecum Nacional de Plantas Medicinales. Guatemala: Universidad de San Carlos de Guatemala; 2006

[2] Cáceres A, Cruz SM, Gaitán I, Guerrero K, Alvarez LE, Marroquín MN. Antioxidant activity and quantitative composition of extracts of Piper species from Guatemala with potential use in natural product industry. Acta Horticulturae. 2012;**964**:77-84

[3] Cruz SM, Cáceres A, Alvarez L, Morales J, Apel MA, Henriquez AT, et al. Chemical composition of essential oils of *Piper jacquemontianum* and Piper variabile from Guatemala and bioactivity of the dichloromethane and methanol extracts. Brazilian Journal of Pharmacognosy. 2011;**21**(4):587-593

[4] Holzmann I, Cechiquel V, Mora T, Cáceres A, Martínez V, Cruz SM, et al. Evaluation of behavioral and pharmacological effects of hydroalcoholic extract of *Valeriana prionophylla* Standl. from Guatemala. Evidence-based Complementary and Alternative Medicine. 2011;**2011**:1-9

[5] Marroquín MN, Cruz SM, Cáceres A. Antioxidant activity and phenolic compounds in three species of Passifloraceae (*Passiflora edulis*, *P. incarnata*, *P. ligularis*) from Guatemala. Acta Horticulturae. 2012;**964**:93-98

[6] Nash DL, Williams LO. Flora of Guatemala. Schlivek LM, editor. Fieldiana: Botany 1976;**24**(12):125-126

[7] Jakovlev V, Isaac O, Flaskamp E. Pharmacologic studies on chamomile compounds. VI. Studies on the antiphlogistic effect of chamazulene and matricine. Planta Medica. 1983;**49**:67-73

[8] Kinghorn A. *Stevia*: The genus *Stevia*. London and New York: Taylor & Francis; 2002. p. 202

[9] King RM, Robinson H. The Genera of the Eupatorieae (Asteraceae). Monographs in Systematic Botany from the Missouri Botanical Garden. Vol. 22. St. Louis: Missouri Botanical Garden; 1987

[10] Robinson H, King RM. Eupatorieae—Systematic review. In: Heywood VH, Harborne JB, Turner BL, editors. The Biology and Chemistry of the Compositae. Vol. 1. New York: Academic Press; 1977. pp. 437-485

[11] Pruski JF, Robinson H. Flora Mesoamericana. In: Davidse G, Sousa Sánchez M, Knapp S, Chiang Cabrera F, editors. Asteraceae Bercht. & J. Presl. 2015;**5**(2):554-555

[12] Karaköse H, Jaiswal R, Kuhnert N. Characterization and quantification of hydroxycinnamate derivatives in *Stevia rebaudiana* leaves by LC-MSn. Journal of Agricultural and Food Chemistry. 2011;**59**:10143-10150. DOI: 10.1021/jf202185m

[13] Cerda-García-Rojas CM, Guerra-Ramírez D, Román-Marín LU, Hernández-Hernández JD, Joseph-Nathan P. DFT molecular modeling and NMR conformation analysis of a new longipinenetriolone diester. Journal of Molecular Structure. 2006;**789**:37-42

[14] Sánchez-Arreola E, Cerda-García-Rojas CM, Román LU, Hernández JD, Joseph-Nathan P. Longipinene derivatives from *Stevia porphyrea*. Phytochemistry. 1999;**52**:473-477

[15] Román LU, Morán G, Hernández JD, Cerda-García-Rojas CM, Joseph-Nathan P. Longipinane derivatives

from *Stevia viscida*. Phytochemistry. 1995;**38**(6):1437-1439

[16] Álvarez-García R, Torres-Valencia JM, Román LU, Hernández JD, Cerda-García-Rojas CM, Joseph-Nathan P. Absolute configuration of the α-methylbutyryl residue in longipinene derivatives from *Stevia pilosa*. Phytochemistry. 2005;**66**:639-642

[17] Amaro JM, Adrián M, Cerda CM, Joseph-Nathan P. Longipinene derivatives from *Stevia lucida* and *S. triflora*. Phytochemistry. 1988;**27**(5):1409-1412. DOI: 10.1016/0031-9422(88)80205-X

[18] Guerra-Ramírez D, Cerda-García-Rojas C, Puentes AM, Joseph-Nathan P. Longipinene diesters from *Stevia lucida*. Phytochemistry. 1998;**48**(1):151-154. DOI: 10.1016/S0031-9422(97)00793-0

[19] Prakash Chaturvedula VS, Prakash I. A new diterpene glycoside from *Stevia rebaudiana*. Molecules. 2011;**16**:2937-2943. DOI: 10.3390/molecules16042937

[20] Prakash I, Prakash Chaturvedula VS. Additional minor diterpene glycosides from *Stevia rebaudiana* Bertoni. Molecules. 2013;**18**:13510-13519. DOI: 10.3390/molecules181113510

[21] Machado KN, Turatti ICC, Lopes NP, do Nascimento A. Essential oil composition of *Stevia urticifolia* growing in ouro preto-mg. Chemistry of Natural Compounds. 2015;**51**(5):985-986. DOI: 10.1007/s10600-015-1471-9

[22] Muanda FN, Soulimani R, Diop B, Dicko A. Study on chemical composition and biological activities of essential oil and extracts from *Stevia rebaudiana* Bertoni leaves. LWT-Food Science and Technology. 2011;**44**:1865-1872. DOI: 10.1016/j.lwt.2010.12.002

[23] Machado KN, Tasco AJH, Salvador MJ, Rodrigues IV, Pessoa C, Sousa IJO, et al. Flavonoids, antioxidant, and antiproliferative activities of *Stevia urticifolia*. Chemistry of Natural Compounds. 2017;**53**(6):1167-1169. DOI: 10.1007/s10600-017-2228-4

[24] Román LU, Guerra-Ramírez D, Morán G, Martínez I, Hernández JD, Cerda-García-Rojas CM, et al. First *seco*-C oleananes from nature. Organic Letters. 2004;**6**(2):173-176. DOI: 10.1021/ol036107j

[25] Oberti JC, Gil RR, Sosa VE, Herz W. A guaianolide from *Stevia breviaristata*. Phytochemistry. 1986;**25**(6):1479-1480

[26] Angeles E, Folting K, Grieco PA, Huffman JC, Miranda R, Salmón M. Isolation and structure of stephalic acid, a new clerodane diterpene from *Stevia polycephala*. Phytochemistry. 1982;**21**(7):1804-1806

[27] Khattab SN, Massoud MI, El-Sayed Jad Y, Bekhit AA, El-Faham A. Production and physicochemical assessment of new *stevia* amino acid sweeteners from the natural stevioside. Food Chemistry. 2015;**173**:979-985. DOI: 10.1016/j.foodchem.2014.10.093

[28] Millones C, Mori G, Bacalla J, Vásquez E, Tafur R. Obtención de un filtrante de anís de monte (*Tagetes filifolia* Lag.) edulcorado con hojas de estevia (*Stevia rebaudiana* Bertoni). Scientia Agropecuaria. 2014;**5**:45-51. DOI: 10.17268/sci.agropecu.2014.01.05

[29] Sánchez-Arreola E, Cerda-García-Rojas CM, Joseph-Nathan P, Román LU, Hernández JD. Longipinene derivatives from *Stevia serrata*. Phytochemistry. 1995;**39**(4):853-857

[30] Calderón JS, Quijano L, Gómez F, Ríos T. Prochamazulene sesquiterpene lactones from *Stevia serrata*. Phytochemistry. 1989;**28**(12):3526-3527

[31] Franke R, Schilcher H, editors. Chamomile Industrial Profiles:

Medicinal and Aromatic Plants-Industrial Profiles. Boca Raton: CRC Press Taylor & Francis Group; 2005. p. 279

[32] Román LU, Mora Y, Hernández JD. *Stevia serrata*, a source of chamazulene. Phytoterapia. 1990;**61**(1):84

[33] Simas DL, Mérida-Reyes M, Muñoz-Wug M, Cordeiro M, Giorno TB, Taracena EA, et al. Chemical composition and evaluation of antinociceptive activity of essential oil of *Stevia serrata* Cav. from Guatemala. Natural Product Research. 2017;**33**(4):577-579. DOI: 10.1080/14786419.2017.1399376

[34] Kohda H, Yamazaki K, Tanaka O. Methylripariochromene a from *Stevia serrata*. Phytochemistry. 1976;**15**:847-848

[35] Ticktin T, Dalle SP. Medicinal plant use in the practice of midwifery in rural Honduras. Journal of Ethnopharmacology. 2005;**96**:233-248

[36] Vibrans H, Alipi AM, Pichardo JM. Malezas de México. 2009. Available from: http://www.conabio.gob.mx/malezasdemexico/asteraceae/stevia-serrata/fichas/ficha.htm

Chapter 13

Durable Woods and Antifungal Activity of Their Essential Oils: Case of *Tetraclinis articulata* (Vahl) Masters and *Cedrus atlantica* Manetti

Abdelwahed Fidah, Mohamed Rahouti, Bousselham Kabouchi and Abderrahim Famiri

Abstract

Cedrus atlantica Manetti and *Tetraclinis articulata* (Vahl) Masters are a resinous species originated from North Africa and well known for their durable and noble timbers. This work was conducted to assess the relationship between natural durability of their woods, assessed in previous works by European standards NF EN 350 and CEN/TS 15083-1, and the bioactivity of essential oils extracted from these woods by hydrodistillation and analyzed by GC-FID and GC-MS. Bioassay of sawdust essential oils, conducted by direct contact technique on agar medium on four wood-decaying fungi strains, revealed strong antifungal inhibition especially by *T. articulata* root burl oil due to its richness in phenols. Natural durability classes of *T. articulata* and *C. atlantica* woods were then positively correlated with antifungal activity levels of their oils.

Keywords: *Cedrus atlantica*, *Tetraclinis articulata*, wood durability, essential oils, wood-decaying fungi

1. Introduction

Tetraclinis articulata (Vahl) Masters (Cupressaceae) and *Cedrus atlantica* Manetti (Pinaceae) are threatened species [1] endemic to the western Mediterranean areas [2]. Moroccan *T. articulata* and *C. atlantica* populations occupy an area of about 500,000 and 130,000 ha, respectively, and satisfy many socioeconomic needs of the human riparian populations for various products. But in the last decades, forests of *T. articulata* were exposed to a significant degradation due to a strong demand by the craft sector for timbers [2, 3]. Even if *C. atlantica* forest provides approximately annually 90,000 m^3 of softwood logs intended for sawing and veneer, they also suffer the same fate [4]. Both species are famous for their durable and noble timbers. *T. articulata* root burl provides good quality woody material (hard, homogeneous, and fine grained) with remarkable flecks and an aesthetic aspect very appreciated in cabinetry and marquetry uses [3].

Laboratory tests showed that *T. articulata* trunk wood and its root burl are durable against wood-decaying fungi [5, 6] as well as *C. atlantica* heartwood [6–8].

A lot of waste as slabs and sawdust results from wood processing of those species timbers. However, only sawdust accounts for about 8% of *C. atlantica* sawn timber that contains appreciable amounts of extractives [9–12]. Many recent works highlighted that essential oils (EOs) extracted from *T. articulata* and *C. atlantica* woods possess numerous biocidal activities [12–15]. Nevertheless, only few attempts to investigate the capability of EOs to protect woods against fungi decay have been previously undertaken [16–18]. The relationship between natural durability of *T. articulata* and *C. atlantica* woods and the bioactivity of their EOs was not yet established. Therefore, the present study is devoted to investigate this relationship.

2. Material and methods

2.1 Material used for EOs extraction and chemical analysis

Trunk wood and root burl samples of *T. articulata* were collected from sweepings of craft processing workshops in Khemisset Region (central plate of Morocco), while samples of *C. atlantica* sawdust were collected from wood sawmill in the region of Azrou (Middle Atlas Mountains of Morocco). Sawdust was then sieved into particles of 1 mm size and triplicate samples of 250 g were subjected to hydrodistillation for 4 hours to obtain pure essential oils (EOs).

The chemical analysis and component identification of EOs were performed by gas chromatography (GC-FID) and by gas chromatography coupled with a mass spectroscopy (GC-MS). The identification of EO components was achieved by comparison of their retention indices (RI) relative to (C8-C22) *n*-alkanes with those of known compounds, and by comparison of similar mass spectra using Wiley/NBS mass spectral library of the GC-MS data system and other published mass spectra [19]. The percentage compositions of samples were calculated according to the area of the chromatographic peaks using the total ion current.

2.2 Material used for bioassay

Bioactivity of EOs extracted from *T. articulata* and *C. atlantica* woods was assessed in bioassay against the following wood-decaying basidiomycetes fungi: *Gloeophyllum trabeum* (BAM Ebw.109 strain), *Oligoporus placenta* (FPRL. 280 strain), *Coniophora puteana* (BAM Ebw. 15 strain), and *Trametes versicolor* (CTB 863 A strain). Fungi strains originated from the mycological collection of the Laboratory of Botany, Mycology and Environment, Faculty of Sciences in Rabat, Morocco.

2.3 Experimental

Bioassays of EOs were performed by direct contact of fungi strains on agar medium according to the method reported by Remmal et al. [20]. Oils were first diluted in a sterile solution of tap water-agar at 0.2% in order to obtain a homogeneous mixture, then distributed in test tubes containing 13.5 ml of sterilized malt-agar medium (20 g/l malt extract and 15 g/l agar), and kept at 45°C in a water bath. To obtain the final oil concentrations in the culture medium ranging from 1/250 to 1/5000 v/v, aseptic volumes of 1.5 ml of different dilutions were then added to those tubes before pouring EO-medium mixtures into Petri dishes. Additional control dishes containing only 13.5 ml of culture medium and agar solution at 0.2% (SA) alone were also prepared. Inoculation of Petri dishes was made by two 0.5 cm^2 fragments of 10 days old fungal culture in malt-agar. For each treatment, three repetitions were prepared and incubated in the dark for 7 days at 22°C. At the end of each bioassay, minimal inhibitory concentration (MIC) [21] was determined for each fungus.

2.4 Results and discussion

According to the bioassay conducted on oils extracted from *T. articulata* and *C. atlantica* woods, a significant inhibitory effect on the four tested wood-decaying fungi was observed (**Table 1**) with different levels of inhibition. *T. articulata* root burl wood EOs showed, however, a strong inhibitory action against those fungi strains with oil dilutions over 1/4000 v/v. *G. trabeum* fungus was the most sensitive to the inhibitory effect of this essential oil since it was inhibited by concentrations between 1/5000 for *T. articulata* root burl oil and 1/1000 v/v for Atlas cedar oil. *O. placenta* was the most resistant strain since its growth inhibition was not reached until 1/400 concentrations for *C. atlantica* oil and 1/800 for *T. articulata* trunk wood oil (**Table 1**).

Previous studies by our team [5, 8] showed that *T. articulata* and *C. atlantica* woods were classified as very durable to durable (DC 1 and 2) and means of mass loss of test specimens was below 5.20% compared to those of Scot pine wood (control) (40.70%). According to their durability indexes (*X*) determined by NF EN 350-1 and CEN/TS 15083-1 standards [22, 23] and the biological risks defined by EN 335-2 standard [24], natural durability levels of those woods against wood-decaying fungi allow them to access high-risk classes of biological attacks 4 and 5 for an end-use without preservative treatment regarding decay fungi [25, 26]. Compared to similar studies on Moroccan coniferous woods (**Figure 1**), the natural durability of native Atlas cedar wood is similar to that of *C. atlantica* heartwood (DC 1 and 2) originated from a south Italian plantation [7], whereas *Pinus halepensis* and *P. pinaster* woods were considered as less durable (DC 4) [6, 27]. Generally, pine woods contain less active extractives than those of *Cupressaceae*. Adamopoulos et al. [28] reported that the weakness of natural durability of both heartwood and sapwood of *Pinus leucodermis* is related to low presence of bioactive extractives that can inhibit the brown-rot fungus, *Coniophora puteana*.

In addition, other works by our team [8, 29] revealed that EOs of *C. atlantica* wood is dominated by ketones (52.05%) and alcohols (26.58%), while those of thuya are dominated by alcohols (about 55–78%) and sesquiterpenes (13–22%) (**Table 2**). Major components of *C. atlantica* oil are, respectively, E-γ-atlantone, E-α-atlantone, 5-isocedranol, 9-iso-thujopsanone, cedranone, Z-α-atlantone, cedroxyde, and 14-hydroxy-δ-cadinene [8] (**Table 3**).

	Essential oils fungal strains	Thuya trunk wood	Thuya root burl	Atlas cedar wood
Specific MIC	*T. versicolor*	1/1000	1/4000	1/800
	C. puteana	1/1000	1/4000	1/400
	G. trabeum	1/1200	1/5000	1/1000
	O. placenta	1/800	1/5000	1/400
Global MIC		1/800	1/4000	1/400

Table 1.
Minimal inhibitory concentrations (MIC) (v/v) determined for essential oils of thuya and Atlas cedar woods by bioassay conducted on malt-agar medium on wood-decaying fungi.

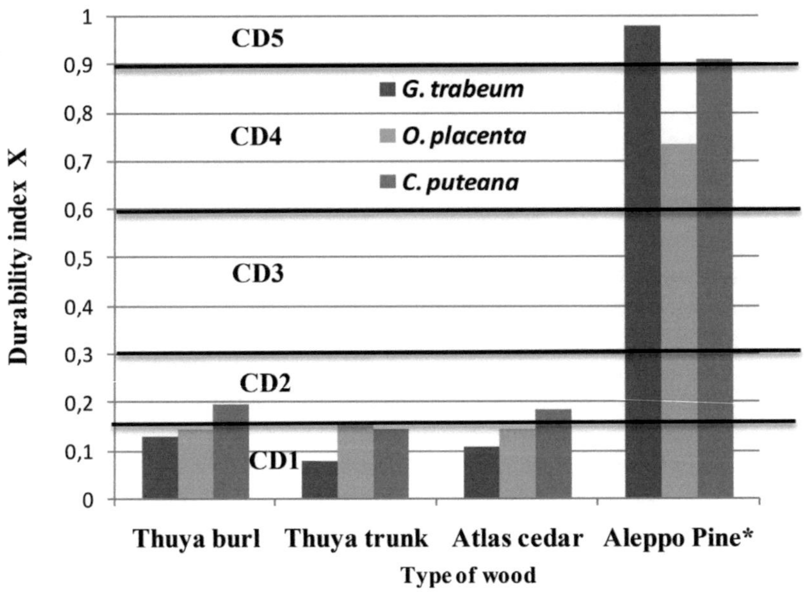

Figure 1.
Natural durability of studied woods against wood-decaying fungi.

Chemical group	Thuya trunk wood	Thuya root burl	Atlas cedar wood
Alcohols	54.80	78.50	26.60
Esters	—	—	5.70
Ketones	1.40	0.30	52.00
Oxydes	0.90	0.60	3.40
Terpenes	31.20	14.10	8.00
Others	1.75	1.15	—
Global	90.04	94.65	95.69

Table 2.
Chemical groups (in%) identified, by GC-MS, in essential oils of Tetraclinis articulata and Cedrus atlantica woods.

EOs of *T. articulata* woods are rich in thymol, 3-tert-butyl-4-methoxyphenol, cedrol, and *α*-cedrene (**Table 3**). The oil of *T. articulata* trunk wood contains phenols such as *α*- and *β*-acorenol, cedrol, and totarol, along with terpenes as *α*-cedrene that can protect this wood from fungi decay. The strong inhibitory effect of *T. articulata* root burl oil is probably due to its oxygenated fraction rich in phenols as thymol, 3-tert-butyl-4-methoxyphenol, and cedrol. A previous work already highlighted this significant antibacterial activity at low concentrations [30]. Abundance of tropolones and phenols in *Cupressaceae* woods may explain their high natural durability grades against wood-decaying fungi as reported by Haluk and Roussel [16] and by more recent works [13, 30, 31]. Early investigations found that thujaplicines (tropolones) of *Thuja plicata* possesses strong inhibitory effect against the wood blue stain fungi and many wood-decaying fungi [32].

Natural durability of *C. atlantica* wood can be correlated to bioactivity of its oils essentially rich in sesquiterpene ketones as atlantones [9, 14]. In the literature,

Chemical group	Component	KI	TTW	TRB	ACW
Alcohols	Thymol	1290	8.20	39.80	—
	3-Tert-butyl-4-methoxyphenol	1491	16.50	24.70	—
	Turmerol	1578	—	—	3.45
	Cedrol	1596	14.80	6.35	1.90
	α-Acorenol	1633	4.20	1.10	—
	β-Acorenol	1637	4.15	1.30	—
	Himachalol	1647	—	0.30	2.45
	5-Isocedranol	1669	—	—	11.70
	β-Santalol	1741	—	—	2.00
	E-Z-Farnesol	1742	—	—	1.10
	Totarol	2314	2.60	1.50	—
Esters	Hexyl isobutyrate	1150	—	—	1.38
	Benzyl benzoate	1762	—	—	1.16
	Z-β-Santalol acetate	1823	—	—	1.15
	Z-Ternine	1838	—	—	1.25
Ketones	Camphor	1143	0.11	—	1.28
	Cedranone	1620	—	—	4.13
	9-Iso-thujopsanone	1637	—	—	4.45
	Deodarone (Dihydro-2,2,6-trimethyl-6-(4-methyl-3-cyclohexen-1-yl)-2H-pyran-3(4H)-one)	1694	—	—	1.07
	E γ-Atlantone	1701	—	—	19.73
	Z α-Atlantone	1713	—	—	4.02
	E α-Atlantone	1773	0.74	—	16.86
Oxides	italicene oxide	1538	0.91	0.59	—
	Cedroxide	1704	—	—	2.38
Terpenes	α-Cedrene	1409	17.59	7.77	—
	β-Cedrene	1434	4.00	1.80	—
	α-Himachalene	1447	—	1.07	—
	Thujupsadiene	1462	—	1.10	—
	α-Acoradiene	1464	2.33	—	—
	β-Acoradiene	1465	2.26	—	—
	β-Himachalene	1499	—	1.25	0.79
	α-Dehydro-ar-himachalene	1511	—	—	1.13
	γ-Cadinene	1513	1.18	0.31	
	γ-Dehydro-ar-himachalene	1529	—	—	1.57
	14 Hydroxy-murolene	1775	—	—	1.00
	14 Hydroxy δ-cadinene	1799	1.21	0.24	1.94

KI, Kovàts Index; TTW, Thuya Trunk Wood; TRB, Thuya Root Burl; ACW, Atlas Cedar Wood.

Table 3.
The main components (in%) identified, by GC-MS, in essential oils of Tetraclinis articulata and Cedrus atlantica woods.

African padauk (*Pterocarpus soyauxii*) medium-heavy heartwood manifests a remarkable decay resistance attributed to its specific extractive compounds [33]. Moreover, extractive compounds obtained from black locust heartwood were able to increase the native durability of European beech against wood-decaying fungi from class 5 (not durable) to class 3 (moderately durable) [34].

The strong antifungal activity of *T. articulata* root burl wood EOs is probably related to their alcohol fraction, rich in thymol and 3-tert-butyl-4-methoxyphenol, which confer them this significant bioactivity. The combined action of two phenolic compounds, such as thymol and carvacrol, was previously reported [17, 35]. Action of phenolic compounds on fungi is primarily based on the inhibition of fungal enzymes containing SH group in their active site [36, 37]. The antifungal activity of the EOs of *C. atlantica* wood can be correlated to its sesquiterpene ketones, which are mainly atlantones (about 40.61%). Bioassays conducted with pure α-atlantones extracted from *Decalepis hamiltonii* revealed great inhibitory activity against pests and molds [38]. In our study, alcohols present in *C. atlantica* oil in significant amount (more than 26%), such as isocedranol, tumerol, himachalol, and cedrol, may also be involved in the inhibitory effect of this oil. Other compounds such as cadinenes (monoterpene hydrocarbon) could also have a great antimicrobial property [39]. However, synergistic action of two or more components of essential oils extracted from thuya and Atlas cedar woods can also be involved in the observed bioactivity reported in our study.

Antifungal activity of several thyme species oils was recently tested successfully against wood-decaying fungi especially those of *Thymus bleicherianus* [17]. Furthermore, EOs of *T. articulata* root burl showed an antibacterial activity two to six times greater compared to that of reference antibiotic and were more effective on *Staphylococcus aureus* (Gram$^+$) and *Escherichia coli* (Gram$^-$) with significant bacteriostatic and bactericidal effects [40]. Regarding inhibition mechanisms of active compounds of EOs, it was reported that volatile alcohols can act on both lytic and synthetic enzyme pathways and growth inhibition consequently occurred after breaking off natural extension hyphae of fungi [41, 42].

Active compounds contained in oils of *T. articulata* and *C. atlantica* woods, which were successfully tested against wood-decaying fungi, may protect and ensure good natural durability levels for these woods ranging from very durable (DC1) to durable (DC2) classes.

Findings of this study could allow us to consider recovering wastes from *T. articulata* and *C. atlantica* wood processing for the extraction of bioactive oils and use them as biocide especially in preservative treatment of less-durable woods such as those of pines. Also, these results are in favor of more protection for these threatened species and for the rehabilitation in their natural environment. Natural compounds of *T. articulata* and *C. atlantica* oils can then replace the use of current petrochemical compounds that are becoming more and more criticized for their harmful effect on human health and the environment.

3. Conclusion

The present study has clarified the relationship between the natural durability of *T. articulata* and *C. atlantica* woods and the antifungal activity of their EOs. According to the bioassay conducted on those oils, a significant inhibitory activity was obtained on the four wood-rotting fungi tested, mainly for oils extracted from *T. articulata* root burl. Stronger inhibitory effect was then reached by dilutions over 1/5000 v/v for this oil rich in phenols. The durability classes of *T. articulata* and *C. atlantica* woods positively correlated with bioactivity of their oils against the three wood-decaying fungi specified by the CEN/TS 15083-1 standard. Active EOs extracted from wastes of these woods can be valorized as wood-preservative treatment.

Acknowledgements

This project is supported by Forest Research Center in Rabat (High Commission for Water and Forests and Fight against Desertification) in collaboration with Faculty of Sciences in Rabat (Mohammed V University), Morocco.

Author details

Abdelwahed Fidah[1*], Mohamed Rahouti[2], Bousselham Kabouchi[2] and Abderrahim Famiri[1]

1 Forest Research Center, Rabat, Morocco

2 Faculty of Sciences, Mohammed V University in Rabat, Morocco

*Address all correspondence to: fidah.abdelwahed@gmail.com

References

[1] IUCN. The IUCN Red List of threatened species, Version 2018-1 [Internet]. Available from: http://www.iucnredlist.org [Accessed 17 July 2018]

[2] M'Hirit O, Blerot P. Le grand livre de la forêt marocaine. 1st ed. Bruxelles, Belgium: Mardaga Editions; 1999. p. 280. ISBN-10: 2870096860

[3] Dakak J. La qualité du bois de Thuya de Maghreb (*Tetraclinis articulata*) et ses conditions de développement sur ses principaux sites phytoécologiques de son bloc méridional au Maroc [thesis]. Nancy, France: Ecole Nationale du Génie Rural, des Eaux et des Forêts; 2002

[4] HCEFLCD. Synthèse de l'Inventaire Forestier National Marocain. Direction de développement forestier. Haut Commissariat aux Eaux et Forêts et à la Lutte Contre la Désertification. Technical Report. Rabat, Maroc; 2013

[5] Fidah A, Rahouti M, Kabouchi B, Ziani M, El Bouhtoury-Charrier F, Famiri A. Natural durability of *Tetraclinis articulata* (Vahl) Masters woods against wood decay fungi: Laboratory test. Wood Research. 2015;**60**(6):865-872. Available form: http://www.centrumdp.sk/wr/201506/02.pdf

[6] Fidah A, Salhi N, Janah T, Rahouti M, Kabouchi B, El Alami A, et al. Comparative natural durability of four Mediterranean softwoods against wood decay fungi. Journal of the Indian Academy of Wood Science. 2016a;**13**(2):132-137. DOI: 10.1007/s13196-016-0174-6

[7] Brunetti M, De Capua EL, Macchioni N, Monachello S. Natural durability, physical and mechanical properties of Atlas cedar (*Cedrus atlantica* Manetti) wood from southern Italy. Annals of Forest Science. 2001;**58**:607-613. Available form: https://www.afs-journal.org/articles/forest/pdf/2001/06/28-00-curseur.pdf

[8] Fidah A, Salhi N, Rahouti M, Kabouchi B, Ziani M, Aberchane M, et al. Natural durability of *Cedrus atlantica* Manetti wood related to the bioactivity of its essential oil against wood decaying fungi. Maderas Ciencia y Tecnologia. 2016b;**18**(4):567-576. DOI: 10.4067/S0718-221X2016005000049

[9] Aberchane M, Fechtal M, Chaouch A. Analysis of Moroccan Atlas cedarwood oil (*Cedrus atlantica* Manetti). Journal of Essential Oil Research. 2004;**16**:542-547. DOI: 10.1080/10412905.2004.9698793

[10] Aberchane M, Fechtal M, Chaouch A, El Abid A. Effet de l'infection du bois de Cèdre de l'atlas (*Cedrus atlantica* M.) par les champignons Trametes pini et *Ungulina officinalis*. Annales de la Recherche Forestière du Maroc. 2006;**37**:105-114

[11] El Bouhtoury-Charrier F, Hakam A, Famiri A, Ziani M, Charrier B. Wood characterization of *Tetraclinis articulata* and evaluation of its resistance against ligninolytic fungi. In: Proceedings of IRG Annual Meeting (IRG/WP 09-10697); 24-28 May 2009; Beijing China; Conference: 09-05

[12] Bourkhiss M, Hnach M, Lakhlifi T, Bourkhiss B, Ouhssine M, Satrani B. Production et caractérisation de l'huile essentielle de la sciure de bois de *Tetraclinis articulata* (Vahl) Masters. Bulletin de la Société Royale des Sciences de Liège. 2010;**79**:4-11

[13] Satrani B, Aberchane M, Farah A, Chaouch A, Talbi M. Composition chimique et activité antimicrobienne des huiles essentielles extraites par hydrodistillation fractionnée du bois de *Cedrus atlantica* Manetti. Acta Botanica

Gallica. 2006;**153**(1):97-104. DOI: 10.1080/12538078.2006.10515524

[14] Derwich E, Benziane Z, Boukir A. Chemical composition and in vitro antibacterial activity of the essential oil of *Cedrus atlantica*. International Journal of Agriculture and Biology. 2010;**12**:381-385. Available from: http://www.fspublishers.org:80/ijab/past-issues/IJABVOL_12_NO_3/12.pdf

[15] Fidah A. Etude de la durabilité naturelle des bois de *Cedrus atlantica* Manetti et de *Tetraclinis articulata* (Vahl) Masters et bioactivité de leurs huiles essentielles vis-à-vis des champignons basidiomycètes lignivores [thesis]. Maroc: Faculté des Sciences de Rabat; 2016

[16] Haluk JP, Roussel C. Caractérisation et origine des tropolones responsables de la durabilité naturelle des Cupressacées. Application potentielle en préservation du bois. Annals of Forest Science. 2000;**57**:819-829. Available from: https://hal.archives-ouvertes.fr/hal-00883441

[17] El Ajjouri M, Satrani B, Ghanmi M, Aafi A, Farah A, Rahouti M, et al. Activité antifongique des huiles essentielles de Thymus bleicherianus Pomel et *Thymus capitatus* (L.) Hoffm. & Link contre les champignons de pourriture du bois d'œuvre. Biotechnologie, Agronomie, Société et Environnement. 2008;**12**(4):345-351. Available from: http://www.pressesagro.be/base/text/v12n4/345.pdf

[18] Hassane SOS, Ghanmi M, Satrani B, Farah A, Mansouri N, Chaouch A. Activité antifongique contre la pourriture du bois de l'huile essentielle de Pelargonium x asperum Erthrt. Ex willd des Îles Comores. Bulletin de la Société Royale des Sciences de Liège. 2012;**81**:36-49. Available from: https://popups.uliege.be/0037-9565/index.php?id=3659&file=1&pid=3650

[19] Adams RP. Identification of Essential Oil Components by Gas Chromatography/Mass Spectrometry. 4th ed. Illinois, USA: Allured publishing corporation, Carol Stream; 2007. p. 804. ISBN-13: 978-1932633214

[20] Remmal A, Tantaoui-Eiaraki A, Bouchikhi T, Rhayour K, Ettayebi M. Lmproved·method for determination of antimicrobial activity of essential oils in agar medium. Journal of Essential Oil Research. 1993;**5**:179-184. DOI: 10.1080/10412905.1993.9698197

[21] Tantaoui-Elaraki A, Lattaoui N, Benjilali B, Errifi A. Antimicrobial activity of four chemically different essential oils. Rivista Italiana EPPOS. 1992;**6**:13-22. DOI: 10.1080/10412905.1993.9698169

[22] NF EN 350-1 Standard. Durability of Wood and Wood-Based Products—Natural Durability of Solid Wood. Part 1: Guide to the Principles of Testing and Classification of the Natural Durability of Wood. Brussels: European Committee for Standarization; 1994. p. 18

[23] CEN/TS 15083-1 Standard. Durability of Wood and Wood-Based Products—Determination of the Natural Durability of Solid Wood against Wood-Destroying Fungi-Test Methods, Part 1: Basidiomycetes. Brussels: European Committee for Standarization; 2005. p. 24

[24] NF EN 335 Standard. Durability of Wood and Wood-Based Products—Use Classes: Definitions, Application to Solid Wood and Wood-Based Products. Brussels: European Committee for Standarization; 2013. p. 18

[25] NF EN 460 Standard. Durability of Wood and Wood-Based Products. Natural Durability of Solid Wood Guide of the Durability Requirements for Wood to be Used in Hazard Classes. Brussels: European Committee for Standarization; 1994

[26] Van Acker J, Stevens M, Carey J, Sierra-Alvarez R, Militz H, Le Bayon I, et al. Biological durability of wood in relation to end-use. Holz als Roh-und Werkstoff. 2003;**61**(1):33-45. DOI: 10.1007/s00107-002-0351-8

[27] Thevenon M-F, Janah T, Rahouti M, Langbour P, Gerard J. Investigations on the durability of two secondary Pine species (*Pinus halepensis*, *Pinus uncinata*) within the scope of the European natural durability standards revision. In: Proceedings of IUFRO conference, Division 5 "Forest Products"; 8-13 July 2012; Lisbon, Portugal

[28] Adamopoulos S, Gellerich A, Mantanis G, Kalaitzi T, Militz H. Resistance of *Pinus leucodermis* heartwood and sapwood against the brown-rot fungus *Coniophora puteana*. Wood Material Science & Engineering. 2012;**7**(4):242-244. DOI: 10.1080/17480272.2012.684705

[29] Fidah A, Salhi N, Rahouti M, Kabouchi B, Ismaili MR, Ziani M, et al. Chemical variability and antifungal activity of *Tetraclinis articulata* (Vahl) masters woods and leaves essential oils against wood decaying fungi. International Journal of Pharmacognosy and Phytochemical Research. 2017;**9**(1):123-128. ISSN: 0975-4873

[30] El Moussaouiti M, Talbaoui A, Gmouh S, Aberchane M, Benjouad A, Bakri Y, et al. Chemical composition and bactericidal evaluation of essential oil of *Tetraclinis articulata* burl wood from Morocco. Journal of the Indian Academy of Wood Science. 2010;**7**(1-2):14-18. DOI: 10.1007/s13196-010-0003-2

[31] El Hanbali F, Amusant N, Mellouki F, Akssira M, Baudasse C. Potentiality of use extracts from *Tetraclinis articulata* like biocide against wood destroying organisms: *Reticulitermes santonensis*. In: Proceedings of IRG/WP 07-30418; 20-24 May 2007; Jackson USA

[32] Rennerfelt E. Investigations of thujaplicine, a fungicidal substance in the heartwood of *Thuja plicata* D. Don. Physiologia Plantarum. 1948;**1**:245-254. DOI: 10.1111/j.1399-3054.1948.tb07128.x

[33] Nzokou P, Kamdem DP. Fungal decay resistance of nondurable aspen wood treated with extractives from African padauk (*Pterocarpus soyauxii*). Journal of Tropical Forest Products. 2003;**9**(1-2):125-133

[34] Sablik P, Giagli K, Paril P, Baar J, Rademacher P. Impact of extractive chemical compounds from durable wood species on fungal decay after impregnation of nondurable wood species. European Journal of Wood and Wood Products. 2015;**74**(2):231-236. DOI: 10.1007/s00107-015-0984-z

[35] Trombetta D, Saija A, Bisignano G. Study on the mechanisms of the antibacterial action of some plant, β-unsaturated aldehydes. Letters in Applied Microbiology. 2002;**35**:285-290. DOI: 10.1046/j.1472-765X.2002.01190.x

[36] Cowan MM. Plant products as antimicrobial agents. Clinical Microbiology Reviews. 1999;**12**(4): 564-582. Available from: http://cmr.asm.org/content/12/4/564

[37] Celimene CC, Micales JA, Ferge L, Young RA. Efficacy of pinosylvins against white-rot and brown-rot fungi. Holzforschung. 1999;**53**:491-497. Available from: https://www.fpl.fs.fed.us/documnts/pdf1999/celim99a.pdf

[38] Thangadurai D, Anitha S, Pullaiah T, Reddy PN, Ramachandraiah OS. Essential oil constituents and in vitro antimicrobial activity of *Decalepis hamiltonii* roots against foodborne pathogens. Journal of Agricultural and Food Chemistry. 2002;**50**(11):3147-3149. DOI: 10.1021/jf011541q

[39] Keawsa-Ard S, Liawruangrath B, Liawruangrath S, Teerawutgulrag A, Pyne SG. Chemical constituents and antioxidant and biological activities of the essential oil from leaves of *Solanum spirale*. Natural Product Communications. 2012;7(7):955-958. DOI: 10.1177/1934578X1200700740

[40] Akbli M, Rhallabi N, Ait Mhand R, Akssira M, Mellouki F. Activité antibactérienne de l'huile essentielle de la sciure du bois de loupe de *Tetraclinis articulata* (Vahl) Masters du Maroc sur des souches d'origine clinique. International Journal of Innovation and Applied Studies. 2016;**16**:314-321. Available from: http://www.ijias.issr-journals.org/

[41] Kahn V, Andrawis A. Inhibition of mushroom tyrosinase by tropolone. Phytochemistry. 1985;**24**:905-908. DOI: 10.1016/S0031-9422(00)83150-7

[42] Takeuchi M, Ichishima E. Inhibition study of tyrosinase from *Aspergillus oryzae*. Agricultural and Biological Chemistry. 1989;**53**:557-558. DOI: 10.1080/00021369.1989.10869309